HOMOGENEOUS CATALYSIS WITH COMPOUNDS OF RHODIUM
AND IRIDIUM

CATALYSIS BY METAL COMPLEXES

RONALD S. DICKSON

*Department of Chemistry, Monash University,
Clayton, Victoria, Australia*

HOMOGENEOUS CATALYSIS WITH COMPOUNDS OF RHODIUM AND IRIDIUM

D. REIDEL PUBLISHING COMPANY

A MEMBER OF THE KLUWER  ACADEMIC PUBLISHERS GROUP

DORDRECHT / BOSTON / LANCASTER

Library of Congress Cataloging in Publication Data

Dickson, Ronald S.
  Homogeneous catalysis with compounds of rhodium and iridium.

  (Catalysis by metal complexes)
  Bibliography: p.
  Includes index.
  1. Catalysis. 2. Rhodium catalysts. 3. Iridium catalysts.
  I. Title. II. Series.
  QD505.D53 1985   541.3'95   85-18311
  ISBN 90-277-1880-6

Published by D. Reidel Publishing Company.
P.O. Box 17, 3300 AA Dordrecht, Holland

Sold and distributed in the U.S.A. and Canada
by Kluwer Academic Publishers
190 Old Derby Street, Hingham, MA 02043, U.S.A.

In all other countries, sold and distributed
by Kluwer Academic Publishers Group,
P.O. Box 322, 3300 AH Dordrecht, Holland.

Printed in The Netherlands

To
two very different sources of inspiration –
my family (at home), and
my research group (in the lab.)

CONTENTS

PREFACE

Some years ago, I agreed to contribute a volume to the Academic Press 'Organo-metallic Chemistry' series — the metals to be covered were *rhodium* and *iridium*. Initially, my plan was to discuss both the fundamental organometallic chemistry and applications in organic synthesis. When the first draft of the manuscript was complete, it was apparent that I had exceeded my allowance of pages by a huge amount. It was then that I decided that the catalysis section warranted separate treatment. I am grateful to Reidel for agreeing to publish this volume on *Homogeneous Catalysis with Compounds of Rhodium and Iridium* as part of their 'Catalysis by Metal Complexes' series.

The material I had for the original Academic Press project covered the litera-ture to the end of 1978. I decided to update this to the end of 1982 with a few key references from 1983. It is some measure of the rate of progress in this field that the number of references almost doubled during this revision.

Students often ask me what is special about rhodium and iridium — why are these particular metals so often used to catalyse organic reactions? An insight into the mechanism of the catalytic reactions, and in particular the role played by the metal, provides the answer. It is now well established that the reaction schemes for homogeneous catalysis are all based on the repetition of a limited set of reaction types. These are coordinative-addition, oxidative-addition plus its reverse, i.e. reductive-elimination, and *cis* migration. For a metal complex to be catalytically active, it must be capable of participating in all of these types of reaction and of exercising kinetic control over the total reaction. Rhodium, and to some extent iridium, fulfills these requirements better than most other metals. The high specificity that is often obtained with these metals offsets their relatively high cost.

The book has been written as a research monograph, and is intended to provide information useful to organic and organometallic chemists in industrial and research laboratories. The first chapter provides a general account of catalysis, contrasting homogeneous and heterogeneous systems, considering the problems associated with recovering soluble catalysts and pointing to the new challenges associated with changing feedstocks. The subsequent chapters deal individually with different classes of reaction — C—H bond activation, C—H bond formation, carbonylation, $C-O$ bond formation, functional group removal and $C-C$ bond formation. Within this organization, the discussion is based predominantly on catalyst type.

In the writing of this book, I have been helped by a number of people. Peter Maitlis suggested the initial contact with Reidel. I am most grateful to Vicki who helped with the more recent literature search and with the organizing of reference material, and to Val who typed the final draft of the manuscript. The accuracy of the textual material and the diagrams are entirely my responsibity – if there are errors of fact or interpretation, I apologize. My research students again tolerated my preoccupation with book activities without complaint, and I appreciate their patience and cooperation. Finally, I promise my family that there are no plans to take on another project of this magnitude in the foreseeable future – thanks June, Mark, Paul, David, Gill and Ibi for your continuing love and friendship.

Ron S. Dickson

April, 1985

ABBREVIATIONS

Chemicals, ligands, radicals, etc.

Ac	acetyl, CH_3CO
acac	acetylacetonate anion
acacH	acetylacetone
Alk	alkyl
Ar	aryl
BINAP	2,2'-bis(diphenylphosphino)-1,1'-binaphthyl
bipy	2,2'-dipyridine, or bipyridine
BMPP	benzylmethylphenylphosphine
BPPFA	α[2,1'-bisdiphenylphosphinoferrocenyl] ethyldimethylamine
BPPM	N-(t-butoxycarbonyl)-4-(diphenylphosphino)-2-[(diphenyl-phosphino)methyl] pyrrolidine
Bu	butyl (superscript i, s, or t refers to iso, secondary or tertiary butyl)
CAMP	cyclohexylanisylmethylphosphine
CHAIRPHOS	1,3-bis(diphenylphosphino)butane
CHIRAPHOS	2,3-bis(diphenylphosphino)butane
COD	1,5-cyclooctadiene
Cy	cyclohexyl
DIOP	2,3-O-isopropylidene-2,3-dihydroxy-1,4-bis(diphenylphosphino)-butane
DIOXOP	bis(diphenylphosphinomethyl)dioxolan
DIPAMP	1,2-bis(phenylanisylphosphino)ethane
DMA	dimethylacetamide, $MeCONMe_2$
DMF	N,N'-dimethylformamide, $HCONMe_2$
DMG	dimethylglyoximato anion
$DMGH_2$	dimethylglyoxime
DMSO	dimethyl sulfoxide, Me_2SO
Et	ethyl, CH_3CH_2
HMDB	hexamethyldewarbenzene
HMPT	hexamethylphosphoric triamide, $OP(NMe_2)_3$

L	ligand
Me	methyl, CH_3
NBD	norbornadiene
NORPHOS	$(-)$-(R,R)-2-*exo*-3-*endo*-bis(diphenylphosphino)bicyclo[2.2.1]-heptene
NPTP	naphthylphenyltolylphosphine
OAc	acetate ion
Ⓟ	polymer support
PAMP	phenylanisylmethylphosphine
PDPE	1,2-bis(diphenylphosphino)-1-phenylethane
Ph	phenyl, C_6H_5
phen	1,10-phenanthroline
PNNP	N,N'-bis(diphenylphosphino)-1,4-diphenyl-2,3-diaminobutane
PPFA	α-[2-diphenylphosphinoferrocenyl] ethyldimethylamine
PPPM	4-pivaloyl-4-(diphenylphosphino)-2-[(diphenylphosphino)-methyl] pyrrolidine
Pr	propyl (superscript i refers to iso)
PROPHOS	1,2-bis(diphenylphosphino)propane
py	pyridine
pz	pyrazolyl
R	alkyl group
S	solvent
[*Si*]	silica support
SKEWPHOS	2,4-bis(diphenylphosphino)pentane
THF	tetrahydrofuran
X	halogen

Units, etc.

Å	Angstrom unit, 10^{-10} m
atm	atmospheres, 1 atm = 101 325 Pa
cm^{-1}	wave number
ee	enantiomeric excess
EPR	electron paramagnetic resonance
ESCA	electron spectroscopy for chemical analysis
(g)	gaseous state
Hz	hertz, s^{-1}
IR	infrared
kJ	kilojoule
MPa	megapascal
nm	nanometres, 10^{-9} m

NMR	nuclear magnetic resonance
Pa	pascal, $1\ Pa = 1\ N\,m^{-2}$
psi	pounds per square inch, $1\ psi = 6894.8\ Pa$
μm	micron, $10^{-6}\ m$

CHAPTER 1

SOME GENERAL COMMENTS ON TRANSITION
METAL CATALYSTS

The making and breaking of metal–carbon bonds has become an important and versatile tool in synthetic organic chemistry [see, for example 50, 179, 339, 402, 766, 981, 1134, 1236, 1251, 1367]. Transition metal assisted reactions used for the manufacture of organic compounds on an industrial scale include the oxidation, hydrogenation, hydroformylation, isomerization and polymerization of alkenes, diene cyclooligomerization and alcohol carbonylation. Other reactions, such as the asymmetric hydrogenation of prochiral alkenes, the activation of C—H bonds for H/D exchange, the reduction of ketones by hydrosilation and the decarbonylation of aldehydes are also catalysed by complexes of transition metals. These reactions have wide application in laboratory-scale preparations, and some are also used in the manufacture of pharmaceuticals.

Reactions of the types just mentioned, and indeed, a majority of all organic reactions, are controlled by kinetic rather than thermodynamic factors. The addition of transition metal complexes that can become intimately involved in the reaction sequence is an effective way of increasing the reaction rates. The transition metal catalyst lowers the energy of activation for the reaction by changing the mechanism [981], and in some cases it relaxes restrictions imposed by orbital symmetry control [649, 1114, 1378].

1.1 Catalysts and Catalyst Recovery

During the last 30 years, industrial organic chemistry has been based largely on petroleum products, and most petrochemical processes use heterogeneous rather than homogeneous catalysts. This is principally because heterogeneous catalysts are generally more stable at higher temperatures and are less troublesome to separate from the substrate phase. However, there is a growing interest in homogeneous catalysts because they often show higher selectivity and greater catalytic activity, and they also provide greater control of temperature on the catalyst site. For some commercial processes it has been determined that these advantages of soluble catalysts outweigh the economic problems associated with catalyst recovery. Examples include the hydroformylation of alkenes specifically to straight-chain aldehydes which is catalysed by $HRh(CO)(PPh_3)_2$, and the carbonylation of methanol to acetic acid with $[Rh(CO)_2I_2]^-$ as the active catalyst; these processes are discussed in detail in later chapters. Some of the difficulties associated with separating homogeneous catalysts from the reaction products (and methods of overcoming them) are indicated in Table 1.I.

TABLE 1.I

Methods used for recovery and regeneration of rhodium catalysts
from carbonylation and hydroformylation processes

Process for separation from reaction mixture	Method [reference]
Reverse osmosis	Separation achieved with silicon latex membranes under pressure – catalyst and other species of molecular diameter >10 Å are separated from organic material such as heptanal [608]; similarly with polyalkene membranes [607].
Adsorption	Adsorption on magnesium silicate, then desorption with THF containing Et_3N [509]; adsorption on refractory support, then volatilization and carbonization [547]; adsorption from acidic solution with activated carbon, and isolation by filtration [1196]; adsorption on 'Sirotherm (TM) ' (a thermally regenerable ion-exchange resin used for desalination) [936].
Precipitation	Treatment of the product mixture (alcohols) with anhydrous gaseous NH_3 to precipitate the rhodium complexes [1450].
Cation-exchange treatment	Oxidation with HNO_3/H_2O_2, followed by isolation of rhodium from aqueous phase with cation exchanger – desorption achieved with aqueous HCl [1017]; see also [1072, 1073].
Extraction: two-phase systems	Treatment of reaction residues with paraffins (e.g., hexane and polar organic solvent such as acetonitrile) – rhodium complexes found in polar phase [1194]; separation into aqueous phase by heating with 10% aqueous solution of organic acid such as HCO_2H or $MeCO_2H$ [498]; amino group included in phosphine ligand, e.g., $P(CH_2CH_2CH_2NMe_2)_3$ – catalyst then soluble in weakly acidic media without decomposition [64]; tarry residue treated with aqueous, alkaline NaCN – rhodium then found in aqueous phase [1329]; extraction with CO_2, C_2–C_4 hydrocarbon or HCl(g) under conditions above critical temperatures and pressures [1553]; other examples [204, 453, 472, 506, 750, 811, 1332].
In situ regeneration of catalyst	Soluble carbonyl–rhodium catalyst recovered by heating residue with H_2 at $100-180\,^\circ C$, 200 atm pressure in presence of carrier such as C/Pd, then treating with CO in methanol containing $MeNH_2$, PPh_3 or other L at $120-160\,^\circ C$, $300-700$ atm pressure [1877]; residue refluxed at high pressure, sediment isolated by decantation and solid residue dissolved in fresh catalyst solution under CO at elevated temperature and pressure [500]; treatment with aqueous acid, H_2O_2, X^-,

Table 1.I (cont'd)

Process for separation from reaction mixture	Method [reference]
	PR_3 and then CO results in 90% regeneration of Rh(CO)-X(PPh$_3$)$_2$ which is converted to HRh(CO)(PPh$_3$)$_3$ [1018]; treatment of nonvolatile residue (bottom product) with large excess of PPh$_3$ (100 mol phosphine per g atom of Rh) regenerates HRh(CO)(PPh$_3$)$_3$ [1354]; other examples [404, 416, 1019].

1.2 Supported Catalysts

Recently, much research effort has been directed at bridging the gap between heterogeneous and homogeneous catalysts [1529]. The active catalytic species involved in both types of process may be of a similar nature, and the detailed mechanisms that have been developed for many catalytic reactions in the homogeneous phase may be useful models for the interpretation of heterogeneously catalysed reactions. The link between the two has been cemented by the realization that 'soluble' catalysts can be anchored to insoluble supports such as silica or crosslinked polystyrene [494, 584, 670, 797, 1179, 1550, 1938]. This provides a system which retains the greater activity and selectivity of the homogeneous catalyst, but with the catalyst present in a heterogeneous state which is easily separated from the substrate. Some examples of the use of supported catalysts are provided in later chapters.

A further development in this area involves the generation of supported metal catalysts by the decarbonylation of metal carbonyl clusters. Generally, the support material is treated with a solution of the cluster compound, and decarbonylation is then achieved thermally — or by other means, such as γ-radiation [669]. With KBr as the support, thermogravimetric and infrared results indicate that the decomposiiton temperature of clusters is raised when they are supported [1439]. On oxide supports, however, decomposition can begin at relatively low temperatures.

Some of the earliest investigations in this area involved the formation of metal catalysts by the dispersion and subsequent thermal decarbonylation of $Co_3Rh(CO)_{12}$, $Co_2Rh_2(CO)_{12}$, $Rh_4(CO)_{12}$, $Ir_4(CO)_{12}$ and $Rh_6(CO)_{16}$ on γ-Al$_2$O$_3$ or SiO$_2$ supports [58, 60, 166]. Infrared results showed that the initial adsorption of $Co_2Rh_2(CO)_{12}$ on γ-Al$_2$O$_3$ occurs with the loss of bridging carbonyls, and that the remaining carbonyls are lost progressively at temperatures above 300 K [58]. With $Ir_4(CO)_{12}$ on SiO$_2$ or Al$_2$O$_3$, the loss of carbonyls begins above *ca.* 350 K *in vacuo* and is complete at *ca.* 620 K [58, 769, 1755]. Iridium is the major product of decomposition on silica, but on alumina a fraction of the metal becomes oxidized [1755]. There is evidence from temperature-programed decompositions that oxidation occurs during the latter stages of decarbonylation [775]. This oxidation process seems to be a general phenomenon when metal carbonyl clusters on alumina are decarbonylated at high temperatures.

A supported iridium catalyst of extremely high dispersion (average particle size <1 nm) has been prepared from $Ir_4(CO)_{12}$ and γ-Al_2O_3 by heating at 623 K in a vacuum for 4 h, and then under hydrogen for a further 15 h [59]. Part of the iridium is retained as Ir_4 clusters, but unaggregated cluster units cannot be preserved on the support at temperatures substantially above room temperature. Further study [1755] of this system indicates that the dispersed iridium may be in the form of two-dimensional rafts containing possibly 20 atoms or more.

When $Rh_4(CO)_{12}$ is chemisorbed onto highly divided SiO_2, the integrity of the Rh_4 cluster is maintained only under an atmosphere of CO [1776]. In the absence of CO, the chemisorbed $Rh_4(CO)_{12}$ transforms to $Rh_6(CO)_{16}$ with the release of CO; this indicates that the cluster species are mobile on the surface. Exposure to water converts the rhodium carbonyl clusters to small metal particles of high nuclearity. Treatment with oxygen gives dicarbonyl–rhodium(I) species which can be converted back to $Rh_4(CO)_{12}$ and then $Rh_6(CO)_{16}$ by reaction with CO [166, 585, 1776]. Similar effects are observed when $Rh_6(CO)_{16}$ and $[Rh(CO)_2Cl]_2$ are supported on inorganic oxides [1623]. These results indicate that the moisture content of the support and the method of impregnation can affect the nature of the species obtained.

A battery of techniques (gas adsorption measurements, isotopic measurements, IR and EPR spectroscopy, X-ray photoelectron spectroscopy and transmission electron microscopy) has been used to characterize the behavior of $Rh_6(CO)_{16}$ on alumina [62, 1902]. Decarbonylation of the cluster was achieved with oxygen, and subsequent recarbonylation to $Rh_6(CO)_{16}$ was effected at room temperature − this cycle required the presence of physically adsorbed water on the support. The cycle was not reversible when the cluster was decarbonylated *in vacuo* at >525 K. A highly dispersed catalyst consisting of metal crystallites was obtained by this process − the crystallites did not aggregate further upon heating *in vacuo* above 725 K.

Zeolites have been used to entrap rhodium clusters. The decarbonylation of $Rh_6(CO)_{16}$ supported on zeolite has been achieved *in vacuo* at 373 K or by treatment with oxygen. There is no substantial loss of the Rh_6 cluster structure [585]. It is possible to prepare zeolite-supported rhodium carbonyls in another way. This involves incorporation of $[Rh(NH_3)_5Cl]^{2+}$ at the cationic sites by ion exchange, followed by decomposition of the complex and subsequent carbonylation of the rhodium [1419].

Heteronuclear cluster catalysts can be prepared from solutions of complexes such as $py_2Pt[Ir_6(CO)_{15}]$ and $py_2Pt[Rh(CO)_2(PPh_3)_2]_2$ by impregnation onto Al_2O_3 and subsequent thermal decomposition [1160]. The metal stoichiometry in the supported catalyst is the same as that in the starting metal cluster. Thus this procedure has considerable potential for the formation of catalysts with a well-defined and uniform distribution of two or more metals.

Chemically modified silica supports have been used in some studies. Typically, the supports have surface-attached pendant ligands of the types $-(CH_2)_3NH(CH_2)_2NH_2$, $-(CH_2)_3NHCy$ or $-(CH_2)_3PPh_2$. The accessibility of silica-bonded phosphine ligands to rhodium(I) can be probed by high-resolution ^{31}P NMR spectroscopy [1593]. Spectroscopic data establish that mononuclear species of the type $L_nRh^I(CO)_2$ are formed when these supports are treated with $Rh_6(CO)_{16}$ [966, 967]. Reduction and aggregation occurs when these species are treated with hydrogen at ca. 1 atm pressure and 377 K [1780]. The rhodium crystallites have been characterized by transmission electron microscopy and found to be nearly uniform in size (ca. 12−15 Å and 16−17 Å, respectively on phosphine- and amine-containing supports).

CHEMISORPTION OF CO ON SUPPORTED RHODIUM

The supported metals generated from metal carbonyl clusters must be able to adsorb small molecules if they are to be effective in catalysis. The chemisorption of CO on supported rhodium has been investigated in considerable detail. One of the more recent studies [1937] shows that three species of adsorbed CO can be detected when rhodium on γ-Al_2O_3 is treated with CO. The species are represented as Rh—CO (with ν(CO) near 2000 cm^{-1}), $Rh(CO)_2$ (2100 and 2030 cm^{-1}) and $Rh_2(\mu$-CO) (1900−1850 cm^{-1}), and the site distribution and molecular mobility of the chemisorbed CO have been determined.

Carbonyl species are also formed by treatment of supported $RhCl_3$ with CO at 373 K [1557]. With silica as the support, a dicarbonyl—rhodium(I) species is formed. The same species, together with a monocarbonyl complex and a small amount of a monocarbonyl—rhodium(III) complex, is detected when the support is magnesium oxide. The extent of formation of the rhodium(III) species is increased by treatment of the surface with oxygen.

1.3 Organic Feedstocks

At the present time, alkenes are the basic feedstock in the organic chemicals industry. The alkenes are obtained by the cracking of light petroleum fractions.

Ethylene is the major product, but significant amounts of propene, butadiene and aromatics are also obtained; acetylene and propyne are minor co-products. As the global reserves of natural gas and oil dwindle, this dependence on petroleum feedstocks must decrease. Hence, there is a need to find appropriate alternative feedstocks. Considerable interest is being shown in the use of '*synthesis gas*' (a mixture of CO and H_2) as a major feedstock.

A number of established processes already depend on synthesis gas. For example, it has been used to hydroformylate alkenes to aldehydes and alcohols since the 1940s. Presently, synthesis gas is obtained most economically by the reforming of natural gas or naphtha. In the near future, it is more likely to be produced by the controlled combustion of coal, shale or organic waste in the presence of steam:

$$[C] + \tfrac{1}{2} O_2 \longrightarrow CO$$

$$CO + H_2O \rightleftharpoons CO_2 + H_2$$

The reaction between CO and H_2O is known as the '*water–gas shift reaction*', and it adjusts the ratio of CO and H_2 in the synthesis gas. It therefore plays an integral role in determining the types of organic products that will be obtained from synthesis gas.

The conversion of synthesis gas to organic products is often referred to as '*Fischer–Tropsch*' chemistry. For a considerable time, it has had limited commercial use for the production of fuels. Recently, there has been a resurgence of interest in Fischer–Tropsch and related chemistry because it is potentially attractive for the large-scale production of alkenes, long-chain alkanes and oxygen-containing compounds including alcohols and acids [469, 953, 1133]. Heterogeneous catalysts such as supported Fe, Co or Ni are preferred for large-scale Fischer–Tropsch processes, but there are problems with poor selectivity in the formation of products and with catalyst deactivation. There is certainly a need to develop a better understanding of the mechanisms of these reactions, and homogeneous catalysis will doubtless play a key role here. The use of soluble rhodium and iridium compounds is discussed in Chapter 4.

1.4 Organic Reactions Catalysed by Rhodium and Iridium Compounds

The range of synthetically important organic reactions is immense. One way to classify these reactions is in terms of the type of C–X bond being formed or broken, and this type of organization is used in this book. The discussion is restricted to reactions that are catalysed by soluble compounds of rhodium and iridium, and consideration is given to the range of catalyst sources, the nature of the active catalysts and the scope and mechanisms of the catalytic reactions. Further information about the synthesis and properties of the catalyst precursors can be obtained from two comprehensive accounts of the organo-metallic chemistry of rhodium and iridium that have been published recently [445, 1924].

CHAPTER 2

THE ACTIVATION OF C—H AND C—C BONDS: DEHYDROGENATION, H/D EXCHANGE AND ISOMERIZATION REACTIONS

The catalytic reforming of petroleum is a major industrial process [1604]. It achieves an increase in the volatility of the paraffins by the hydrogenolysis of C—C bonds (Equation (2.1a)). The isomerization of linear paraffins to branched-chain species also occurs (Equation (2.1b)) and these have higher octane ratings than their linear analogs.

$$RCH_2CH_2CH_2R \longrightarrow \begin{cases} \xrightarrow{H_2} RCH_3 + CH_3CH_2R & (2.1a) \\[2ex] \xrightarrow{} RCH_2\overset{\displaystyle CH_3}{\underset{\displaystyle |}{C}}HR & (2.1b) \end{cases}$$

An improvement in octane rating is also achieved by the dehydrogenation of cycloalkanes to aromatics. The catalysts used for these C—C and C—H bond-cleavage reactions are heterogeneous systems such as platinum metals on oxide supports [583]. Generally, they operate under severe conditions and they are not very selective. Similar bond-activation reactions are implicated in the conversion of coal to liquid fuels, and again heterogeneous catalysts and severe conditions are needed [391].

Relatively few reactions of saturated hydrocarbons have been catalysed by soluble metal complexes, but there is a growing interest in this field [1236, 1366, 1368, 1589, 1590, 1905]. Further investigations are warranted because alkanes would provide a much cheaper feedstock than alkenes for the organic chemical industry. More far-reaching investigations have been conducted with sterically strained systems such as cyclopropanes, and with unsaturated systems where the activation of C—H and/or C—C bonds in the substrate is readily achieved by coordination. With suitable catalysts, these systems readily undergo reactions such as dehydrogenation, H/D exchange, double-bond migration and ring-opening.

The catalysts for these reactions are often formed from readily available compounds, including rhodium and iridium species such as $RhCl_3 \cdot 3H_2O$, $RhCl(PPh_3)_3$ and $Ir(CO)Cl(PPh_3)_2$. However, less usual compounds may provide more active catalysts. For example, $[Rh(Ph_2PCH_2CH_2PPh_2)_2]$, which is obtained by electrochemical reduction of $[Rh(Ph_2PCH_2CH_2PPh_2)_2]^+$ in benzonitrile, is particularly effective for the activation of C—H bonds [1634]. When the reduction is performed in the presence of cyclohexane, there is H-abstraction to generate cyclohexyl radicals which couple to form bicyclohexyl and disproportionate to give cyclohexene plus cyclohexane. In a similar manner, H-abstraction from t-butylbenzene yields isobutylbenzene. The factors that seem to enhance the cleavage of C—H bonds in reactions between hydrocarbons and soluble metal complexes have been summarized by Muetterties [1236].

2.1 The Activation of C—H Bonds in Saturated Hydrocarbons

The activation of C—H bonds in alkyl–transition metal complexes is a common phenomenon. Indeed, β-hydride elimination is one of the well-established pathways for the decomposition of transition metal alkyls, and it occurs readily provided there is a vacant site on the metal [339]:

$$M-CH_2CH_2R \longrightarrow M-H + H_2C=CHR$$

This reaction plagued the development of transition metal alkyl chemistry until strategies were developed [401] to inhibit the decomposition pathway.

In contrast, the C—H bond in saturated hydrocarbons is especially resistant to activation by metal complexes. Bond-energy considerations [55] provide one explanation for the paucity of reactions of the type:

$$RCH_3 + M \rightleftharpoons RCH_2-M-H$$

The promotion of this type of reaction by soluble metal complexes was first reported as recently as 1979, and rhodium and iridium complexes figure prominently in this pioneering work.

Crabtree and co-workers [366] have shown that the solvated iridium complexes $[H_2Ir(PR_3)_2S_2]^+$ (S = H_2O or acetone) react with cyclopentane at 80 °C in the presence of 3,3-dimethyl-1-butene (which acts as an H-acceptor) to give $[HIr(\eta-C_5H_5)(PR_3)_2]^+$. Under the same conditions, cyclooctane gives $[Ir(1,5-COD)(PMe_3)_2]^+$, but an analogous benzene compound was not obtained with cyclohexane.

Pentamethylcyclopentadienyl complexes have been used in other investigations. Under photolysis conditions, the iridium compound $H_2Ir(\eta-C_5Me_5)(PMe_3)$ inserts directly into the C—H bond of normal, iso and cyclic alkanes [862]. The reaction with CMe_4, for example, gives $HIr(\eta-C_5Me_5)(CH_2CMe_3)(PMe_3)$. The alkane can be eliminated from these complexes by heating at 110 °C. Similar reactions occur with $H_2Rh(\eta-C_5Me_5)(PMe_3)$ in liquid propane; here, the primary C—H bond is involved in the addition reaction [885]. The rhodium complex also activates arene C—H bonds, but in these reactions there is evidence for precoordination of the arene. The stoichiometric oxidative-addition of alkane C—H bonds also occurs with [Ir(η-C$_5$Me$_5$)(CO)]. When a solution of Ir(η-C$_5$Me$_5$)(CO)$_2$ in neopentane or cyclohexane is irradiated at room temperature, there is a loss of CO and the formation of HIr(η-C$_5$Me$_5$)(CO)R (R = neopentyl or cyclohexyl) [771]. These products are stable in dilute solutions only.

Although the activation of C—H bonds is achieved under mild conditions with these electron-rich complexes, the reactions are stoichiometric. A further understanding of the detailed mechanisms must be developed if catalytic applications based on soluble metal complexes are to emerge. Continued investigations may also establish if there is a link between the behavior of these homogeneous systems and the interactions between saturated hydrocarbons and metal surfaces.

2.2 Dehydrogenation

A few rhodium and iridium complexes have been used as catalysts for the dehydrogenation of saturated hydrocarbons. The conversion of dodecane to dodecene, and of 1-methyl-1-cyclohexene to toluene, has been achieved at 350–385 °C in the presence of Ir(CO)Cl(PPh$_3$)$_2$ [137]. The same catalyst, as well as the rhodium complexes RhCl(PPh$_3$)$_3$ and RhCl$_3$(AsPh$_3$)$_3$, assist the dehydrogenation of cyclic hydrocarbons at 225–265 °C; for example, 10,11-dihydro-5-H-dibenzo[a,d]cycloheptene is converted to 5-H-dibenzo[a,d]cyclo-heptene [185]. As mentioned previously, the formation of cyclohexene together with bicyclohexyl from cyclohexane is promoted by [Rh(Ph$_2$PCH$_2$CH$_2$PPh$_2$)$_2$] [1634]; this reaction occurs at room temperature.

The dehydrogenation of alcohols can also be accomplished with rhodium and iridium catalysts. For example, refluxing 2-propanol is converted slowly to dimethylketone in the presence of [Rh(SnCl$_3$)$_2$(μ-Cl)]$_2$ (see Equations (2.2) and (2.3)) [311]:

$$Me_2CHOH + [Rh^{III}\!-\!Cl] \longrightarrow Me_2CO + [Rh^{III}\!-\!H] + HCl \qquad (2.2)$$

$$[Rh^{III}\!-\!H] + HCl \longrightarrow H_2 + [Rh^{III}\!-\!Cl] \qquad (2.3)$$

The dehydrogenation of 2-propanol has also been achieved photocatalytically in the presence of RhCl(PPh$_3$)$_3$ [76], and with Rh$_2$(OAc)$_4$ immobilized on a Ph$_2$P-modified silica surface [1592]. The use of a supported catalyst seems to eliminate an induction period and leads to a higher initial rate of reaction than observed with the analogous soluble catalyst. In other examples of the dehydrogenation of 2-propanol there is H-transfer to an unsaturated substrate – these types of reactions are discussed further in Chapter 3.

The conversion of benzylic alcohols to carbonyl compounds has been achieved in a phase-transfer system (NaOH/benzene/$PhCH_2N^+Et_3Cl^-$) with $[Rh(CO)_2Cl]_2$ as the catalyst [53]. With the chiral alcohol PhMeCHOH, enantioselection in the dehydrogenation reaction has been achieved with catalysts prepared *in situ* from $[Rh(alkene)_2Cl]_2$ complexes and the chiral phosphines, neomenthyl-diphenylphosphine [1289, 1290, 1291] or DIOP [1291];* there is some pre-ference for dehydrogenation of the (R)-(+)-conformer of the alcohol.

The catalytic dehydrogenation of formic acid to give CO_2 and H_2 has been accomplished with various complexes including $Rh(CO)ClL_2$ (L = PPh_3, $P(OPr^i)_3$) [1950], $H_2IrCl(PPh_3)_3$ [331] and $[M(CO)_2I_4]^-$ (M = Rh or Ir) [530]. A mechanistic study [1668] of the catalytic decomposition of formic acid by $Rh(C_6H_4PPh_2)(PPh_3)_2$ has identified the following steps in the reaction – an initial oxidative-addition of formic acid to produce $Rh(HCO_2)(PPh_3)_3$, then β-hydride elimination to form $HRh(PPh_3)_3$ and CO_2, and finally reaction between $HRh(PPh_3)_3$ and formic acid to give $Rh(HCO_2)(PPh_3)_3$ and H_2. The kinetics of the reaction with $Ir(CO)Cl(PPh_3)_2$ are first order in formic acid and in the iridium complex, and are consistent with this type of mechanism [1949]. In many other reactions with rhodium and iridium halides, there is decarbonylation of the formic acid and the formation of carbonyl–metal complexes [445].

The dehydrogenation of benzylamine to benzonitrile has been achieved with $RhCl(PPh_3)_3$ as the catalyst [1940]. In toluene at 110 °C, the conversion is about 27%.

2.3 H/D Exchange

Iridium complexes are particularly effective for the activation of C—H bonds in benzene and substituted benzenes. Thus exchange between C_6H_6 and D_2, or C_6D_6 and H_2, is achieved at 100 °C in the presence of H_5IrL_2 (L = PEt_2Ph, PEt_3, PMe_3) [114], and alkylbenzenes such as PhMe and PhEt undergo exchange with D_2O at 130 °C with Na_3IrCl_6 as the catalyst [568]. The relative rates for the $H_5Ir(PMe_3)_2$-catalysed exchange of mono- and di-substituted benzenes with D_2 increase in the order p-Me_2 < Me ≃ OMe < H < CF_3 ≃ F < p-F_2 [960]. The mechanism for these reactions presumably involves a series of alternating reductive-elimination and oxidative-addition steps.

The formation of deuterated alkenes (and alkanes) is observed in some reactions of alkenes with D_2. For example, the reaction between ethylene

---

* Catalysts of the type $RhClL_a$ (L = appropriate phosphine) are formed in these systems; DIOP = 2,3-O-isopropylidene-2,3-dihydroxyl-1,4-bis(diphenylphosphino)butane – see Structure 3-45 in Chapter 3.

and D_2 in the presence of $Ir(CO)Cl(PPh_3)_2$ gives d_1- and d_2-ethylenes as well as d_0-, d_1-, d_2-, d_3- and d_4-ethanes [481]. Distribution of deuterium onto all carbon atoms occurs when vinyl- and allyl-deuterated 1-alkenes such as $CH_3CH_2CH_2CD{=}CDH$ are treated with $RhCl_3$ in 2-propanol [664]. The H/D exchange in these reactions probably involves $H-M-D$ complexes and the reversible addition of $M-H(D)$ to the coordinated alkene. The observation [1547] of exchange between D_2 or C_2D_4 and the $M-H$ bond in $HIr(CO)(PPh_3)_3$ or $HRh(Ph_2PCH_2CH_2PPh_2)_2$ supports this suggestion, as does the formation of C_2H_4, C_2H_3D ... C_2D_4 from the reaction between $HRh(CO)(PPh_3)_3$ and *trans*-$CHD{=}CHD$ in benzene [777].

H/D exchange between butane and D_2 is catalysed by silica-bound complexes of the type $[Si]$-$ORh(allyl)H$ [1893]. At 100 °C and 1 atm pressure of D_2, a mixture of deuterated butanes $(d_0{-}d_9)$ is obtained.

Exchange between D_2 and the $O-H$ and $N-H$ bonds in compounds such as $MeCO_2H$, MeOH and morpholine has also been achieved. The catalysts for these reactions include $Ir(CO)Cl(PPh_3)_2$, $H_2IrCl(PPh_3)_3$, $H_3Ir(PPh_3)_3$ and $RhCl(PPh_3)_3$ [480]. With tertiary amines, H/D exchange has been catalysed by metal carbonyl clusters, including $Ir_4(CO)_{12}$ [1939] and $Rh_6(CO)_{16}$ in the presence of D_2O [1038]. The proposed mechanism [1595] involves metal cluster intermediates with $R_2C{=}NR_2^+$, μ_3-R_2NCH_2C and η^2-R_2NCH_2CHR ligands. Unusual stereoselectivities are observed in some of these reactions. Catalytic D for H exchange in tertiary amines has been used in a modeling study of hydro-denitrogenation reactions. Soluble catalysts derived from $Rh_6(CO)_{16}$ display a reactivity pattern very similar to that shown by heterogeneous $Co-Mo$-supported catalysts [1037].

2.4   Unsaturated Alcohol to Aldehyde Conversions

There are several examples of the conversion of unsaturated alcohols to aldehydes (Equation (2.4)) catalysed by rhodium(I) complexes:

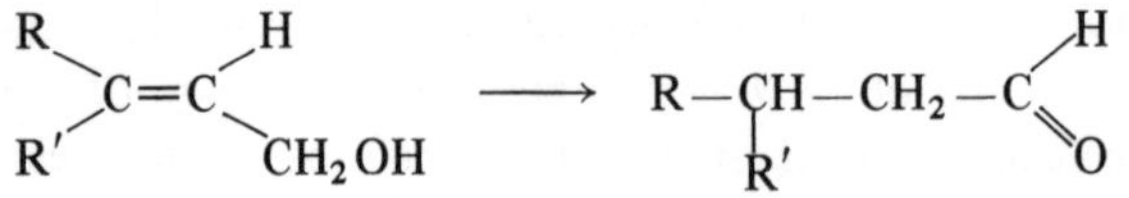

$$R_2C{=}CH\text{-}CH_2OH \longrightarrow R\text{-}CH\text{-}CH_2\text{-}CHO \qquad (2.4)$$

These include the formation of isobutyraldehyde from methylallyl alcohol in the presence of $HRh(CO)(PPh_3)_3$ and CF_3CH_2OH [1719], of 2,3-dihydro-2-furfuraldehyde (2-1) from furfuryl alcohol with $[Rh(COD)Cl]_2$ as the catalyst [1476], and of $HOCH_2CH_2CH_2CHO$ and BuCHO from the unsaturated alcohols $HOCH_2CH{=}CHCH_2OH$ and $H_2C{=}CHCH_2CH_2OH$, respectively, with carborane–rhodium catalysts of the type 3,1,2-$[(PPh_3)_2RhHC_2B_9H_{11}]$ [1959].

Another catalyst obtained by treating the iridium complex $[Ir(COD)(PMePh_2)_2]^+$ with H_2 in tetrahydrofuran (presumably $[H_2Ir(PMePh_2)_2(THF)_2]^+$) promotes the conversion of $H_2C=CHCH_2OH$ to CH_3CH_2CHO [125].

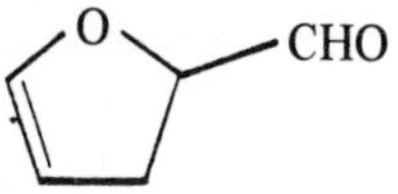

(2-1)

Some other conversions use $[Rh(CO)_2Cl]_2$ as the catalyst source. The transformation of Structure 2-2 to 2-3 occurs at 120 °C, and it has been established that this reaction proceeds via a *homo*-1,5-H shift [157]. Much milder conditions (25–30 °C) can be used [51] with a two-phase system (8 M NaOH/CH_2Cl_2) to convert 5-hexen-1-ol and 4-phenyl-3-buten-2-ol to the appropriate carbonyl compounds in high yields (100% and 94%, respectively). With $[Rh(CO)_2Cl]_2$ on a polybenzimidazole fiber support, allyl alcohol is transformed to Me_2CHCHO (84%) and $EtCH(OH)CHO$ (16%); the catalyst is completely ineffective without the support [778].

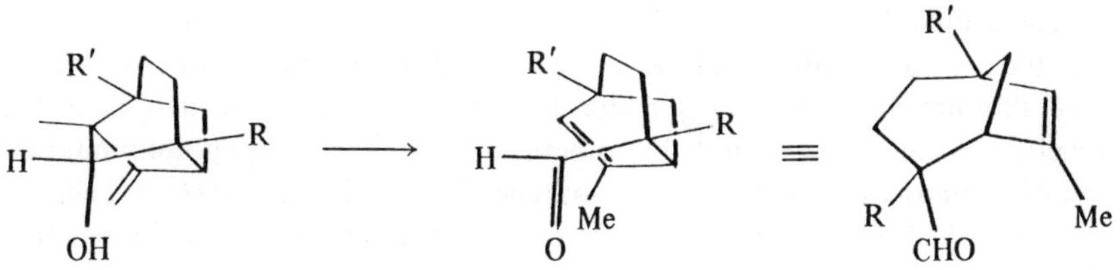

(2-2)   (2-3)

The formation of β-keto esters from α-diazo-β-hydroxy esters is a similar type of reaction:

$$RCH(OH)C(=N_2)CO_2Et \xrightarrow{-N_2} [RCH(OH)\ddot{C}CO_2Et]$$

$$\longrightarrow RCOCH_2CO_2Et$$

Nearly quantitative conversions can be obtained with $Rh_2(OAc)_4$ as the catalyst [1379].

2.5 Double-bond Migration

A variety of rhodium and iridium compounds will catalyse the isomerization of alkenes. Some of these catalysts (e.g., $RhCl(PPh_3)_3$) are also effective for the hydrogenation of alkenes, and this can lead to complications in the interpretation of hydrogenation results. In this section, isomerization reactions that occur in the absence of hydrogen will be emphasized; there will, however, be some consideration of isomerizations that occur under hydrogenation conditions.

CATALYST SOURCES

RhCl(PPh₃)₃

The rhodium(I) complex $RhCl(PPh_3)_3$ is an effective isomerization catalyst. It promotes double-bond migration in a range of alkenes, including the following: 1-butene [1807], *cis-* and *trans*-2-butene [1808], phenylbutenes [191]; *cis*-2-pentene (under hydrogenation conditions — the rate of isomerization is one third that of hydrogenation) [199]; 1- and 2-heptene (isomerization is dependent on the presence of O_2 and on the solvent used) [93]; 1-nonene (gives 2- and 3-nonene oxide in the presence of O_2) [1811]; allylbenzene, *p*-allylanisole [191], and allyl ethers such as $ROCH_2CH{=}CH_2$ [350]; 1-methoxy-1,4-cyclohexadiene and related dienes [172].

If the reactions with simple alkenes are allowed to proceed to completion, an equilibrium mixture of all possible isomers is formed. For example, in the isomerization of 1-butene in dichloromethane at 25 °C, the equilibrium mixture consists of *trans*-2-butene (69%), *cis*-2-butene (25%) and 1-butene (6%). There is, however, a high *cis-* to *trans*-2-butene ratio in the early stages of the reaction [1807, 1808].

In some cases, the effectiveness of the catalyst is improved by the addition of a metal halide. Thus, {RhCl(PPh₃)₃ + SnCl₂} catalyses the conversions of 1,4,9-decatriene to the 1,6,8-conformer, $H_2C{=}CHCH_2CO_2H$ to $MeCH{=}CHCO_2H$ [1608] and $MeO_2CCH_2C({=}CH_2)CO_2Me$ to (E)- and (Z)-$MeO_2CC(Me){=}CHCO_2Me$ [1511]. The last of these isomerization reactions occurs at 100 °C and shows second-order kinetics in the ester and rhodium complex. The isomerizations of 4-vinylcyclohexene and of 1,5-cyclooctadiene are catalysed by {RhCl(PPh₃)₃ + EtAlCl₂, AgBF₄, AlCl₃ or ZnCl₂} [784], and double-bond migration within diene fatty acids and esters, including linoleic acid and methyl linoleate, is catalysed by {RhCl(PPh₃)₃ + HCl, SnCl₂, FeCl₃, CrCl₃ or LiCl} [1606, 1607].

Catalysts have been formed by supporting $RhCl(PPh_3)_3$ on silica gel with attached groups of the types $-OSi(OSiMe_3)(CH_2PPh_2)OSiMe_2(CH_2)_nPPh_2$ or $-OSiMe_2(CH_2)_nPPh_2$ (n = 1, 3, 7, 9). The activity of these catalysts for the isomerization of 1-pentene increases with increasing values of n [1854].

M(CO)XL$_2$

The carbonyl complexes M(CO)Cl(PPh$_3$)$_2$ (M = Rh or Ir) are effective iso-merization catalysts only if they are activated in certain ways. They have been used as the catalyst source with terminal alkenes such as 1-butene, 1-pentene, 1-heptene and 1-nonene. The isomerization of these alkenes can be initiated by ultraviolet irradiation [1670], or by the addition of oxygen [170] or hydrogen [481, 1681]. Under hydrogenation conditions, the complexes Ir(CO)ClL$_2$ (L = PCy$_3$ or P(OPh)$_3$) are less active than Ir(CO)Cl(PPh$_3$)$_2$ [1681], but a marked increase in activity is achieved by replacing PPh$_3$ with bidentate ligands of the type Ph$_2$PCH$_2$CH$_2$NMe$_2$ or Ph$_2$PC$_6$H$_4$CH$_2$NMe$_2$ [1494]. The observa-tion that the isomerization of alkenes in the presence of Ir(CO)XL$_2$ complexes is very slow under N$_2$ (if it occurs at all) but fast under H$_2$, implies that the active catalyst is an hydrido—metal species. Other substrates that isomerize in the presence of Ir(CO)Cl(PPh$_3$)$_2$ include vinylcycloalkenes [1092] (the double bond is retained in the *exo* position), phenylbutenes, allylbenzene, *p*-allylanisole [191], and acyclic enones and dienones [26, 933, 1388] (e.g., **2-4** → **2-5** at 250 °C).

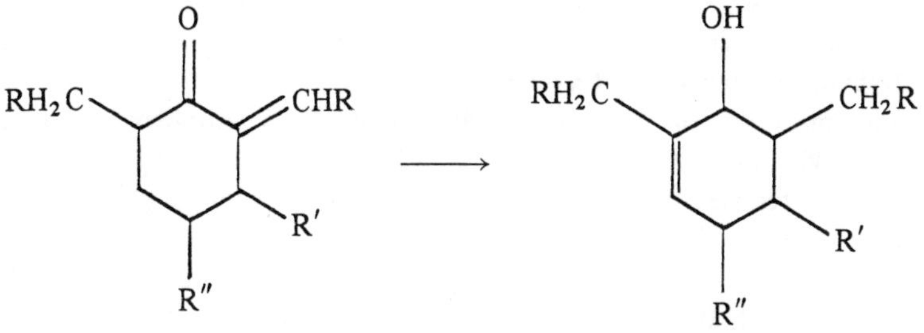

(2-4)   (2-5)

Other compounds

Some double-bond migration reactions have been achieved with the hydrido compounds HM(CO)L$_3$ as the catalyst source. The isomerization of 1-pentene to *cis*- and *trans*-2-pentene occurs with {HIr(CO)(PPh$_3$)$_3$ + O$_2$} [170] and with {HRh(CO)(PPh$_3$)$_3$ + H$_2$} [1930]. In the latter case, isomerization and hydrogenation proceed at about the same rate at 27 °C and 1 atm H$_2$ pres-sure. The conversion of 1-heptene to *cis*- and *trans*-2-heptene is catalysed by HRh(CO)(PPh$_3$)$_3$ [1670, 1709], and the rate is enhanced by ultraviolet irradia-tion. HRh(CO)(PPh$_3$)$_3$ is also effective [191] for the isomerization of phenyl-butenes, allylbenzene and *p*-allylanisole at the boiling temperature of the alkene, and HM(CO)(PPh$_3$)$_3$ (M = Rh or Ir) achieves the conversion of RCHClCH=CH$_2$

to $RCH=CHCH_2Cl$ by a [1,3]-sigmatropic shift of the chloro group [1674].
Another hydrido complex, $HRh(PPh_3)_4$, has been used to catalyse the preparation of substituted aliphatic N-(1-propenyl)amides from N-allylamides by
double-bond migration [1661].

Several other isomerization catalysts have been derived from alkene
or diene complexes of rhodium(I) and iridium(I). The ethylene complexes
$[Rh(C_2H_4)_2X]_2$ (X = Cl, $SnCl_3$ or acac*) are effective under mild conditions
in the presence of acids such as HCl [370, 377] and acetic acid [379]; presumably the active catalysts are formed by oxidative-addition of the acid to
give hydrido–rhodium(III) species. With $[Rh(C_2H_4)_2Cl]_2$, added ethylene has
an abnormal effect on the rate constant for the isomerization of butenes [1326].
The activity of the ethylene–rhodium catalysts is increased about 100 times
when they are supported on silica [1742]. Catalysts have also been formed from
a range of diene complexes including $[Rh(COD)Cl]_2$ [1516], Rh(diene)acac
[80], $[Rh(diene)(PF_3)Cl]_x$ [1268], Rh(η-C_5H_5)(diene) [80], Rh(COD)L
(L = 5-ethylcyclohexenyl) [1726], $[M(COD)(MeCN)_2]^+$ (M = Rh, Ir) [616],
$[Ir(COD)(PMePh_2)_2]^+$ [124, 1330] and $\{[Ir(COD)Cl]_2 + NaBPh_4\}$ supported
on phosphinated polystyrene [285]. With $[Ir(COD)(PMePh_2)_2][PF_6]$ as the
catalyst, the isomerization can be carried out under neutral and mild conditions;
this is a good catalyst for the isomerization of O-allyl protective groups in
carbohydrates such as 2-6 (R = CH_2Ph) [1330].

Incorporation of the chiral phosphine 2,2′-bis(diphenylphosphino)-1,1′-
binaphthyl (binap; 2-7) into a cyclooctadiene–rhodium complex gives a catalyst
which is effective for asymmetric isomerization reactions [1767]. Thus, with
$[Rh(COD)(binap)][ClO_4]$, N,N-diethylnerylamine (2-8) or N,N-diethylgeramyl-
amine (2-9) is converted to optically active N,N-diethylcitronellal-(E)-enamine
(2-10, R,R′ = H, Me or Me, H; >95% ee**, >95% chemoselectivity).

(2-6)   (2-7)

* Hacac = acetylacetone. Systematic name; 2,4-pentanedione.
** See Chapter 3, p. 87 for definition.

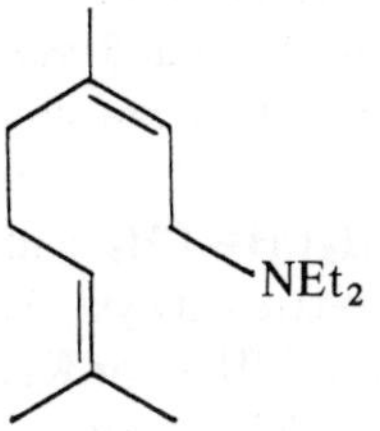

(2-8) 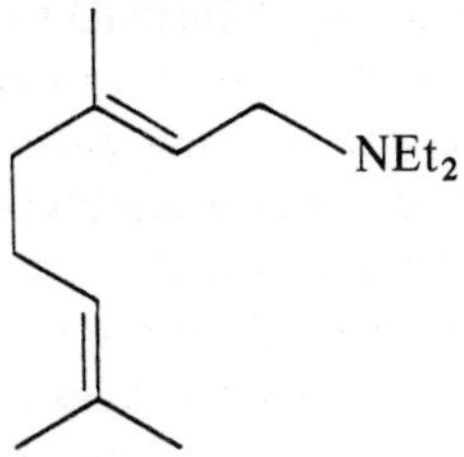   (2-9)

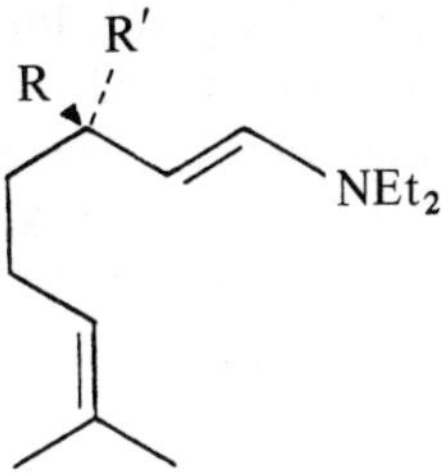

(2-10)

In other investigations, the catalysts have been obtained from $Rh(CO)_2(L-L)$ (e.g., HL_2 = bis(salicylidene)diamine or bis(salicylidene)ethylenediamine) [1653], $[Rh(CO)_3(PPh_3)_2][BPh_4]$ supported on phosphinated polystyrene [285], $[Rh(NO)(MeCN)_4]^+$ [344], and carborane complexes such as 3,1,2-$[(PPh_3)_2RhHC_2B_9H_{11}]$ and 2,1,7-$[(PPh_3)_2RhHC_2B_9H_{11}]$ [1957]. Allyl aromatics and allylamines have been isomerized under phase-transfer conditions with $[Rh(CO)_2Cl]_2$ as the catalyst [52]. The organic phase can be C_6H_6 or CH_2Cl_2, and aqueous NaOH plus a quaternary ammonium salt is used as the phase-transfer catalyst.

In some cases, catalysts have been derived from complexes of rhodium(III) and iridium(III). For example, $RhCl_3 \cdot 3H_2O$ is a convenient catalyst to use for the isomerization of 1-hexene and various heptenes [663, 1968] and pentenes [18, 880]. In a typical reaction, an alcohol such as C_2H_5OH is used as the co-catalyst, and isomerization is achieved at reflux temperatures under N_2. The $RhCl_3 \cdot 3H_2O$ system has also been used for various enone transpositions [620, 621], in the synthesis of some difficultly accessible unsaturated steroids [68] and for the conversion of N-allyl-amides and -imides to aliphatic enamides [1661]. The catalyst can be supported on 'Silocel' or a porous solid support, but it is more active in solution [1488]. It has been suggested that the $RhCl_3$-catalysed isomerization of cyclodecadiene proceeds via the $[Rh(diene)Cl]_2$ complex [1532], but the corresponding isomerization of 1,5-cyclooctadiene to the 1,3-isomer is inhibited by [Rh(COD)Cl]$_2$ [1258]. Treatment of RhCl$_3$ or IrCl$_3$ with CO yields catalysts for the isomerization of *cis*- and *trans*-2-butene and 2-pentene, but the rates are not particularly fast; they are slower, for instance, than those with Pd or Pt on carbon [497].

Water-insoluble allylic compounds such as PhCH$_2$CH=CH$_2$ and PhOCH$_2$-CH=CH$_2$ have been isomerized with a two-liquid phase catalyst. The catalyst system consists of aqueous metal halides (e.g., RhCl$_3 \cdot 3$H$_2$O or K$_3$IrCl$_6$) plus quaternary ammonium or phosphonium salts of lipophilic anions. It has been used to convert PhCH$_2$CH=CH$_2$ to PhCH=CHCH$_3$ (70% after 3 h; *trans/cis* ratio = 5.5) [1525].

Various derivatives of RhCl$_3$, including RhCl$_3$(PPh$_3$)$_3$ [776], RhCl$_3$(NCPh)$_3$ [1968], HRhCl$_2$(PPh$_3$)$_3$ [776], H$_2$RhCl(PPh$_3$)$_2$ [160] and {RhCl$_3$py$_3$ + NaBH$_4$/DMF} [3] are also effective isomerization catalysts. The iridium compounds H$_2$IrCl$_6$ and HIrCl$_2$(PEt$_2$Ph)$_3$ assist the conversion of 1,5-cyclooctadiene to the 1,3-isomer [1258], and 1-pentene isomerizes to *cis*- and *trans*-2-pentenes in the presence of HIrCl$_2$(PEt$_2$Ph)$_3$ with added oxygen [170].

RELATIVE ACTIVITIES AND RATES

Several comparative studies have revealed trends in the behavior of these systems. Isomerizations with RhCl(PPh$_3$)$_3$ are much faster than with RhCl$_3 \cdot 3$H$_2$O [350]. In the isomerization of 1-heptene, the following orders of catalyst activity have been established [1683]: Ir(CO)ClL$_2$, L = PPh$_3$ > PCy$_3$ > P(OPh)$_3$; HRh(CO)(PPh$_3$)$_3$ ≫ HIr(CO)(PPh$_3$)$_3$ > Ir(CO)Cl(PPh$_3$)$_2$ ≫ Rh(CO)Cl(PPh$_3$)$_2$. With catalysts of the type [Me$_4$N]$_n$[MX(SnX$_3$)$_5$]$^{n-}$, the catalytic activity in the selective isomerization of PhCH$_2$CH=CH$_2$ increases in the order: M = Os < Ru < Ir < Rh < Pt; X = Cl < Br [207].

The following order of relative rates has been established for the isomerization of heptenes with Ir(CO)XL$_2$ complexes as the catalyst [1681]: 1-heptene ≃ *cis*-2-heptene > *trans*-3-heptene > *trans*-2-heptene. In general, terminal alkenes isomerize more rapidly than internal alkenes, and this is attributed to weaker coordination of the internal alkenes [374, 375]. Another study of the isomerization of 1-heptene with RhCl$_3$ or RhCl$_3$(NCPh)$_3$ as the catalyst in acetic acid has shown that the yield decreases in the order *cis*-2- > *trans*-2- ≫ *trans*-3- > *cis*-3-heptene [1968].

MECHANISMS

Cramer [374, 375] has suggested the mechanism shown in Scheme 2.1 for the isomerization of alkenes catalysed by an HML$_a$ complex. The observation of extensive H/D exchange in systems of the type {HIr(CO)(PPh$_3$)$_3$ + C$_2$D$_4$}

[1547] and {HRh(CO)(PPh$_3$)$_3$ + *trans*-CDH═CHD} [777] supports the general concepts embodied in this mechanism. The mechanism also probably operates when the catalyst source is RhCl(PPh$_3$)$_3$ [1807, 1808] or [RhCl(PPh$_3$)$_2$]$_2$ [394], and the isomerization is achieved under hydrogenation conditions.

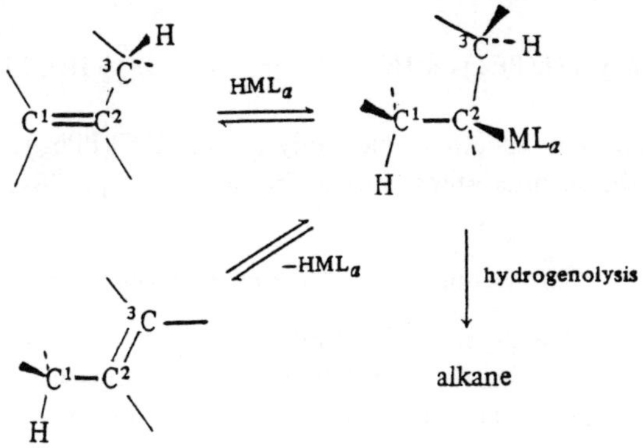

Scheme 2.1

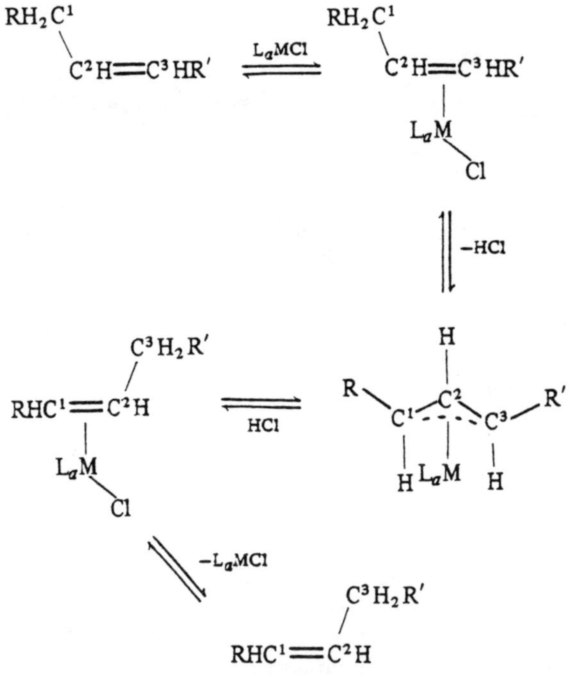

Scheme 2.2

Isomerization reactions promoted by metal catalysts with no hydrido–metal bonds probably follow the 1,3-hydrogen shift mechanism shown in Scheme 2.2. This involves π-allyl–metal complexes as intermediates; the ability of HCl to add to a π-allyl–metal complex has been demonstrated [1267] for the system:

$$(\eta\text{-allyl})Rh(PF_3)_3 + HCl \longrightarrow \text{alkene} + \tfrac{1}{2}[Rh(PF_3)_2Cl]_2 + PF_3$$

A further kinetic and spectroscopic study of the $RhCl(PPh_3)_3$–butene system has confirmed the essential steps outlined in the scheme [1808].

2.6 Geometric *cis* $\rightleftharpoons$ *trans* Isomerization

In some instances, it is possible to get *cis* $\rightleftharpoons$ *trans* stereomutation of an alkene. This occurs, for example, with *cis*-2-pentene in the presence of $RhCl(PPh_3)_3$ as the catalyst [199]. The reaction occurs under hydrogenation conditions (30 °C, 1 atm H_2 pressure, benzene/ethanol) and is accompanied by double-bond migration (Scheme 2.3) and hydrogenation.

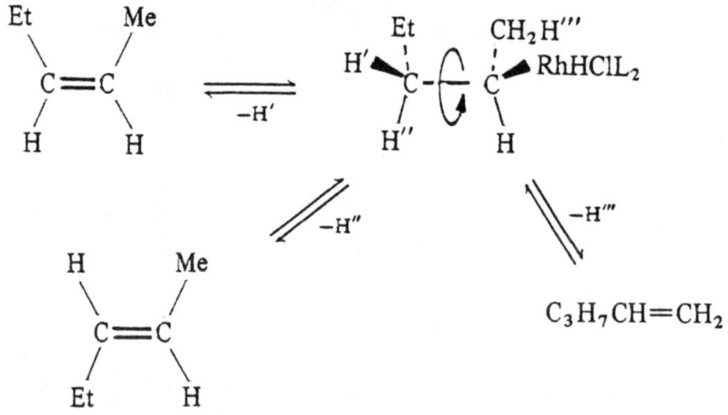

Scheme 2.3

The stereomutation *cis*-2-heptene $\rightleftharpoons$ *trans*-2-heptene with $HRh(CO)(PPh_3)_3$ as the catalyst has also been observed [1709]. Scheme 2.4 shows a possible pathway for this type of isomerization reaction. It is based on intermediates identified in the reactions of the unsaturated diester $Me(CO_2Me)C=CH(CO_2Me)$ with rhodium compounds [666].

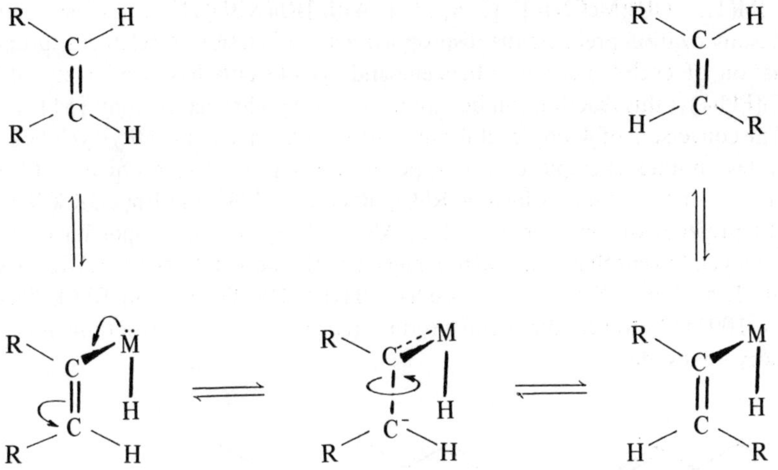

Scheme 2.4

The geometric isomerization of epoxy alcohols has also been accomplished with $HRh(CO)(PPh_3)_3$ and related compounds. For example, the conversion of *cis*-1,2-epoxy-3-hydroxyhexane to the *trans* conformer is achieved with $HRh(CO)(PPh_3)_3$ at 110 °C (24% conversion after 4 h) [1095].

2.7 Intermolecular Disproportionation Reactions

ALKENES

Some double-bond migration reactions occur intermolecularly. This type of reaction is illustrated by the conversion of 1,4-cyclohexadiene into benzene and cyclohexene as shown in Equation (2.5).

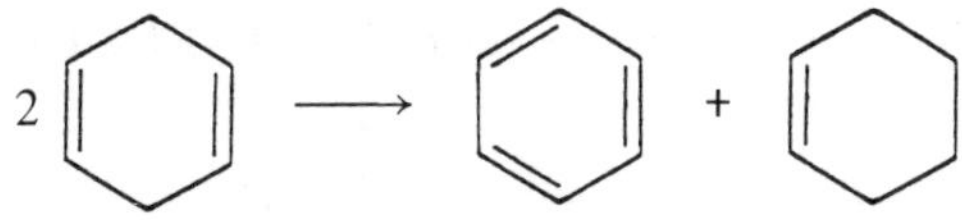

$$2 \bigcirc \longrightarrow \bigcirc + \bigcirc \qquad (2.5)$$

trans-$Ir(CO)Cl(PPh_3)_2$ is an effective catalyst for the reaction (60% disproportionation after 180 h at 80 °C), but $HIr(CO)(PPh_3)_3$ only induces isomerization to 1,3-cyclohexadiene [1090, 1091]. Although $Ir(CO)Cl(PPh_3)_2$ will not catalyse the disproportionation of 1,3-cyclohexadiene into benzene and cyclohexene, this transformation has been achieved with other catalyst systems including $Rh(\eta\text{-}C_5Me_5)(1,3\text{-cyclohexadiene})$ [1229, 1230], $[Rh(NBD)_2]^+$ [617] and $[M(1,5\text{-COD})(MeCN)_2]^+$ [616, 617]. With $[Rh(NBD)_2]^+$, it has been shown that isomerization precedes the disproportionation reaction. A related disproportionation of cyclohexene into benzene and cyclohexane has been catalysed by $RhCl(PPh_3)_3$; this reaction can be monitored by gas chromatography [1442].

The conversion of 4-vinylcyclohexene into ethylbenzene and ethylcyclohexene provides another example of this type of rearrangement (see Equation (2.6)). Catalysts for this reaction include $RhCl_3$ in ethanol [89] and $Rh(CO)Cl(PPh_3)_2$ in the presence of Lewis acids such as $AlCl_3$ [783]. The disproportionation of 4-vinylcyclohexene has also been achieved with the catalysts $Ir(CO)XL_2$ (X = Cl, I; L = PPh_3: X = Cl; L = $AsPh_3$), '$H_2Ir(CO)(PPh_3)_2$' and $RhCl(PPh_3)_3$ [784, 1091]; however, the major product (ca. 80%) of these reactions is ethylidenecyclohexene.

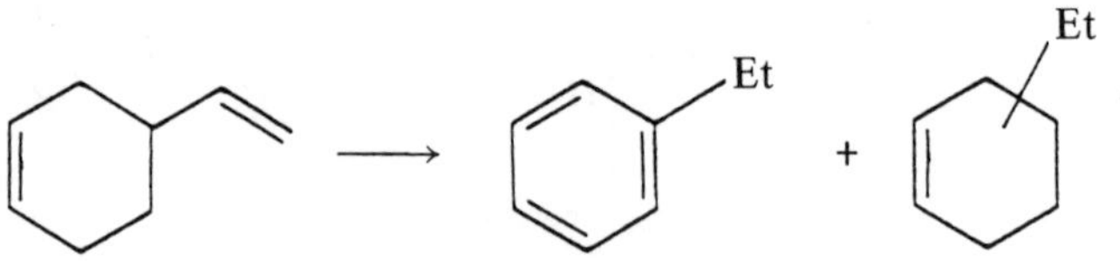

$$ (2.6) $$

Another example of alkene disproportionation is provided by the transformation of an acyclic alkene into an alkane and an alkadiene. The formation of hexane and trans,trans-2,4-hexadiene from 2-hexene, and of pentane and isoprene from $MeCH\!=\!CMe_2$, illustrate this type of reaction. The hydrido complexes H_5IrL_2 (L = PPh_3, PPr^i_3, PMe_3) and mer-$H_3Ir(PPh_3)_3$ are suitable catalysts for these disproportionations [330, 1131]. Related reactions with smaller alkenes led to the proposed reaction pathway shown in Scheme 2.5; isolation of the complexes $Ir(C_2H_4)_2(PPh_3)\{PPh_2(C_6H_4)\}$, $Ir(C_2H_4)_2(PPr^i_3)\text{-}$ $\{PPr^i_2(i\text{-}C_3H_6)\}$, $Ir(\pi\text{-}C_3H_5)L_2$ (L = PPh_3, PPr^i_3) and $HIr(C_4H_6)(PPr^i_3)_2$ provide examples of the types of intermediates involved in the reaction scheme [330].

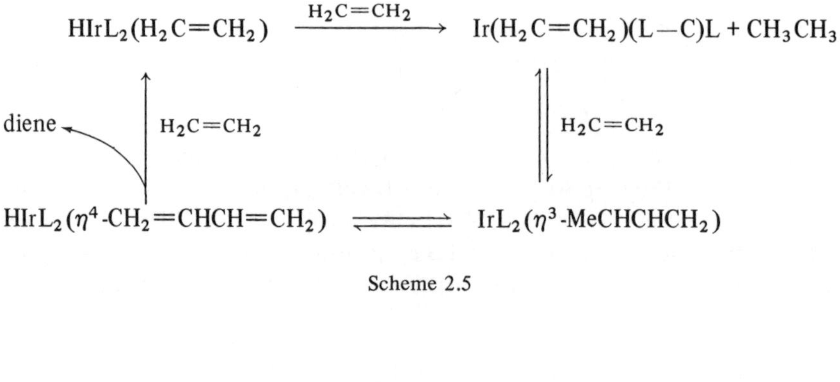

Scheme 2.5

ALDEHYDES, ETHERS AND ESTERS

The redox disproportionation of an aldehyde into an alcohol plus a carboxylic acid occurs in neutral or alkaline aqueous medium with catalysts of the type $M_2(\eta\text{-}C_5Me_5)_2Cl_4$ (M = Rh, Ir) or $Rh_2(\eta\text{-}C_5Me_5)_2(OH)_3Cl \cdot 4H_2O$ [347, 348] :

$$2\ RCHO \xrightarrow{\ H_2O\ } RCH_2OH + RCO_2H$$

A kinetic study of the conversion of acetaldehyde into ethanol and acetic acid has established that the reaction is first order in the aldehyde and one-half order in catalyst. In the proposed mechanism [348], hydride transfer occurs in a stepwise manner *via* a metal-hydride species (– cf. the Cannizzaro and Tischenko reactions). For aldehydes which are poorly soluble in water, an aqueous/organic solvent can be used.

The conversion of an allyl ether into an aldehyde and an alkene (Equation (2.7)) is also promoted by rhodium compounds.

$$(RCH{=}CHCH_2)_2O \longrightarrow RCH{=}CHCHO + RCH{=}CHMe \qquad (2.7)$$

Suitable catalysts for the reaction include $RhCl(PPh_3)_3$, $Rh(CO)Cl(PPh_3)_2$ and $\{RhCl_3 \cdot 3H_2O + PPh_3$ in ethanol$\}$ [1903].

In refluxing tetrahydrofuran containing $RhCl(PPh_3)_3$, 2-propenylchloro-acetate ($ClCH_2CO_2CH_2CH{=}CH_2$) is converted into $ClCH_2CO_2H$ and $ClCH_2\text{-}CO_2CH_2C({=}CH_2)CH_2CH{=}CH_2$ [1203]. The latter product is obtained by the formal elimination of allene and its subsequent insertion into the CO bond of the ester function of the starting chloroacetate.

2.8 Valence Isomerization; Ring-opening Reactions

Although the ring-opening of cyclopropanes and of bicyclo[n.1.0]-alkanes and -alkenes is formally an orbital symmetry-forbidden reaction, it is often feasible through thermal processes because of high ring strain [571]. These isomerizations are also promoted in the presence of certain metal complexes, and rhodium(I) compounds are among the most effective catalysts [178, 220, 1114].

$[Rh(CO)_2Cl]_2$ is an effective catalyst for the opening of substituted cyclopropanes [327, 1417, 1868], and there has been particular interest in the rearrangements of vinylcyclopropanes [1524]. An example is shown in Equation (2.8).

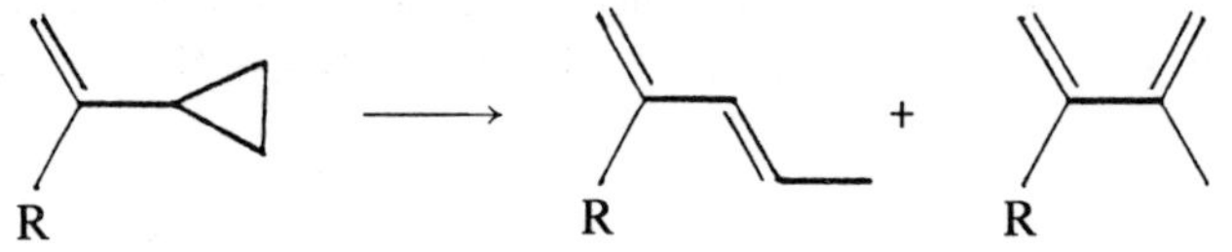

$$\tag{2.8}$$

In a typical reaction, vinylcyclopropane is heated in chloroform at 88 °C in the presence of $[Rh(CO)_2Cl]_2$ to give *trans*-1,3-pentadiene as the major product. A study of the conversion of stereospecifically deuterium-labelled vinylcyclopropanes to 1,3- and 1,4-dienes has shown that ring-cleavage is accompanied by *trans* to *cis* isomerization [1517]. The mechanism for ring-cleavage in these reactions involves the initial coordination of the vinyl group with rhodium(I), the subsequent formation of a metallocyclic allyl–Rh^{III}–alkyl intermediate and stereospecific *cis*-β-hydride elimination [1515]. The rearrangements of oxocyclopropanes to vinyl ethers (Equation (2.9)) are catalysed by several transition metal compounds, including $[Rh(CO)_2Cl]_2$ [461]. The absence of either the alkoxy or carbalkoxy group renders the cyclopropane derivatives inert to ring-cleavage.

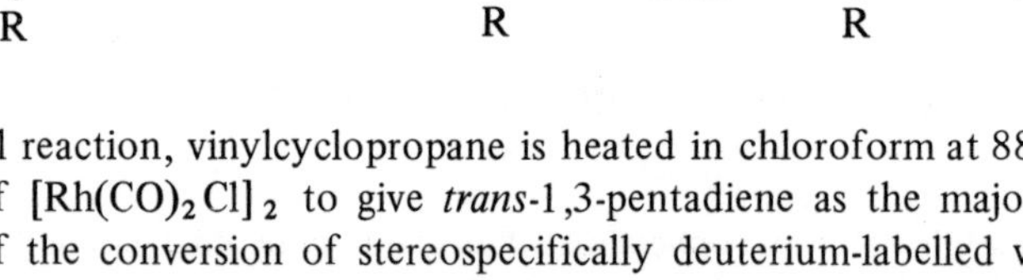

$$\tag{2.9}$$

Bicyclic compounds have been studied more extensively, and there has been particular interest in the behavior of bicyclo[1.1.0]butane, which is the simplest bicyclic system. Although bicyclo[1.1.0]butane has a high strain energy of almost 270 kJ mol^{-1} [1911], it does not rearrange at temperatures below 150 °C. When thermal rearrangement does occur, there is cleavage of two of the non-bridged C–C bonds to form 1,3-butadienes [582]. The same rearrangement is catalysed by $[Rh(CO)_2Cl]_2$ in chloroform at room temperature, and the results of various studies are summarized in Table 2.I; in some cases, additional products are identified. The formation of 1,3-butadienes occurs by stepwise cleavage of one side and then the center bond, and the reaction has been interpreted as a transition metal retrocarbene addition [582]. The reaction pathway is depicted in Scheme 2.6.

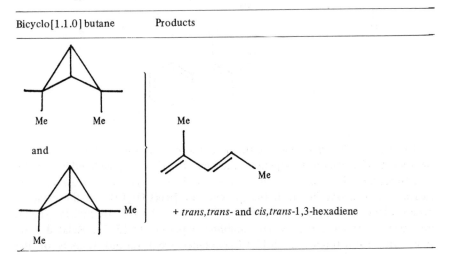

Scheme 2-6

The behavior of these systems is affected by variation of the metal, the attached ligands [577, 582] and the solvent [579].

TABLE 2.I

Products formed from the [Rh(CO)$_2$Cl]$_2$-catalysed rearrangement of substituted bicyclo[1.1.0]butanes [576, 578, 581, 582, 1512]

Bicyclo[1.1.0]butane	Products
(structures)	Me / Me structure + *trans,trans*- and *cis,trans*-1,3-hexadiene

TABLE 2.I (cont'd.)

Bicyclo[1.1.0]butane	Products

The size of the ring which is fused to the cyclopropyl system can have a significant influence on ring-opening reactions. The C—C bridge bond in bicyclo-[2.1.0]pentane has an exceedingly high strain energy, but it rearranges by cleavage of just one bond; in the presence of $[Rh(CO)_2Cl]_2$ as the catalyst, cyclopentene is formed quantitatively (Equation (2.10)). An hydrido—metal intermediate is indicated by D-scrambling experiments [575]. Related reactions with substituted bicyclo[2.1.0]pentanes yield various isomers of the substituted cyclopentenes [1912]. Alternative catalysts for these reactions include RhCl(PPh$_3$)$_3$ and Ir(CO)Cl(PPh$_3$)$_2$.

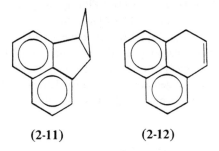

$$\hspace{10cm} (2.10)$$

$$\hspace{10cm} (2.11)$$

The rearrangement of 6-substituted bicyclo[3.1.0]hex-2-enes to 1- or 6-substituted cyclohexadienes (Equation (2.11)) is achieved readily in the presence of RhCl(PPh$_3$)$_3$ or Rh(CO)Cl(PPh$_3$)$_2$ [115, 130, 131]. The rates of these reactions are accelerated dramatically by the presence of O$_2$, and it seems likely that [RhCl(PPh$_3$)$_2$(O$_2$)]$_2$ is the active catalyst precursor. The nature and orientation of the substituents also have a significant effect on the rearrangement of bicyclohexenes. The rates are increased by the introduction of methyl or carboxymethyl substituents at C-6, with the effect being maximized when the methyl group is *exo* and/or the carboxymethyl group is *endo* [130]. The degree of unsaturation of the fused ring does not appear to affect this rearrangement reaction. Thus, cycloprop[*a*]acenaphthalene (**2-11**) transforms to phenalene (**2-12**) in the presence of [Rh(CO)$_2$Cl]$_2$ [1363].

(2-11)   **(2-12)**

This reaction is believed to involve oxidative-addition of Rh^I to the central C—C bond from above the flap, followed by shifting of the C-7 (*exo*)-H to yield an η^3-allyl—Rh^{III} complex, and subsequent rearrangement of this intermediate around the periphery of the phenalene ring. It is interesting that the

Cope rearrangement of *endo*-6-vinylbicyclo[3.1.0]hex-2-ene (**2-13**) to bicyclo[3.2.1]octa-2,6-diene is inhibited by coordination to rhodium(I). Thus treatment of **2-13** with $Rh(C_2H_4)_2(acac)$ forms {*endo*-6-vinylbicyclo[3.1.0]hex-2-ene}Rh(acac) which yields mainly the bicyclo[3.3.0]octa-2,6-diene complex upon thermolysis [78]. This reaction has been followed by NMR spectroscopy and the intermediates **2-14** to **2-16** in Scheme 2.7 have been identified [35]. The proportion of **2-15** to **2-16** is influenced by the solvent; the ratio changes from 88:12 in C_6D_6 to 33:67 in CD_2Cl_2.

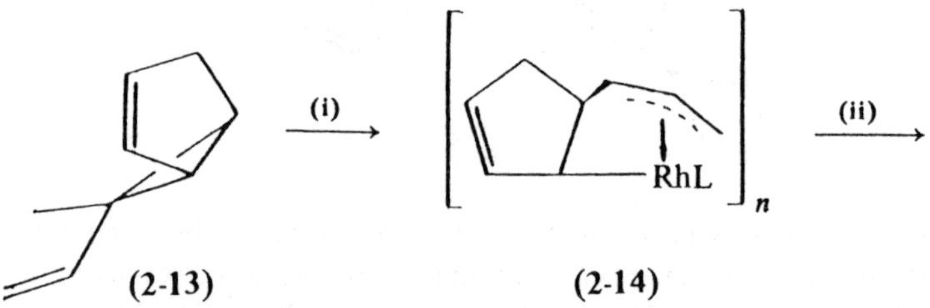

(i) $Rh(C_2H_4)_2L$ where L = acac
(ii) thermolysis

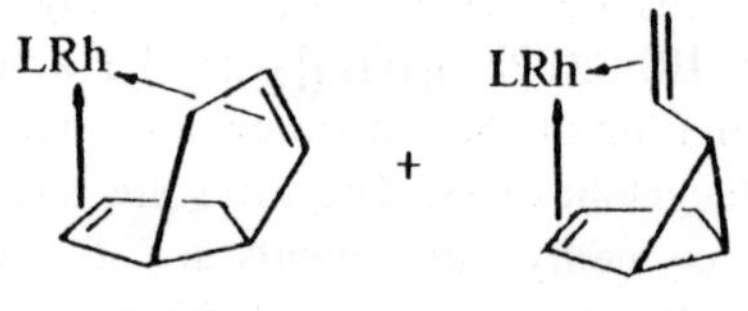

Scheme 2.7

In the presence of $RhCl(PPh_3)_3$ or $Rh(CO)Cl(PPh_3)_2$, bicyclo[4.1.0]hept-2-ene (**2-17**) is less reactive than bicyclo[3.1.0]hex-2-ene, but it forms an analogous product [130, 131]. Bicyclo[6.1.0]non-2-ene (**2-18**) is unreactive with these catalysts, but it does rearrange in the presence of $[Rh(CO)_2Cl]_2$ to give *cis*-8,9-dihydroindene [623] or *cis,cis*-1,4-cyclononadiene [1517]. The $[Rh(CO)_2Cl]_2$-catalysed rearrangement of bicyclo[6.1.0]nona-2,4,6-triene (**2-19**) at 35 °C gives the cyclopentene product **2-20** exclusively [623].

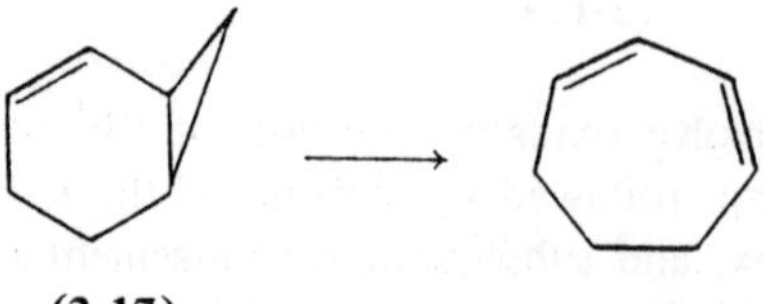

(2-17)

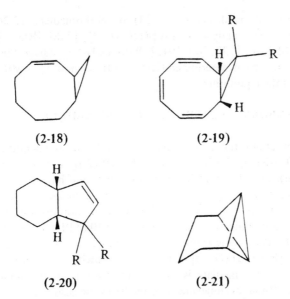

(2-18)   (2-19)

(2-20)   (2-21)

Cyclopropane ring-cleavage has been achieved with a number of tricyclic substrates. The rearrangement of tricyclo[4.4.0.0^{2,7}]heptane (2-21) and various methyl-substituted derivatives to the appropriate methylencyclohexene (2-22) is catalysed by [Rh(CO)$_2$Cl]$_2$, [Ir(CO)$_3$Cl]$_x$, Rh(CO)Cl(PPh$_3$)$_2$ [573, 574, 576] or [Rh(NBD)Cl]$_2$ [1913]; greater than 90% conversions are achieved. Different products are obtained with other catalysts; with Ag$^+$ for example, 1,3-cyclo-heptatriene is obtained quantitatively. The ethers 2-23 and 2-24 are sometimes formed from 2-21 when [Rh(CO)$_2$Cl]$_2$ is used in methanol, and the product composition is determined by the rate of addition of the tricycloheptane [580] and not by the ageing of the catalyst [397, 398].

(2-22)   (2-23)   (2-24)

(2-25)   ⟶   (2-26)

The conversion of quadricyclane (2-25) to norbornadiene (2-26) occurs readily in the presence of catalysts such as $[Rh(CO)_2Cl]_2$ [738], $[Rh(C_2H_4)_2Cl]_2$ [740] or $[Rh(NBD)Cl]_2$ [738, 740, 1913]. With the latter catalyst, the enthalpy of isomerization has been determined ($\Delta H_r \simeq 110$ kJ mol^{-1}). NMR measurements [1769] on the equilibrium:

$$[Rh(NBD)Cl]_2 + 2\,NBD \quad \rightleftharpoons \quad 2\,Rh(NBD)_2Cl$$

provide evidence about the active catalyst at different temperatures in this system. At 40 °C, the equilibrium is far to the left ($K_{40\,°C} \simeq 3.5 \times 10^{-6}$ M^{-1} by extrapolation), and this explains why norbornadiene has no effect on the catalytic isomerization of quadricyclane. A marked increase in the rate of isomerization is observed at 1 °C in the presence of norbornadiene, and this is attributed to the presence of greater amounts of $Rh(NBD)_2Cl$ which is apparently a more efficient isomerization catalyst. Kinetic studies [315] with the related catalyst $[Rh(NBD)(OAc)]_2$ indicate a stepwise mechanism in which a rhodocyclobutane intermediate is formed by insertion of rhodium(I) into a cyclopropyl ring. Rhodocyclohexane intermediates are also formed, and these lead to two *endo*-bis(norbornadienes).

Interest in these systems has been maintained because of their possible use for the storage of energy derived from sunlight. The principles are illustrated by the following example. Upon exposure to sunlight, the norbornadiene derivative 2-27 is readily and quantitatively converted to the corresponding quadricyclane (2-28). The back-isomerization also proceeds quantitatively in the presence of $[Rh(CO)_2Cl]_2$ as a catalyst [1128]. Thus, a cyclic system can be established.

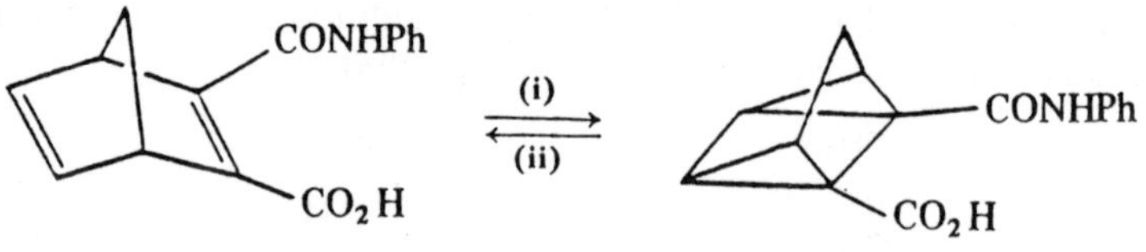

(2-27)	(2-28)

(i) sunlight
(ii) $[Rh(CO)_2Cl]_2$

In the $[Rh(CO)_2Cl]_2$-catalysed reactions of the 3-oxaquadricyclanes (2-29; R = Me, R$'$ = H, Me or RR$'$ = $(CH_2)_8$), the metal complex reacts preferentially at the C—O—C fragment to give 2-30 [739]; different products are formed when 2-29 is heated or is treated with other catalysts. Three products (2-32 to 2-34) can be obtained from the valence isomerization of the tricyclo[4.1.0.0^{2,7}]hept-

4-en-3-one (**2-31**). Formation of **2-32** predominates (9:1:1) when the catalyst is $[Rh(CO)_2Cl]_2$ [1732], but initiation of the isomerization reaction with ultraviolet irradiation or $AgClO_4$ gives mainly **2-34**.

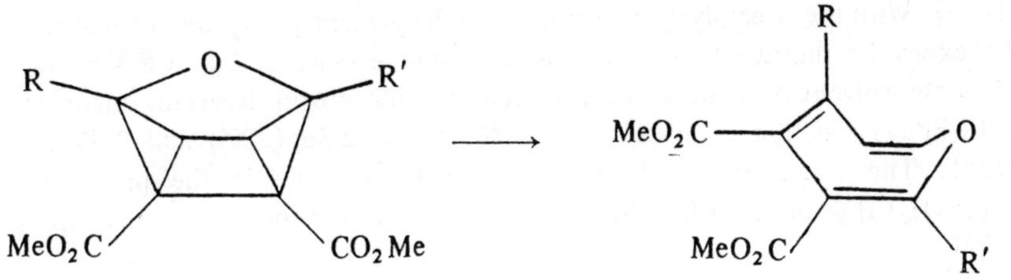

(2-29)   (2-30)

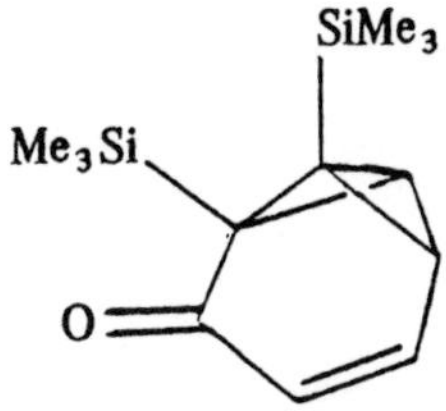

(2-31)

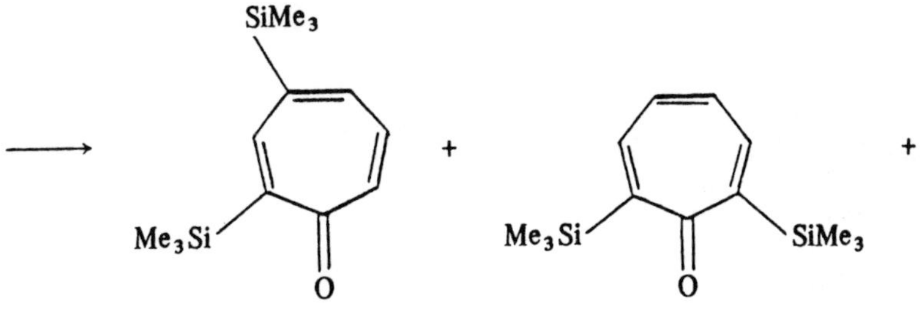

(2-32)   (2-33)

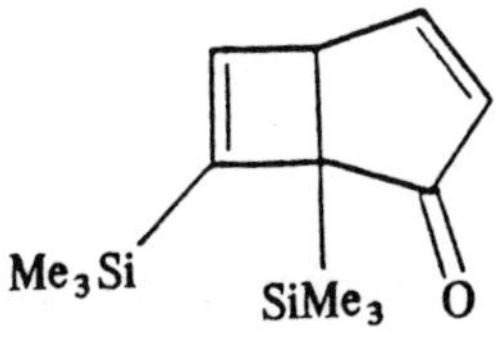

(2-34)

Various products have been obtained from the rearrangement of tricyclo-[3.2.1.0^{2,4}]oct-6-ene (2-35). With [Rh(CO)$_2$Cl]$_2$ as the catalyst, tetracyclo-[3.3.0.0^{2,8}.0^{4,6}]octane (2-36) is the major product [1871], but 5-methylene-bicyclo[2.2.1]hept-2-ene (2-37) is obtained in the presence of Ir(CO)Cl(PPh$_3$)$_2$ [1872]. With the Ir catalyst oxygen needs to be present [273], and formation of the exocyclic methylene product occurs more efficiently but less selectively when the solvent is changed from benzene to chloroform. Rearrangement with RhCl(PPh$_3$)$_3$ as the catalyst yields 2-38 (62%), 2-36 (32%) and 2-39 (6%) [934]. The ring-opening of tricyclo[3.2.1.0^{1,5}]octane in the presence of [Rh(CO)$_2$Cl]$_2$ or [Ir(CO)$_3$Cl]$_x$ produces 4- and 5-methylenecycloheptene [572].

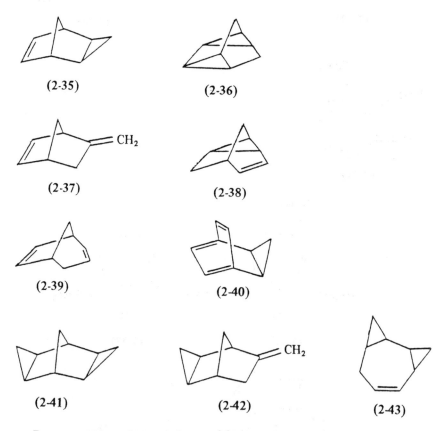

(2-35)   (2-36)

(2-37)   (2-38)

(2-39)   (2-40)

(2-41)   (2-42)   (2-43)

Rearrangements of tricyclo[3.2.2.0^{2,4}]nonatriene (2-40) have been achieved with [Rh(CO)$_2$Cl]$_2$ and RhCl(PPh$_3$)$_3$ as catalysts [935]; bicyclo[4.2.1]non-atriene is formed in each case. When the catalyst is RhCl(PPh$_3$)$_3$, the additional products are barbaralane (51%) and bicyclo[3.2.2]nonatriene (9%). The conversion of 2-41 to 2-42 is achieved in the presence of Ir(CO)Cl(PPh$_3$)$_2$ [1872], and *syn*-1,3-bishomocycloheptatriene (2-43) is converted to bicyclo[6.1]nona-3,6-diene and cyclonona-1,3,6-triene on heating in benzene with [Rh(CO)$_2$Cl]$_2$ [1362]. In a similar manner, 1,3-dehydroadamantane (2-44) isomerizes to 3-methylenebicyclo[3.3.1]non-6-ene (2-45) [1636]. The ring-opening of thiatetra-cyclo-undecatriene (2-46) to 1-benzothiepin (2-47) is catalysed by [Rh(CO)$_2$Cl]$_2$ or Rh(CO)$_2$(acac) [1241, 1242].

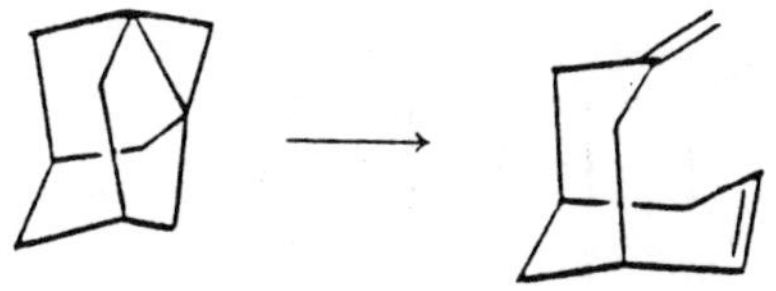

(2-44)   (2-45)

(2-46)   (2-47)

Ring-opening reactions involving four-membered rings are of interest also, and the conversion of hexamethylbicyclo[2.2.0]hexa-2,5-diene (HMDB; 2-48) to hexamethylbenzene provides an example of this type of reaction. This conversion is catalysed by [Rh(HMDB)Cl]$_2$ at 65—70 °C [1869, 1870]. The effectiveness of other catalysts decreases in the order [Rh(CO)$_2$Cl]$_2$ > Rh$_4$(CO)$_{12}$ ≃ RhCl$_3$/EtOH > RhCl(PPh$_3$)$_3$ [88]. With [Rh(diene)Cl]$_2$ (diene = NBD or COD) as the catalyst, the unsubstituted bicyclo[2.2.0]hexane is converted to cyclohexene [1635]; the steps in this conversion are recognized as oxidative-addition by rhodium(I) across the strained 1,4-C—C bond (rate-determining), H-migration from a β-C to rhodium to give a metallocyclic intermediate and reductive-elimination.

The products obtained from the rearrangement of *syn*-tricyclooctane (2-49) are 2-50 (70%), 1,5-cyclooctadiene (28%) and 2-51 (2%). [Rh(NBD)Cl]$_2$ is an effective catalyst for this reaction [1635], but [Rh(CO)$_2$Cl]$_2$ is not because it forms a stable acyl—rhodium complex [1384]. *anti*-tricyclooctane does not rearrange in the presence of [Rh(NBD)Cl]$_2$, but it also forms an acyl complex with [Rh(CO)$_2$Cl]$_2$.

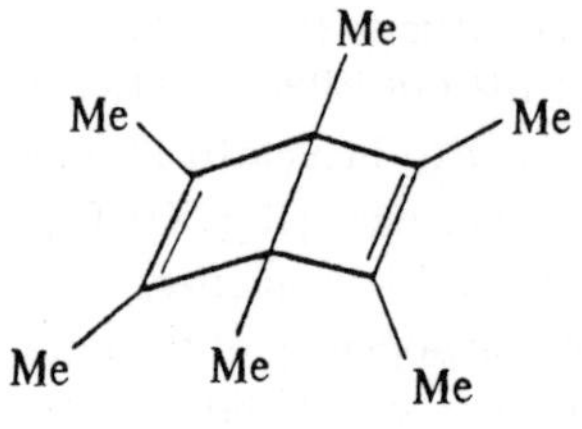

(2-48)

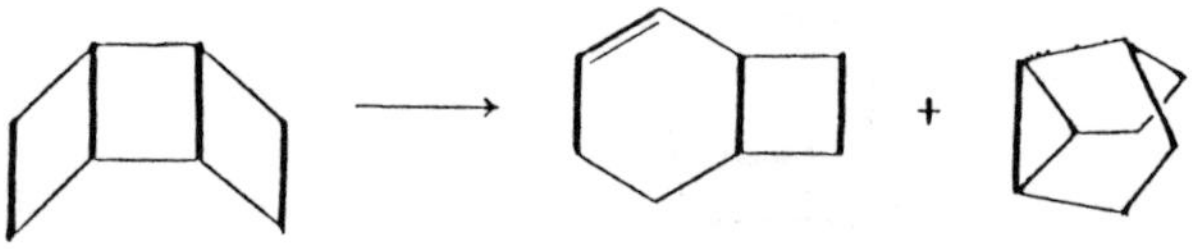

(2-49)   (2-50)   (2-51)

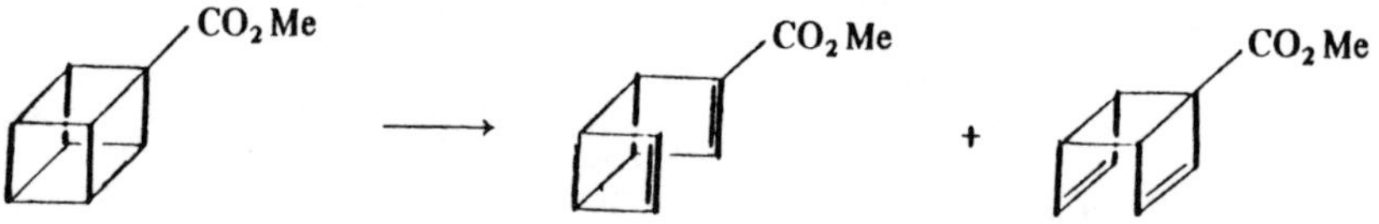

(2-52)   (2-53)   (2-54)

There has been considerable interest in the valence isomerization of cubane and homocubanes. The conversion of **2-52** to **2-53** (71%) and **2-54** (29%) is catalysed by diene–rhodium complexes such as $[Rh(NBD)Cl]_2$, $[Rh(COD)Cl]_2$ and $Rh(COD)Cl(PPh_3)$ [292]. It has been suggested that this reaction proceeds through a nonconcerted oxidative-addition, and this is supported by the isolation of the complex **2-55** from the reaction of cubane with $[Rh(CO)_2Cl]_2$.

The rearrangement of 4-substituted homocubanes (**2-56**) gives **2-57** and **2-58**, with the product distribution depending on the catalyst used [1361]. Both products are obtained with $[Ir(CO)_3Cl]_x$, but only the diene **2-57** is formed in the presence of $[Rh(NBD)Cl]_2$. With $[Rh(NBD)Cl]_2$ as the catalyst, 1,8-bishomocubane rearranges to **2-59** and **2-60**; **2-61** is obtained with some other catalysts but not in the presence of $[Rh(NBD)Cl]_2$ [1360]. Isolation of the metallocyclic intermediate **2-62** from the reaction between 1,3-bishomocubane and $[Ir(CO)_3Cl]_x$ [195] is again consistent with a nonconerted oxidative-addition mechanism for these rearrangements.

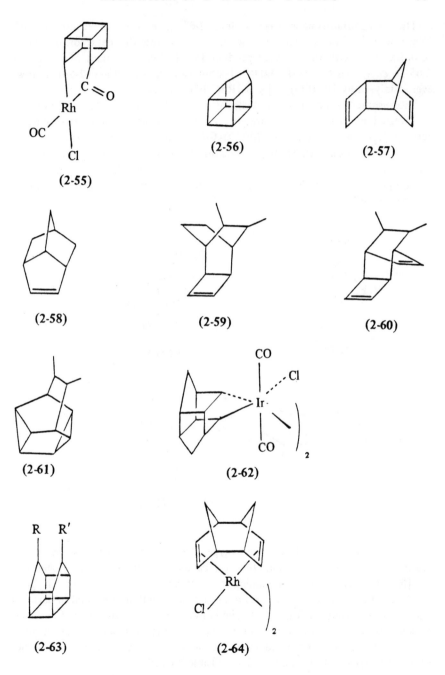

(2-55)

(2-56)

(2-57)

(2-58)

(2-59)

(2-60)

(2-61)

(2-62)

(2-63)

(2-64)

The secopentaprismane cage system **2-63** is cleaved in reactions with [Rh(NBD)Cl]$_2$, but the reaction is noncatalytic because the metal forms a stable complex **2-64** with the cleavage product [479]. Rearrangement of the dione **2-65** to *cis,syn,cis*-tricyclo[5.3.0.0^{2,6}]-4,8-decadiene-3,10-dione (**2-66**) is, however, catalysed by [Rh(CO)$_2$Cl]$_2$ or [Rh(NBD)Cl]$_2$ [478].

Photoisomerization of the tricyclic ketone **2-67** gives **2-68**. This reaction can be reversed by heating in the presence of a rhodium(I) catalyst; the catalytic activity decreases in the order [Rh(CO)$_2$Cl]$_2$ > [Rh(NBD)Cl]$_2$ > RhCl(PPh$_3$)$_3$ [884]. As mentioned previously, reversion of the photoisomerization is potentially useful for energy storage. This system is less effective than other substrates; the rough order of activity is quadricyclanes $\simeq$ prismanes $\simeq$ cubanes > homocubanes > **2-67**.

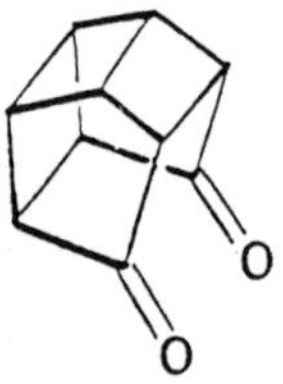

(**2-65**) 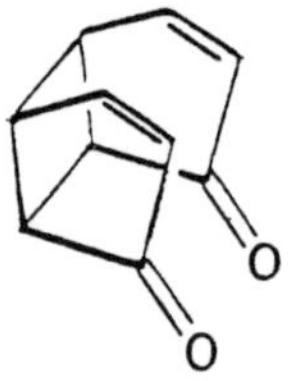   (**2-66**)

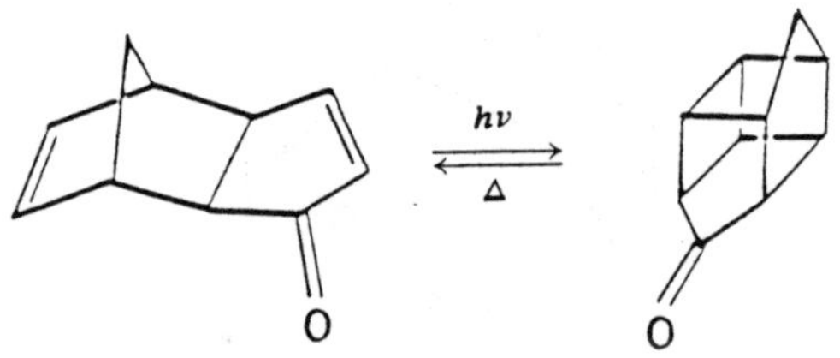

(**2-67**)   (**2-68**)

The cleavage of C—O bonds in oxygen heterocycles provides some related rearrangement reactions. Again, the best catalyst for these reactions seems to be [Rh(CO)$_2$Cl]$_2$. The ring-opening of epoxides and oxetans generally gives aldehydes and/or ketones, and examples of these reactions are listed in parts (a) and (b) respectively of Table 2.II. The conversion of aryl-substituted epoxides to ketones is also catalysed by RhCl(PPh$_3$)$_3$ [1184]. The solvent has an influence on the products obtained from the ring-opening of 7-oxanorbornadienes (Table 2.II(b)) and 3-oxaquadricyclanes (Table 2.II(c)).

TABLE 2.II
[Rh(CO)$_2$Cl]$_2$-catalysed ring-opening reactions involving C—O bond rupture

Reactant	Products	Reference
(a) 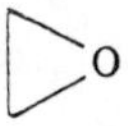		
vinyl epoxides, 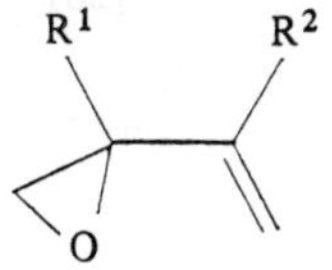	α,β-unsaturated aldehydes 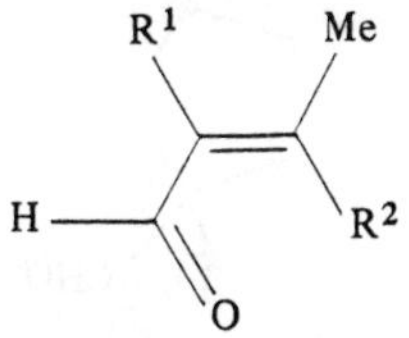	[20]
butadiene epoxide	2-butenal	[1082]
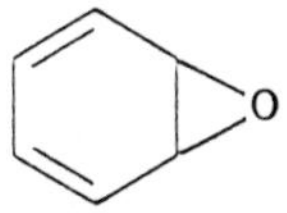	phenol + benzene (1:1)	[84]′
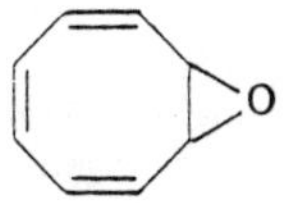	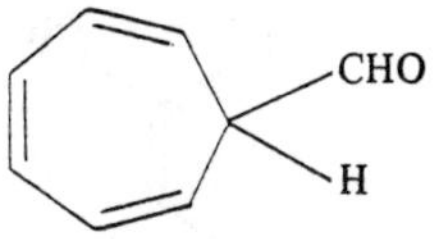	[623]
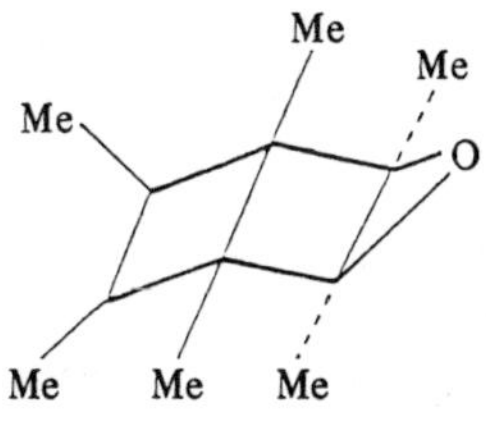	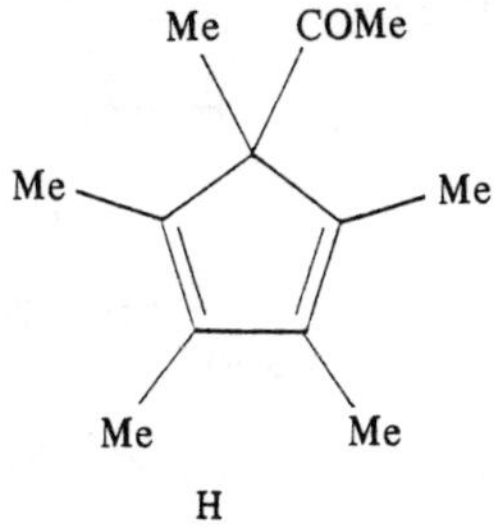	[630]
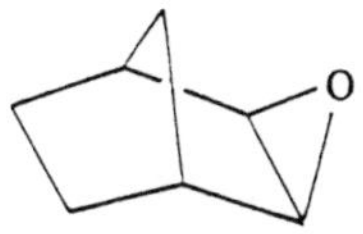	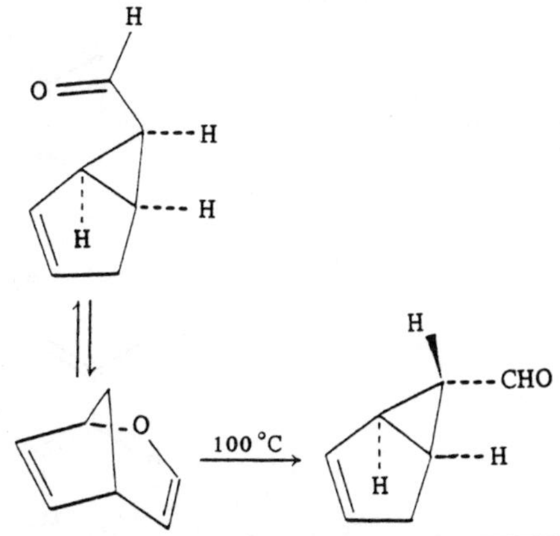	[630]

TABLE 2.II (cont'd.)

Reactant	Products	Reference
(b) 		
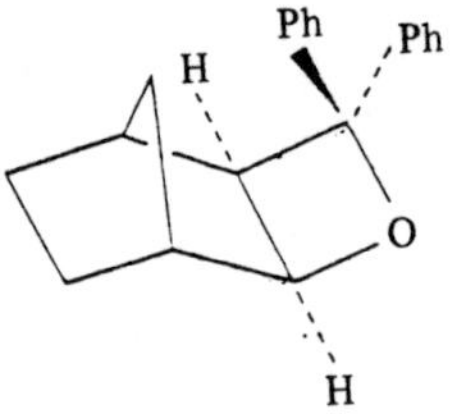	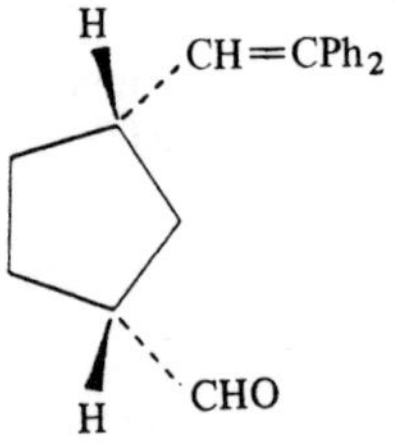	[20]
Me$_2$- and Me$_3$-oxetans	but-2-ene (highly stereoselective ring-cleavage)	[282]
(c) 		
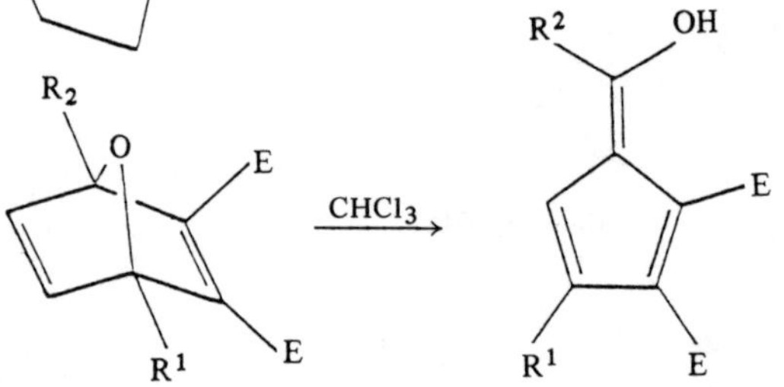		[250, 736]
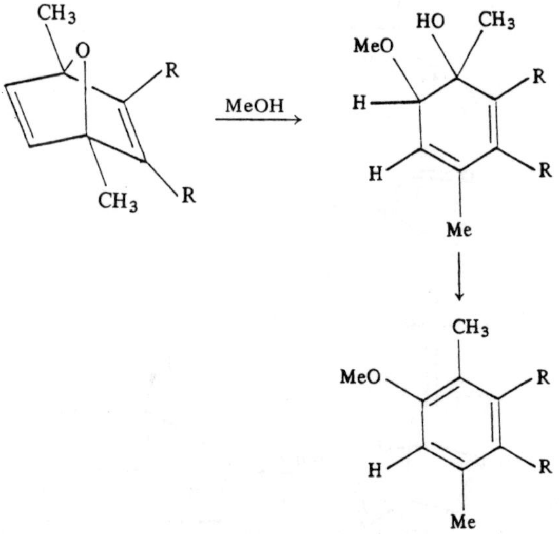		[737]

TABLE 2.II (cont'd.)

Reactant	Products	Reference

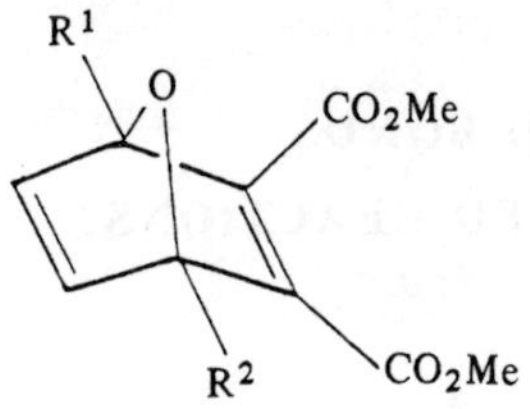

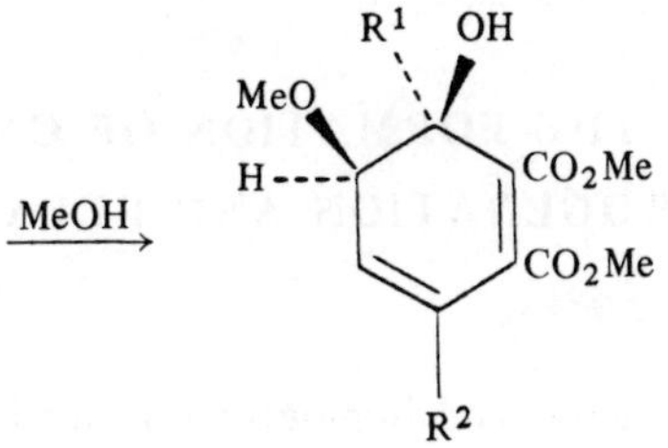

[83]

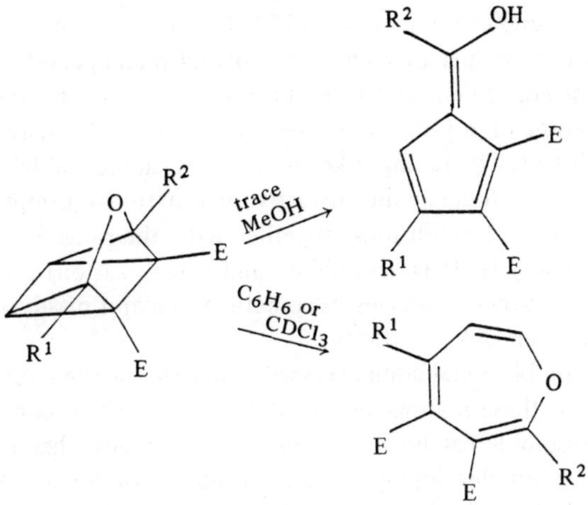

[250, 1493]

+ other products

In the $[Rh(CO)_2Cl]_2$-catalysed rearrangement of (norborn-2-eno)[d]-(3,6-dihydro-1,2-dioxine) (2-69) there is cleavage of the O—O bond to give the aldehyde 2-70 [648]. This is a minor product in the thermal rearrangement.

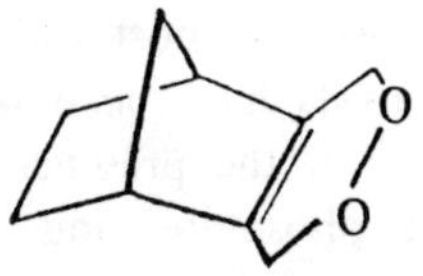

(2-69)

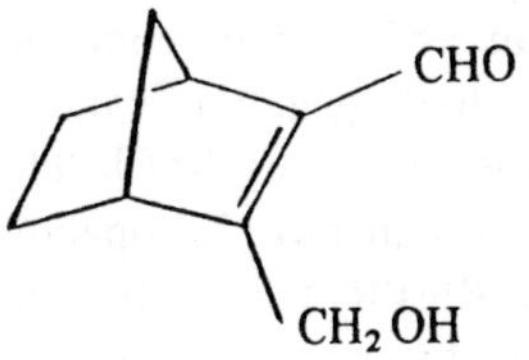

(2-70)

CHAPTER 3

THE FORMATION OF C—H BONDS:
HYDROGENATION AND RELATED REACTIONS

Diverse heterogeneous and homogeneous metal systems are known to catalyse hydrogenation reactions. The general field of hydrogenation in organic chemistry is well surveyed in recent reference books [851, 1154, 1503] and review articles [177, 450, 852]. Heterogeneous species such as *Raney nickel* and *palladium on carbon* are often the most convenient and sometimes the most active catalysts to use for the hydrogenation of organic molecules. They assist the reduction of a wide variety of substrates including alkenes, alkynes, arenes, aldehydes, ketones and nitro compounds. Indeed, the diversity of functional groups that can be hydrogenated under mild conditions, together with the range of other reactions (such as hydrogenolysis, H/D scrambling and isomerization) that are catalysed by the same heterogeneous species, is sometimes a major disadvantage — heterogeneous catalysts often lack specificity.

An impressive array of soluble metal complexes will also catalyse the reduction of organic compounds, and these homogeneous catalysts often show enhanced selectivity compared to insoluble catalysts. This has induced many chemists to expend considerable effort on developing an understanding of homogeneous hydrogenation reactions. It is clear that the stability of the catalyst—substrate transition state in some of these systems is greatly influenced by steric and electronic effects, and it is this sensitivity that generates the enhanced regio- and stereo-specificity. The knowledge gained from these widespread studies is now being turned to considerable synthetic advantage, particularly in individual steps for the preparation of natural products and related compounds.

Rhodium and iridium compounds are amongst the most widely used homogeneous hydrogenation catalysts. Such catalysts have been formed from such diverse species as $RhCl(PPh_3)_3$, $RhCl_3py_3$ in the presence of $NaBH_4$, $[IrCl_6]^{2-}$ in aqueous 2-propanol containing a phosphite, and the solvated cations $[H_2Rh(PR_3)_2S_2]^+$. In general, compounds of rhodium are more active than the analogous compounds of iridium [1678], but the reverse order is sometimes observed. Related complexes with chiral ligands have been used for the asymmetric hydrogenation of prochiral substrates, and many of the catalytic systems have been attached to polymer or other supports.

Although hydrogenation is generally achieved with molecular H_2, other sources of [H] are sometimes used. In some cases, reduction is best accomplished with synthesis gas (a mixture of H_2 and CO). The reduction of alkenes and alkynes by means of formic acid in the presence of complexes such as $RhCl(PPh_3)_3$ and $Ir(CO)Cl(PPh_3)_3$ has also been described [1873]:

$$ML_a + HCO_2H \longrightarrow H_2ML_a + CO_2$$

The reduction of some ketones has been achieved by H-transfer from alcoholic solvents such as 2-propanol. Ketone reduction can also be achieved by hydrosilation followed by hydrolysis.

The detailed discussion in this chapter is organized according to catalyst type. For each major catalyst system, a consideration is given to the nature of the active catalyst, typical conditions for the hydrogenation reaction, the types of substrate that can be reduced, any special features such as selectivity, and the mechanism of the hydrogenation reaction. Separate sections are devoted to asymmetric hydrogenation reactions, supported metal catalysts and hydrosilation reactions.

3.1 $RhCl(PPh_3)_3$ and $HRh(PPh_3)_a$

THE CATALYST SYSTEMS

The sterically crowded molecule $RhCl(PPh_3)_3$ is probably the most extensively studied of all the known homogeneous hydrogenation catalysts [411, 865]. It is often referred to as *Wilkinson's catalyst* in recognition of Wilkinson's extensive investigations of the scope, selectivity and mechanism of the hydrogenation of alkenes in the presence of $RhCl(PPh_3)_3$. The complex is most easily prepared by treating $RhCl_3 \cdot 3H_2O$ with triphenylphosphine in ethanol [1343]; it is available commercially (1983 price, approx. \$8−12 per gram depending on the quantity ordered). Various samples of $RhCl(PPh_3)_3$ from laboratory and commerical preparations have been analysed by X-ray photoelectron spectroscopy, and this has revealed the presence of substantial amounts of an Rh^{III} species [291].

The catalyst is widely and routinely used for the hydrogenation of a range of unsaturated organic molecules. Generally, reduction is achieved under mild conditions − room temperature and partial pressures of 1 atm or less of H_2 are sufficient to achieve good rates of reaction. Aromatic nuclei are inert, and hence benzene and toluene are useful solvents in homogeneous hydrogenations with $RhCl(PPh_3)_3$. The addition of a polar co-solvent such as ethanol or acetone results in a significant increase in the rate of hydrogenation. The rate is doubled, for example, when a benzene/ethanol mixture is used rather than pure benzene.

The catalyst is slowly deactivated by chlorinated hydrocarbons such as $CHCl_3$ or CCl_4 due to oxidation to form $RhCl_3(PPh_3)_3$.

This catalyst has some important advantages over other hydrogenation catalysts. In terms of the rate of hydrogenation, for instance, there is greater discrimination between structure types than with heterogeneous catalysts. Thus, $RhCl(PPh_3)_3$ catalyses the hydrogenation of alkenes specifically in the presence of other easily reduced groups such as $-CH{=}O$ and $-NO_2$, terminal alkenes in the presence of internal alkenes and less substituted alkenes in the presence of more substituted alkenes. Moreover, there is only a minor degree of scrambling when D_2 or T_2 is added to the double bond of n-alkenes [22, 85, 1217] or cyclohexene [85]; the extent of H/T exchange that accompanies the reduction reaction is conveniently monitored by ^3H NMR spectroscopy [487]. This specificity of addition is especially useful in the preparation of deuterium- or tritium-labelled natural products such as fatty acids [21], prostaglandins [69, 980] or triterpenes [1424], and drugs including methadone [1565]. Extensive H/D scrambling has been reported, however, during the deuteration of large cycloalkenes such as cyclooctene [85, 127]. A further advantage of the $RhCl(PPh_3)_3$ catalyst is a lack of disproportionation or hydrogenolysis during hydrogenation. Thus, the hydrogenation of 1,4-dihydrobenzenes gives cyclohexanes without disproportionation to benzene and cyclohexane [1603]. Further, the hydrogenation of cyclopropylalkenes (equation 3.1) gives mainly cyclopropyl alkanes with only minor amounts of pentanes being formed [688]. In the reduction of arene-diazonium salts, there is hydrogenolysis, and benzenes and naphthalenes are formed [1129].

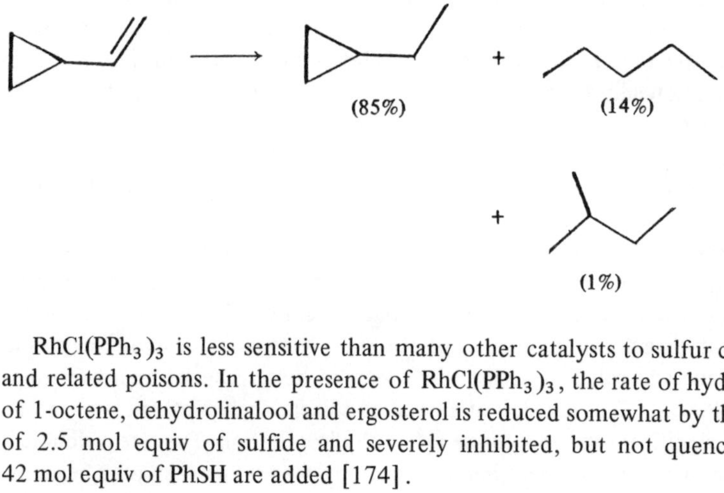

(85%)   (14%)

(3.1)

(1%)

$RhCl(PPh_3)_3$ is less sensitive than many other catalysts to sulfur compounds and related poisons. In the presence of $RhCl(PPh_3)_3$, the rate of hydrogenation of 1-octene, dehydrolinalool and ergosterol is reduced somewhat by the addition of 2.5 mol equiv of sulfide and severely inhibited, but not quenched, when 42 mol equiv of PhSH are added [174].

The hydrogenation of alkenes with $RhCl(PPh_3)_3$ as the catalyst is inhibited by PPh_3. When this problem occurs, it can be overcome by the addition of silver polystyrene sulfonate which acts as a selective phosphine absorber [146]. The rate of hydrogenation of several alkenes has been increased by treating $RhCl(PPh_3)_3$ with the phosphine absorber in the presence of ethylene, and then adding the second alkene. Various other '*additives*' affect the rate of hydrogenation favorably. Phenols accelerate the uptake of H_2 by $RhCl(PPh_3)_3$ and lithium halides in an alcohol greatly accelerate the reaction [758].

Triphenylphosphine dissociation from $RhCl(PPh_3)_3$ is induced by weak ultraviolet irradiation, and this leads to an increase in the rate of hydrogenation of cyclooctene by a factor of 2.5 [1687]. The enhancement factor is 1.7 for the corresponding hydrogenation of cyclohexene [1381].

The presence of a small amount of O_2 or H_2O_2 also activates $RhCl(PPh_3)_3$. Thus the hydrogenation of cyclohexene is accelerated by a factor of 1.3 at an O_2/Rh ratio of 0.7 [1842]. However, enhancement of the rate of hydrogenation of alkenes by O_2 is often accompanied by double-bond migration. This isomerization is inhibited, and the catalytic activity is still enhanced, when O_2 is added to $RhCl(PPh_3)_3$ in the presence of PPh_3 [92]. Thus the rate enhancement is probably not due to the oxidation of dissociated PPh_3, but to the reaction of $RhCl(PPh_3)_3$ with O_2. The oxygenation of $RhCl(PPh_3)_3$ in solution has been studied spectroscopically; ESR results are consistent with an $Rh^{II}O_2^-$ interaction in the oxygenated complex [1848]. In another study, crystals of $RhCl(PPh_3)_2(O_2) \cdot CH_2Cl_2$ were isolated when O_2 was added to a solution of $RhCl(PPh_3)_3$ in CH_2Cl_2, and an X-ray structure determination [143] has revealed the dimeric molecule illustrated in 3-1.

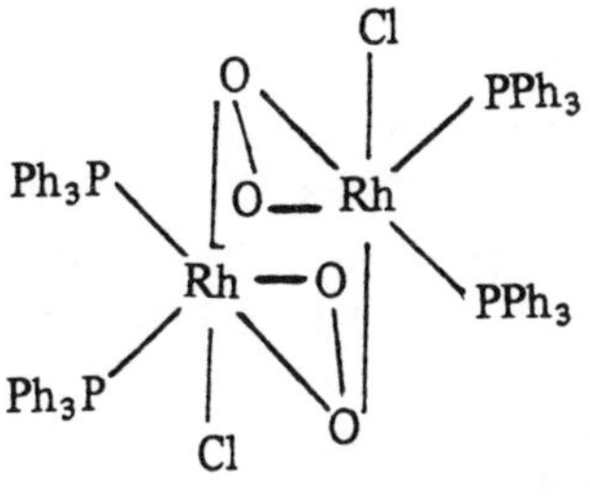

(3-1)

For the system $RhCl(PPh_3)_3$ + $AlEt_3$, a rapid Cl/Et metathesis produces $HRh(PPh_3)_3$ which is 39 times more active than $RhCl(PPh_3)_3$ for the hydrogenation of ethylene in benzene at 20 °C [1667]. The addition of Et_3N to $RhCl(PPh_3)_3$ produces a catalyst which is very active for the hydrogenation of aromatic nitro compounds to amines [1025]. Presumably the amine abstracts HCl from the $\{RhCl(PPh_3)_3 + H_2\}$ system to give $HRh(PPh_3)_3$. The related coordinatively saturated hydrido complex $HRh(PPh_3)_4$ gives a catalyst which is effective for the reduction of 1,3-dienes to terminal monoalkenes [1418] and for the hydrogenation of unsaturated nitriles [434].

REACTION PATHWAYS

Two modes of reaction are possible in metal-catalysed hydrogenation reactions. The formation of a hydride can be followed by coordination of an alkene or, alternatively, the formation of the alkene complex can precede the reaction with H_2. The former pathway is followed when $RhCl(PPh_3)_3$ is the catalyst. This was first established by the observation that ethane is obtained when $H_2 RhCl(PPh_3)_2$ is treated with ethylene, but not when $RhCl(H_2C{=}CH_2)(PPh_3)_2$ is treated with H_2 [1342].

Many groups have studied the kinetics of hydrogenation of alkenes and cycloalkenes in the presence of $RhCl(PPh_3)_3$ as the catalyst [407, 419, 420, 650, 651, 866, 1162, 1497, 1600, 1654, 1785]. This is a kinetically complex system, and there is still uncertainty about some aspects of the reaction pathway. The initial reaction between $RhCl(PPh_3)_3$ and H_2 can be represented by two sets of equilibria:

$$RhCl(PPh_3)_3 + H_2 \;\rightleftharpoons\; H_2 RhCl(PPh_3)_3 \;\overset{-PPh_3}{\rightleftharpoons}\; H_2 RhCl(PPh_3)_2$$

$$RhCl(PPh_3)_3 \;\overset{-PPh_3}{\rightleftharpoons}\; RhCl(PPh_3)_2 \;\overset{H_2}{\rightleftharpoons}\; H_2 RhCl(PPh_3)_2$$

The system is further complicated by the presence of dimeric species such as $[RhCl(PPh_3)_2]_2$ and $[HRhCl(PPh_3)_2]_2$ in the solution. Generally the concentrations of these dimeric species are not large because the rate of the reaction between $RhCl(PPh_3)_2$ and H_2 is so much faster than the dimerization reactions. Although the equilibrium between $RhCl(PPh_3)_3$ and $RhCl(PPh_3)_2$ lies far to the left ($\beta_1 = 1.4 \times 10^{-4}$ at 25 °C), the dissociated species reacts very rapidly (at least 10^4 times faster than $RhCl(PPh_3)_3$) with H_2 [653]. The solvated dihydrido species, $H_2 RhCl(PPh_3)_2 S$ has been isolated and fully characterized [1342]. In solution, the weakly held solvent molecule is readily displaced by alkene, and the two hydrogens can then be transferred sequentially in a two-step process. Scheme 3.1 presents an overall picture of the hydrogenation process.

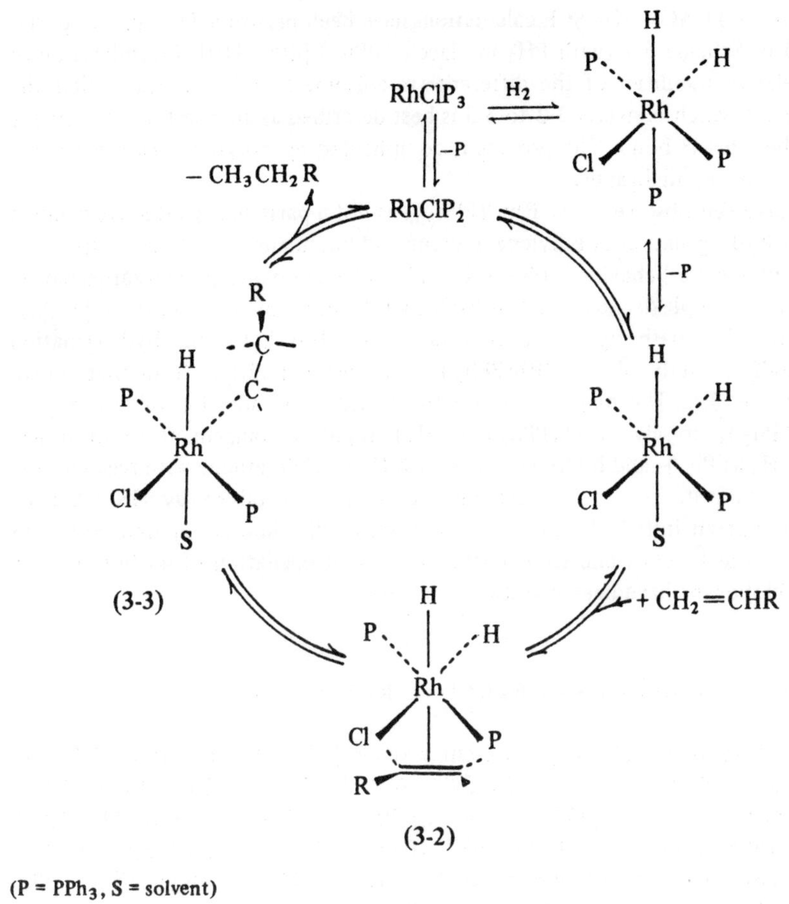

(P = PPh$_3$, S = solvent)

Scheme 3.1

Kinetic and equilibrium data are available for each key step in the catalytic cycle. The initial hydrogen transfer yields an alkylhydrido–rhodium(III) intermediate (3-3), and in most systems this is regarded as the rate-determining step. However, for the hydrogenation of cyclohexene in benzene, the formation of 3-2 is reported [407] to be the rate-determining step. The final step, in which H migrates to the *cis*-alkyl group, is fast. The reversibility of the first hydrogen transfer (i.e. 3-2 ⇌ 3-3) is not important in the hydrogenation process, because it is considerably slower than the final product-forming step. If the formation of 3-2 from 3-3 occurred rapidly, it would result in isomerization (and H/D scrambling) reactions of the alkenes (see Chapter 2).

ab initio LCAO–MO–SCF calculations have been performed for all the species shown in Scheme 3.1 (with PH_3 in place of PPh_3) [409, 410]. Consideration of the relative stabilities of the different stereoisomers of 3-3 indicates that the H-transfer which converts 3-2 to 3-3 is best described as an insertion of ethylene into the Rh—H bond. The process is complicated by polytopal rearrangements of the nonreacting ligands.

It has been observed that $RhCl(PPh_3)_3$ is not a particularly effective catalyst for the hydrogenation of ethylene, styrene and butadiene. For these substrates, a different kinetic behavior is observed, and a scheme involving the hydrogenation of alkene complexes such as $RhCl(PPh_3)_2$(alkene) may be important [1294].

A reaction pathway has been established [1667] for the hydrogenation of ethylene catalysed by $HRh(PPh_3)_3$, and again it differs from that shown in Scheme 3.1. The major steps are the coordinative-addition of ethylene to $HRh(PPh_3)_3$ to give $HRh(PPh_3)_3(C_2H_4)$, rapid rearrangement to form Rh-$(CH_2CH_3)(PPh_3)_3$ and hydrogenolysis with H_2 to yield ethane with regeneration of the catalyst. The same mechanism applies for catalysis by $HRh(PPh_3)_4$, except that an initial phosphine dissociation equilibrium is required. With this catalyst, the kinetics and stereochemistry of hydrogenation of 4-*t*-butylmethy-lenecyclohexane have been investigated [1599].

RANGE OF ALKENES AND RELATIVE RATES

The hydrogenation of many different alkenes [85, 159, 162, 199, 277, 436, 688, 758, 868, 883, 1217, 1342, 1598, 1624, 1625, 1676, 1677, 1915, 1947], cycloalkenes [85, 127, 173, 277, 419, 420, 688, 788, 789, 868, 946, 1279, 1497, 1568, 1600, 1603, 1676, 1677, 1685, 1687], polyenes [159, 176, 277, 333, 868], unsaturated polymers [882], allenes [154] and alkynes [277, 333, 867, 1342, 1676, 1677, 1685, 1915, 1947] has been catalysed by $RhCl(PPh_3)_3$. There have been various studies of the relative rates of hydrogenation of many of these substrates. With some systems, analysis has been assisted by gas chromatographic reaction control [875, 1440]. Differences in the reactivity of alkenes are due primarily to the effect of alkene stereochemistry on the forma-tion constant of H_2RhClL_2(alkene) [1277], and some clear trends are apparent.

Terminal alkenes such as 1-hexene hydrogenate more rapidly than internal alkenes such as 2-hexene [1676, 1977]. With internal alkenes, the *cis*-isomer reacts more rapidly than the *trans* form [199]. The rate of hydrogenation of cycloalkenes decreases somewhat with increasing ring size. Thus, cyclopentene hydrogenates more rapidly than cyclooctene [277]. Chelating dienes such as norbornadiene, 1,5-cyclooctadiene and 1,3-pentadiene hydrogenate very slowly [277, 868], and indeed, can deactivate the catalyst. However, the nonchelating ligand 1,3-cyclooctadiene reacts at about the same rate as cyclohexene. The polycyclic compound hexacyclo[9.3.2.2^{4,7}.0^{2,9}.0^{3,8}.0^{10,12}]octadeca-5,13,15,17-tetraene (3-4), which incorporates diastereotopic double bonds, is hydrogenated specifically at the C(5)—C(6) double bond [396].

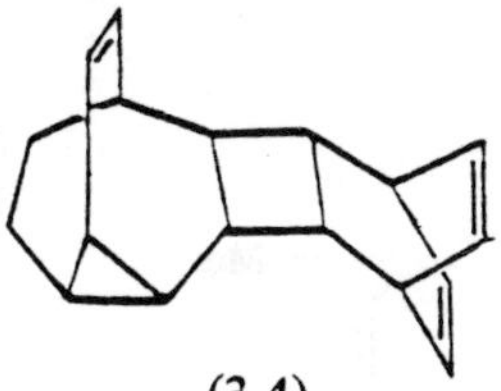

(3-4)

Electron-withdrawing substituents can also affect the rate of hydrogenation. Thus, for the substituted alkenes H$_2$C=CHR, the ease of hydrogenation is in the order R = CN > CO$_2$Me > Ph > Alk [277]. Such effects can lead to specific hydrogenations. Thus, the spiro ether (−)-dehydrogriseofulvin (3-5) gives griseofulvin (3-6), indicating that hydrogenation at MeC=C(CO) is preferred relative to MeOC=C(CO) [176]. It is worth noting that ethers of this type are degraded with heterogeneous catalysts.

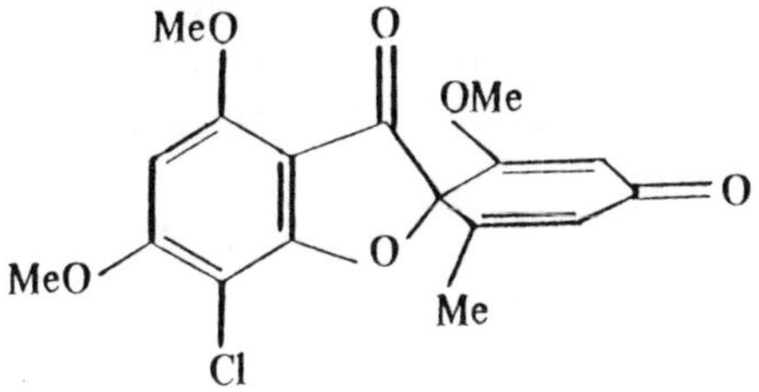

(3-5)

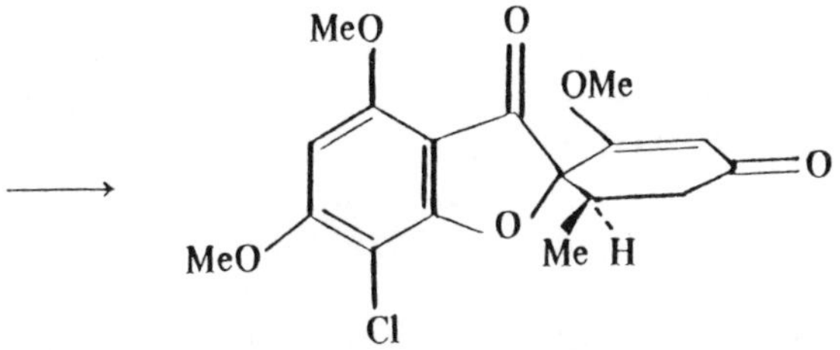

(3-6)

Increasing the degree of substitution at the alkene center generally impedes hydrogenation [868, 1676, 1677]. This is well illustrated by the relative rates of hydrogenation for the methyl-substituted cyclohexenes **3-7** to **3-10** [788, 789]. The hydrogenation of 1-methoxycyclohexa-1,4-diene (**3-11**) to 1-methoxycyclohex-1-ene (**3-12**) [176] and of **3-13** to **3-14** provide other examples. In the latter case, reduction of the cyclohexene is preferred with heterogeneous catalysts such as Pd.

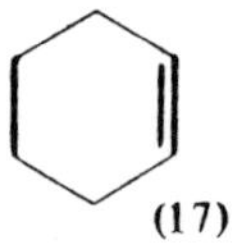

(17)   (0.49)   (0.0)   (0.0)

(3-7) 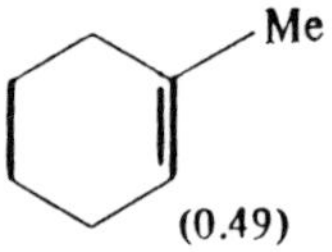   **(3-8)** 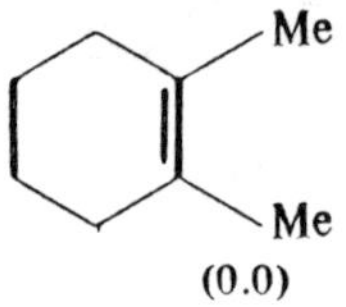   **(3-9)** 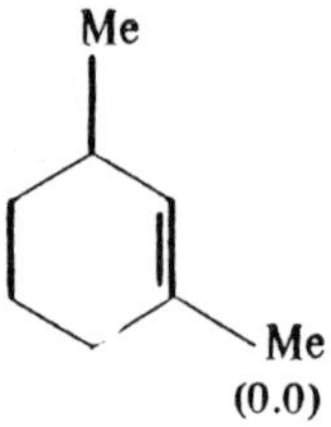   **(3-10)**

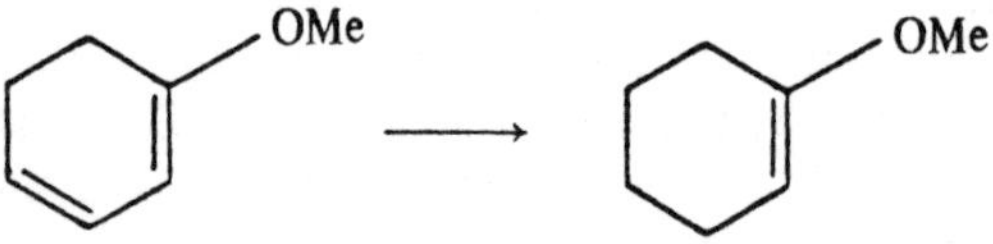

(3-11)   **(3-12)**

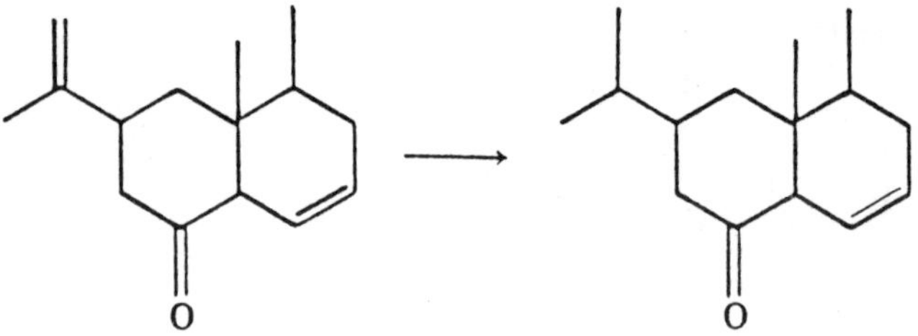

(3-13)   **(3-14)**

The presence of various functional groups in the vicinity of the alkene generally has little effect on the hydrogenation reaction. Reactions with unsaturated aldehydes [554, 868, 1744], ketones [175, 248, 657, 1172, 1327, 1569, 1602], carboxylic acids/esters [173, 657, 1676, 1677, 1689, 1690], nitro compounds [659], nitriles [1569], epoxides [1207], furans, thiophenes and other unsaturated heterocyclic compounds [175, 562, 765, 1594] have been studied, and the alkene is reduced selectively in these systems. Reduction of the α,β-unsaturated nitro compound **3-15** provides a good example (see Equation (3.2)).

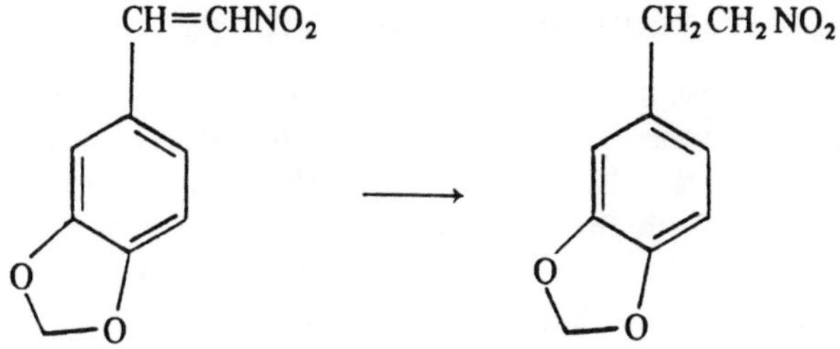

CH=CHNO$_2$ CH$_2$CH$_2$NO$_2$

$\longrightarrow$ (3.2)

(3-15)

It is clear that conjugation of an alkene with a carbonyl does not impede hydrogenation of the alkene. For example, 4-methyl-3-penten-2-one (mesityl oxide) is hydrogenated selectively to methylisobutylketone [1688], 3-16 is readily hydrogenated at the C=C linkage (Equation (3.3)) [1390] and 1,4-naphthaquinone give 1,2,3,4-tetrahydro-1,4-dioxonaphthalene [175, 176].

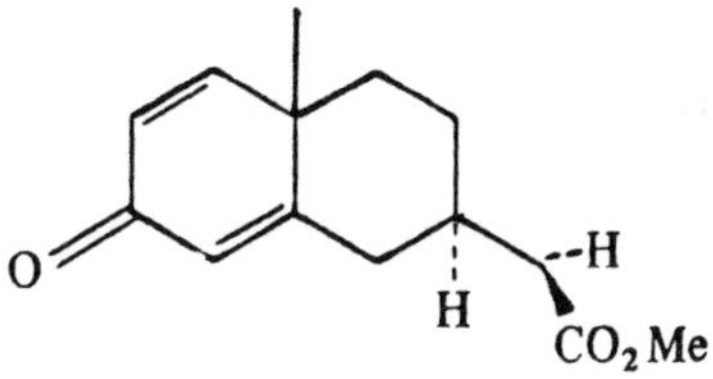

O

--H

H

CO$_2$Me

(3-16)

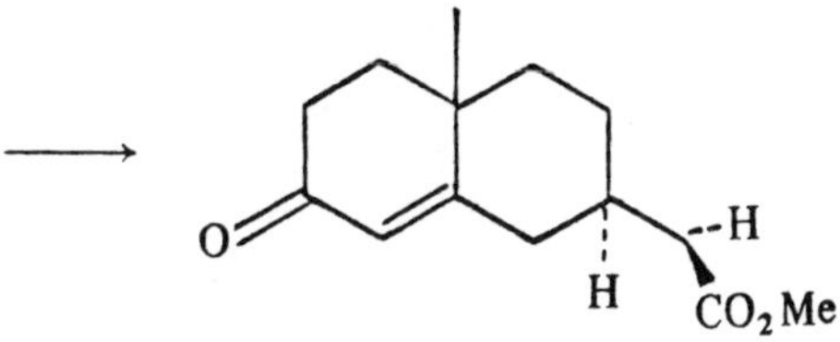

$\longrightarrow$ (3.3)

O

--H

H   CO$_2$Me

However, with quinones of high oxidation potential such as β-naphthaquinone, the rhodium complex is destroyed. The hydrogenation of unsaturated aldehydes is often complicated by the tendency for RhCl(PPh$_3$)$_3$ to achieve decarbonylation (see Chapter 6). When this occurs, there is formation of Rh(CO)Cl(PPh$_3$)$_2$ which is less effective than RhCl(PPh$_3$)$_3$ as a hydrogenation catalyst (see Section 3.2 below). However, some aldehydes, including but-2-enal and trans-2-methylpent-2-enal, have been reduced to the corresponding saturated aldehydes [868, 1206].

STEREOCHEMISTRY AND REGIOSELECTIVITY OF ADDITION

There has been some interest in the stereochemistry of addition during the hydrogenation of alkenes, alkynes and dienes. The results of such studies should be viewed with caution because the stereoselectivity of these reactions can be affected significantly by impurities in the catalyst and/or the solvent. With alkenes, the initial formation of an alkyl—rhodium intermediate (**3.3** in Scheme 3.1) occurs by *cis* addition of H—Rh to the alkene, and subsequent hydrogenolysis to give the alkane occurs with retention of stereochemistry. The overall stereochemistry of these reactions is illustrated in Equations (3.4) and (3.5), where D_2 is added to maleic and fumaric acids to form 1,2-dideutereosuccinic acids [1342].

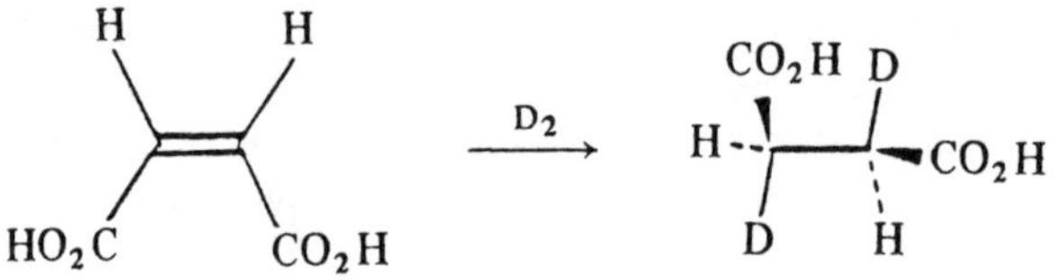

$$\text{(3.4)}$$

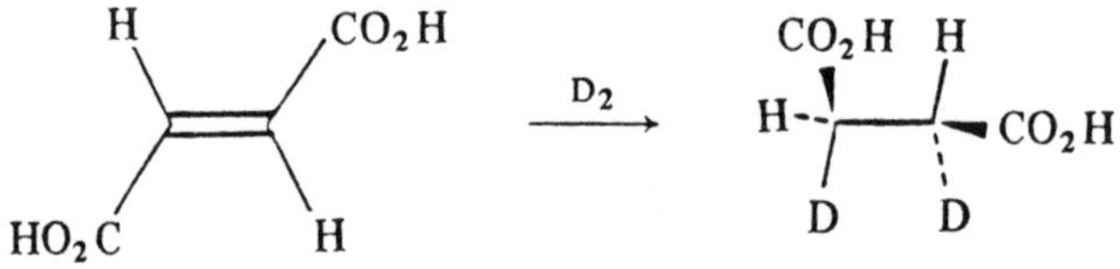

$$\text{(3.5)}$$

The stereochemistry of hydrogenation has also been followed with 4-*t*-butyl-methylenecyclohexane [91, 1193, 1601], and this reaction gives 90% *cis*- and 10% *trans*-4-*t*-butyl-methylcyclohexane.

A study of the reduction of 2-methylenebicyclo[2.2.1]heptane and its methylated derivative has established that the product stereochemistry depends on the equilibrium:

$$H_2 Rh(PPh_3)_2 (S)X + \text{alkene} \rightleftharpoons H_2 Rh(PPh_3)_2 (\text{alkene})X + S$$

Two π-alkene complexes can be formed, and these are precursors of the *endo*- and *exo*-2-methylbicyclo[2.2.1]heptanes [1498].

The addition of H_2 to acyclic dienes occurs as follows: predominantly in the 1,2-position for 2-methyl-1,3-butadiene, exclusively in the 1,2-position for 2,3-dimethyl-1,3-butadiene, in the 1,4-positions for 1,3-butadiene to give a product with a predominantly *cis* configuration and in the 1,4-positions for the isomeric 1,3-pentadienes to give a product with a *trans* configuration [545].

The hydrogenations of steroids and related natural products [161, 163, 173, 174, 299, 448, 609, 1280, 1389, 1390, 1569, 1779, 1867] provide some beautiful illustrations of the delicate selectivity of $RhCl(PPh_3)_3$ compared to many other catalysts, including heterogeneous catalysts. In general, the alkene centers at the 1,2-, 2.3-, 3,4- or 11,12-positions (see **3-17**) hydrogenate without difficulty [448, 1867]. In the conversion of ergosterol acetate (**3-18**) into $5\alpha,6\alpha$-D_2-ergost-7-en-3-β-ol (**3-19**), the least hindered C=C bond is selectively reduced from the less crowded face of the steriod.

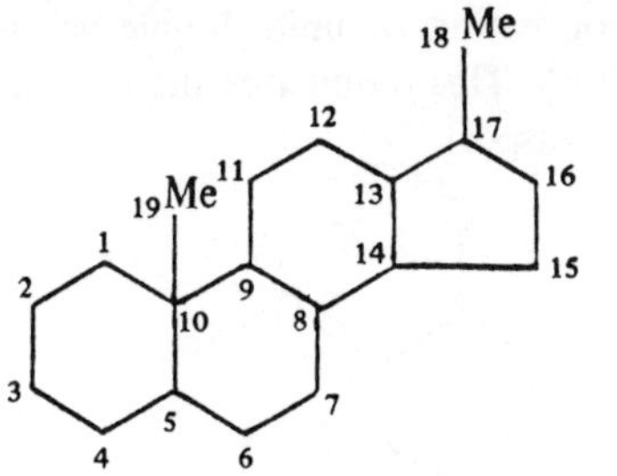

(3-17)

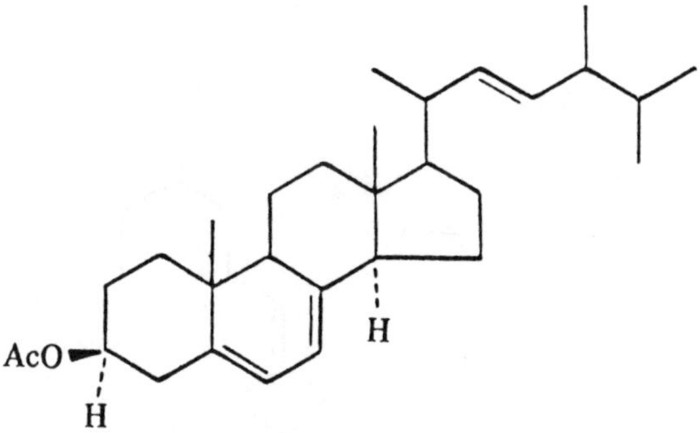

(3-18)

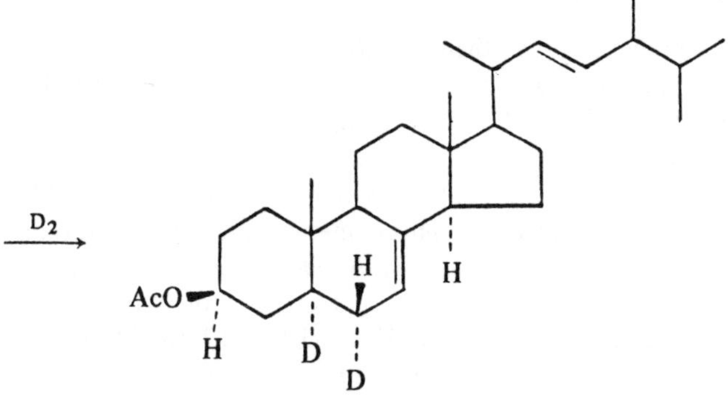

(3-19)

There is no reduction of the trisubstituted 7-ene or the disubstituted 21-ene, and no H/D scrambling is detected [177]. The ability to hydrogenate less-substituted alkenes in the presence of more heavily substituted alkene centers is also an important aid in sesquiterpene synthesis, and this is illustrated by the hydrogenation of linalool (Equation (3.6)) [159] and of carvone (Equation (3.7)) [179]. The direct hydrogenation of lipid double bonds in spinach chloroplasts has been achieved [1468]. This modulates the membrane fluidity and affects photosynthetic electron transport.

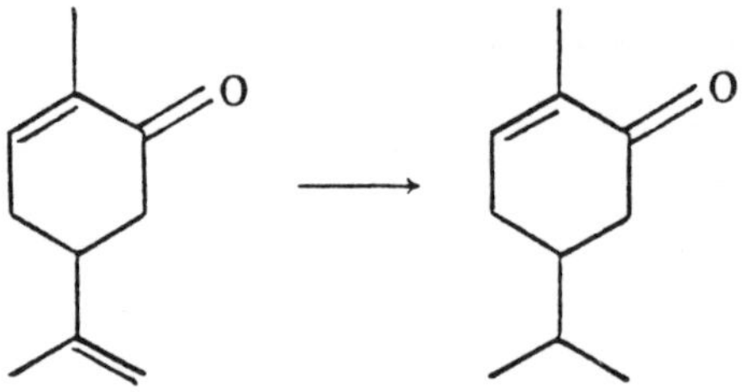

(3.6)

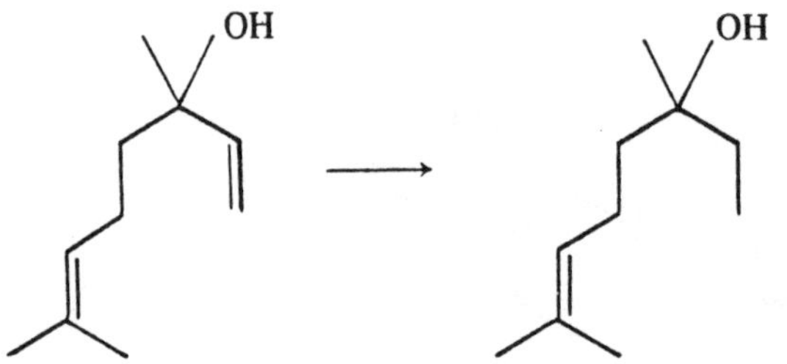

(3.7)

CATALYSTS RELATED TO RhCl(PPh$_3$)$_3$

MXL$_3$ complexes other than RhCl(PPh$_3$)$_3$ have been used as catalysts for the hydrogenation of alkenes, cycloalkenes and alkynes [814, 1916]. Studies of the effect of varying L in the complexes RhClL$_3$ show that RhCl(AsPh$_3$)$_3$ and RhCl(SbPh$_3$)$_3$ are less effective than the triphenylphosphine complex as hydrogenation catalysts [1110]. With the complexes RhCl(PR$_3$)$_3$, the rate of reaction is dependent on the phosphine, but there is little variation in specificity towards a given alkene [1277]. Replacement of PPh$_3$ by P(p-MeOC$_6$H$_4$)$_3$ increases the rate of H$_2$ uptake and the rate of hydrogenation, whereas the rates are decreased with P(p-FC$_6$H$_4$)$_3$ [1215, 1916]. The complexes with more basic tertiary phosphines such as PEt$_3$ or PEt$_2$Ph are catalytically inactive. Studies of the hydrogenation of cyclohexene in benzene with RhX(PPh$_3$)$_3$ as catalysts show that the rate increases in the order X = Cl < Br < I [199, 1342].

Some mixed ligand complexes have been obtained by the reduction of $[RhCl(O_2)(PPh_3)_2]_2$ with ethanol in the presence of 1 equiv of L (e.g. L = $P(C_6H_4X)_3$ with X = p-F, p-OMe, m-Me) [90]. When these systems are used in the hydrogenation of 4-t-butylmethylenecyclohexane, there is kinetic and stereochemical evidence that the species 3-2 in Scheme 3.1 has L exclusively at the site *trans* to the alkene.

The sulfonated derivative, chlorotris(sodium diphenylphosphinobenzene-m-sulfonate)rhodium(I) has been used for hydrogenation reactions in aqueous media. Reduction of the fatty acyl chains of soybean phosphatidylcholine single bilayer vesicles [1103], and of C=C bonds within phospholipid liposomes [1102, 1452] has been achieved. Such reactions may be valuable aids in studies of cellular functions which are dependent on membrane fluidity. Related catalysts have been prepared *in situ* from $RhCl_3 \cdot 3H_2O$ and sodium diphenylphosphinobenzene-o-sulfonate. With this catalyst, the hydrogenation of alkenes such as cyclohexene is governed by the solubility of the alkene in the aqueous phase [467].

The treatment of $RhCl(PPh_3)_3$ in methanol with $Cs^+[7\text{-butenyl-}7,8\text{-}C_2B_9H_{11}]^-$ gives $[closo\text{-}1,3\text{-}\mu\text{-}(\eta^2\text{-}3,4\text{-}H_2C=CHCH_2CH_2)\text{-}3\text{-}H\text{-}3\text{-}PPh_3\text{-}3,1,2\text{-}RhC_2B_9H_{10}]$, which is an extremely reactive hydrogenation catalyst [413]. The related metallocarborane catalyst $[closo\text{-}1,3\text{-}\mu\text{-}2,3\text{-}\mu\{1,2\text{-}\mu\text{-}(\eta^2\text{-}3,4\text{-}CH_2CH_2C(Me)=CHCH_2CH_2CH_2)\}\text{-}3\text{-}H\text{-}3\text{-}PPh_3\text{-}3,1,2\text{-}RhC_2B_9H_9]$ is prepared similarly from $RhCl(PPh_3)_3$ in ethanol and $Cs^+[7,8\text{-bis(butenyl)-}7,8\text{-}C_2B_9H_{10}]$ [414].

The hydrido complexes $HRh(DBP)_a$ (DBP = dibenzophosphole; a = 2 or 3) [260, 742], $HRh\{PhP(CH_2CH_2CH_2PPh_2)_2\}$ [469, 1259, 1260] and $HRh\{PhP(CH_2CH_2CH_2PCy_2)\}_2$ [1259, 1260] catalyse the homogeneous hydrogenation of terminal alkenes. With the latter two catalysts, the hydrogenation reaction occurs about 25 times more rapidly than with $RhCl(PPh_3)_3$. The related complex $(BH_3)Rh\{PhP(CH_2CH_2PPh_2)_2\}$ is also an effective hydrogenation catalyst [469]. The mixed ligand species $HRh(PF_3)(PPh_3)_3$ and $HRh(PF_3)_2(PPh_3)_2$ have been used as hydrogenation catalysts, but there is extensive isomerization of the alkenes [1266].

Polynuclear hydrido—rhodium complexes have been used for the hydrogenation of alkenes. The catalysts include $[HRh\{P(NMe_2)_3\}_2]_2$ [1164], $[HRh\{P(OPr^i)_3\}_2]_2$, $[HRh\{P(OMe)_3\}_2]_3$, $H_4Rh_2\{P(OPr^i)_3\}_4$ and $H_5Rh_3\{P(OMe)_3\}_6$ [405, 1237, 1618]. Particularly high rates of hydrogenation of 1-hexene are achieved with the dimeric systems $[HRhL_2]_2$ (L = $P(OMe)_3$ or $P(OPr^i)_3$). With $[HRh(PPh_3)_2]_2$ as the catalyst, *trans*-alkenes are formed selectively from alkynes [262]. Some related complexes, $HRh(PPr^i_3)_3$ and $H_2Rh_2(\mu\text{-}N_2)(PCy_3)_4$, catalyse the reduction of nitriles to primary amines selectively [1940]. For example, a quantitative yield of $Me(CH_2)_5NH_2$ is obtained from $Me(CH_2)_4CN$ in the presence of $HRh(PPr^i_3)_3$ at 20 °C.

Although IrCl(PPh$_3$)$_3$ readily forms a dihydride, there is no detectable phosphine dissociation and H$_2$IrCl(PPh$_3$)$_3$ is not an active hydrogenation catalyst [140]. The 5-coordinate species H$_2$IrCl(PPh$_3$)$_2$ can be formed by the selective oxidation of one triphenylphosphine ligand in H$_2$IrCl(PPh$_3$)$_3$ with H$_2$O$_2$ [1847] or by the addition of H$_2$ to IrCl(PPh$_3$)$_2$ which is formed *in situ* from [Ir(C$_8$H$_{14}$)$_2$Cl]$_2$ and triphenylphosphine [1845]. Although this compound will catalyse the hydrogenation of alkenes, it also leads to extensive double-bond migration. The bidentate phosphine complex IrCl(Ph$_2$PCH$_2$CH$_2$PPh$_2$) has also been used as a catalyst for the hydrogenation of alkenes and alkynes [816].

3.2 M(CO)XL$_2$

The iridium complex Ir(CO)Cl(PPh$_3$)$_2$, which is often referred to as *Vaska's compound*, can be prepared from halo-iridium compounds and triphenylphosphine in oxygen-containing solvents such as diethyleneglycol or dimethylformamide [1886]. Many related complexes of the type Ir(CO)XL$_2$ are known, and they are noted for their propensity to undergo coordinative- and oxidative-addition reactions. They react reversibly with hydrogen and with alkenes, and this has stimulated interest in the compounds as possible hydrogenation catalysts.

The reaction with H$_2$ forms the dihydrido species H$_2$Ir(CO)XL$_2$, and the magnitude of the equilibrium constant has been determined as a function of X and L [1669, 1699]. Kinetic studies have established that the rate of the reaction with H$_2$ increases in the order X = Cl < Br < I, and that the reaction is faster when L = P(p-MeC$_6$H$_4$)$_3$ or P(p-MeOC$_6$H$_4$)$_3$ rather than PPh$_3$ [856, 1682, 1699, 1700, 1701, 1703, 1852].

The formation of the complexes Ir(CO)Cl(PPh$_3$)$_2$(alkene) has been studied by stopped-flow techniques, and the kinetics are consistent with nucleophilic attack of the metal complex on the alkene to give a dipolar activated complex [645]. The equilibrium constant for the reaction of ethylene with Ir(CO)Cl-(PPh$_3$)$_2$ is about 100 times less than that for the ethylene + RhCl(PPh$_3$)$_3$ system [1130].

The hydrogenation of alkenes in the presence of catalytic amounts of Ir(CO)Cl(PPh$_3$)$_2$ has been achieved in various hydrocarbon solvents [481, 535, 1093, 1345, 1346, 1669, 1680, 1693, 1694, 1714, 1716, 1851, 1932] or even in the absence of solvent [1689]. However, the complex is not a particularly effective catalyst; the hydrogenation of simple alkenes such as ethylene and propylene requires many hours in toluene at 70–80 °C and 1 atm pressure of H$_2$. Moreover, there is competition from isomerization and H/D exchange reactions. The complex Ir(CO)Cl(PCy$_3$)$_2$ does show high selectivity in the conversion of α,β-unsaturated ketones to saturated ketones [1698].

Kinetic studies of hydrogenation reactions in the presence of $Ir(CO)XL_2$ complexes as catalysts indicate that the rate law is complex [265, 856, 1679, 1702, 1704], but it has been interpreted in terms of Scheme 3.2. As with the $RhCl(PPh_3)_3$ system, the mechanism of hydrogen transfer is thought to be stepwise and to involve an alkylhydrido—metal intermediate [1673]. The proposed mechanism involves coordination of the alkene prior to addition of the H_2, this sequence of events being opposite to that for the $RhCl(PPh_3)_3$ reactions.

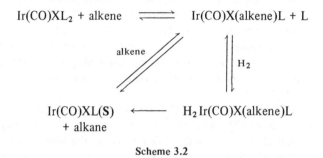

Scheme 3.2

To explain an induction period which is sometimes observed, it is suggested that the alkene—iridium intermediate is involved in a radical chain process which is dependent on the solvent or its impurities [265]. The induction period in DMF is removed by the addition of p-$MeOC_6H_4OH$. It has been suppressed with dibenzoyl peroxide, azobisisobutyrodinitrile or tetrachloro-1,4-benzoquinone, but triethylamine and $SnCl_2 \cdot 2H_2O$ are not suitable [727]. Higher temperatures are also effective.

Other means of improving the catalytic activity of $Ir(CO)Cl(PPh_3)_2$ have been explored. The hydrogenation rate is accelerated by traces of O_2 [856, 1690], and the catalytic activity is also enhanced in coordinating solvents such as dimethylacetamide [856]. The addition of H_2O_2 to selectively oxidize one triphenylphosphine ligand also improves the activity of $Ir(CO)Cl(PPh_3)_2$ toward the hydrogenation of terminal alkynes [1847]. Weak ultraviolet irradiation seems to have the greatest effect, and this increases the activity of the catalyst by a factor of up to 40 [1672, 1686, 1717, 1718, 1720, 1721, 1722]. The influence of quantum fluxes on this photoactivation has been investigated. In studies with the catalyst $Ir(CO)Cl(PPr^i_3)_2$, the rate of hydrogenation of ethyl acrylate is enhanced by irradiation with monochromatic light of wavelength 407 nm (or 578 nm if pentane is incorporated as a photosensitizer) [1715]. Replacement of the triphenylphosphine ligands by bidentate o-diphenylphosphino-N,N-dimethylaniline or o-diphenylphosphino-N,N-dimethylbenzylamine also produces a more effective catalyst for the hydrogenation of 1-hexene [1458].

There is no evidence for hydride formation when a solution of Rh(CO)Cl-(PPh$_3$)$_2$ is treated with H$_2$ at 1 atm pressure [1851]. Nonetheless, Rh(CO)Cl-(PPh$_3$)$_2$ and related Rh(CO)XL$_2$ complexes will act as hydrogenation catalysts [785, 1129, 1579, 1706, 1707, 1708, 1711, 1713, 1714, 1770, 1920]. The effect of changes in X and L have been investigated for the reaction with 1-heptene in toluene at 70 °C, and both hydrogenation and isomerization can occur depending on X and L. The selectivity for hydrogenation decreases in the order: X = Cl, L = PPh$_3$ > PCy$_3$ > P(OPh)$_3$; L = PPh$_3$, X = Cl > Br > I. There is no isomerization in the absence of H$_2$ [1707, 1710, 1711]. The conditions which favor the hydrogenation reaction, and the kinetics of the system, have been investigated [1706, 1707, 1708]. As with the analogous iridium complexes, the hydrogenation reactions require elevated temperatures and are subject to an induction period [1713].

The hydroxy complex trans-Rh(CO)(OH)(PPr^i$_3$)$_2$ has been obtained by treatment of [Rh(NBD)Cl]$_2$ with PPr^i$_3$ and BuLi. It is an active catalyst for the reduction of methyl crotonate with {CO + H$_2$O} as the source of H$_2$ [1323]. The incorporation of other phosphines gives less active catalysts; the order is PPr^i$_3$ > PBu$_3$ > PPhPr^i$_2$ > PPh$_2$Pr^i > PPh$_3$.

Some substituted rhodium carbonyl halides of higher nuclearity have been used as catalyst precursors. The tetrameric complexes Rh$_4$(CO)$_4$Cl$_4$(O$_2$)$_2$L$_2$ (L = tertiary phosphine or phosphinite), which are formed when Rh(CO)ClL$_2$ is treated with oxygen in dimethylacetamide, are effective catalysts for the hydrogenation of alkenes [384]. The catalytic activity of these systems is probably due to dissociation to give dimeric species and then RhClL$_2$.

A polymeric matrix of formula [Rh(CO)Cl(NCC$_6$H$_4$CN)]$_x$ is formed when [Rh(CO)$_2$Cl]$_2$ is treated with 1,4-(NC)$_2$C$_6$H$_4$. This is a heterogeneous system which catalyses the hydrogenation and isomerization of 1-hexene, and the hydrogenation of 2-hexene. These reactions are photoretarded [482].

The cationic species [Rh(CO)(PPh$_3$)$_3$]$^+$ and [Rh(CS)(PPh$_3$)$_3$]$^+$ will catalyse the hydrogenation of cyclic and terminal alkenes at 1 atm H$_2$ pressure and 20 °C. The carbonyl compound is more active than its thiocarbonyl analog [1831].

3.3 HM(CO)(PPh$_3$)$_3$

Treatment of M(CO)XL$_2$ with various reductants in the presence of L gives the hydrido complexes HM(CO)L$_3$. For example, HIr(CO)(PPh$_3$)$_3$ can be obtained from Ir(CO)Cl(PPh$_3$)$_2$ by treatment with sodium borohydride and excess triphenylphosphine [1923]. The analogous hydrido—rhodium complex is most conveniently prepared from RhCl$_3$ · 3H$_2$O in ethanol by reaction with triphenylphosphine, then aqueous formaldehyde and finally potassium hydroxide [25].

The hydrido—rhodium complex $HRh(CO)(PPh_3)_3$ catalyses the hydrogenation of alkenes and alkynes under mild conditions ($25\ °C, < 1$ atm pressure H_2) [918, 1276, 1278, 1598, 1691, 1712]. Measurement of the rates of H_2 uptake by {alkene + catalyst} systems shows similar results for $HRh(CO)(PPh_3)_3$ and $RhCl(PPh_3)_3$ [1276]. $HRh(CO)(PPh_3)_3$ is highly selective for the reduction of 1-alkenes and this is attributed to steric factors caused by the bulky triphenylphosphine groups [1276]. The compound is also a good catalyst for the hydroformylation (Chapter 4) and isomerization (Chapter 2) of alkenes. Groups such as RCHO, RCN and RCO_2R' are not reduced under mild conditions in the presence of $HRh(CO)(PPh_3)_3$.

In solution the behavior of $HRh(CO)(PPh_3)_3$ is consistent with the reversible lons of one triphenylphosphine ligand to give the coordinatively unsaturated species $HRh(CO)(PPh_3)_2$. This accounts for such observations as rapid H/D exchange when D_2 is added to a benzene solution of $HRh(CO)(PPh_3)_3$ (Equation (3.8)) [493, 1276].

$$HRh(CO)(PPh_3)_3 \underset{}{\overset{-PPh_3}{\rightleftharpoons}} HRh(CO)(PPh_3)_2$$

$$\overset{D_2}{\rightleftharpoons} HD_2Rh(CO)(PPh_3)_2$$

$$\overset{-HD}{\rightleftharpoons} DRh(CO)(PPh_3)_2$$

$$\overset{PPh_3}{\rightleftharpoons} DRh(CO)(PPh_3)_3 \qquad (3.8)$$

The proposed pathway for the hydrogenation of alkenes with $HRh(CO)-(PPh_3)_3$ as the catalyst is shown in Scheme 3.3, and again it is suggested that alkene coordination precedes the oxidative addition of H_2 [493, 1276, 1712]. The observation that added PPh_3 inhibits the reduction of terminal alkenes is consistent with $HRh(CO)(PPh_3)_2$ being the active catalyst. The catalytic activity of $HRh(CO)(PPh_3)_3$ for the hydrogenation of alkenes decreases with time, and this may be due to the slow, irreversible loss of H_2 (Equation (3.9)) [1929].

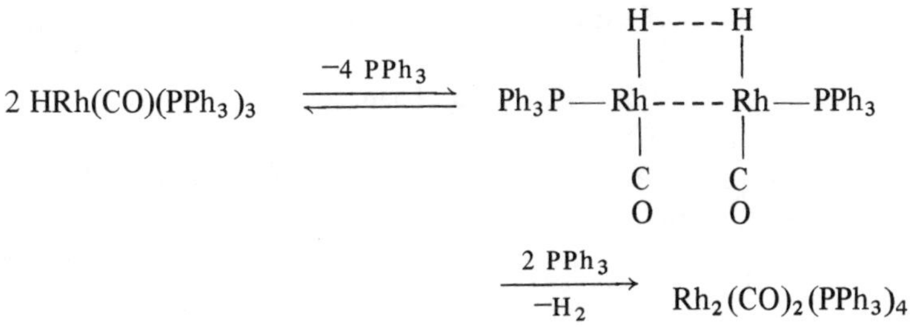

$$2\ HRh(CO)(PPh_3)_3 \underset{}{\overset{-4\ PPh_3}{\rightleftharpoons}} \begin{array}{c} H\text{----}H \\ | \qquad | \\ Ph_3P\text{---}Rh\text{----}Rh\text{---}PPh_3 \\ | \qquad | \\ C \qquad C \\ O \qquad O \end{array}$$

$$\xrightarrow[-H_2]{2\ PPh_3} Rh_2(CO)_2(PPh_3)_4 \qquad (3.9)$$

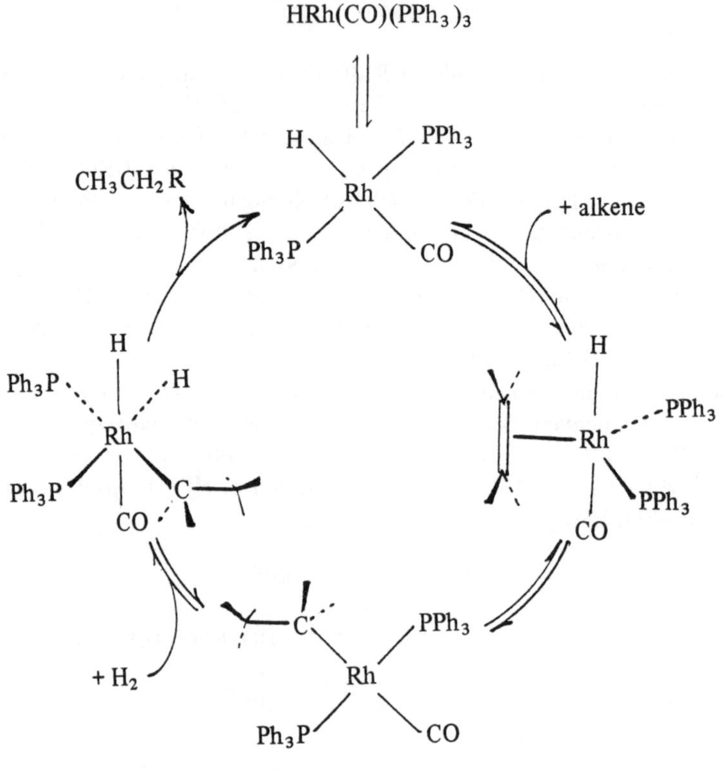

Scheme 3.3

The catalyst is reactivated by irradiation under weak ultraviolet light [1671].

Replacement of the PPh$_3$ ligands in HRh(CO)(PPh$_3$)$_3$ by ionic sodium diphenylphosphinobenzene-m-sulfonate groups gives a water-soluble hydrogenation catalyst. This is active in phospholipid bilayers in water for the hydrogenation (and hydroformylation) of substrates such as 9-decen-1-ol and 10-undecen-1-ol [1451].

The iridium complex HIr(CO)(PPh$_3$)$_3$ has been used as a hydrogenation catalyst [263, 267, 1691, 1716], but it is less active and less selective than its rhodium analog. Studies of the kinetics of the reaction between HIr(CO)(PPh$_3$)$_3$ and H$_2$ have indicated the equilibrium sequence shown in Equation (3.10) [263, 264, 432].

$$\text{HIr(CO)(PPh}_3)_3 \xrightleftharpoons[\;]{-\text{PPh}_3} \text{HIr(CO)(PPh}_3)_2 \xrightleftharpoons[\;]{-\text{PPh}_3} \text{HIr(CO)(PPh}_3)$$

$$\text{H}_3\text{Ir(CO)(PPh}_3)_2 \xrightleftharpoons[\;]{\text{PPh}_3,\text{fast}} \text{H}_3\text{Ir(CO)(PPh}_3) \quad (3.10)$$

The kinetics for the hydrogenation of ethylene and butadiene have been inter-preted in terms of $\text{HIr(CO)(PPh}_3)_2$ being the active catalyst, with the overall reaction pathway being analogous to that shown for the rhodium system in Scheme 3.3 [263, 264, 266]. However, a more recent study of the hydrogena-tion of dimethyl-maleate and -fumarate indicates that the reaction depends on a system of competing reactions involving the complexes $\text{HIr(CO)(PPh}_3)_3$, $\text{H}_3\text{Ir(CO)(PPh}_3)_2$, $\text{HIr(CO)(alkene)(PPh}_3)_2$, $\text{Ir(CO)}(\sigma\text{-alkyl)(PPh}_3)_2$, Ir(CO)-$(\text{alkene})(\sigma\text{-alkyl)(PPh}_3)_2$ and $\overline{\text{Ir(CO)}\{\text{P(C}_6\text{H}_4)\text{Ph}_2\}}(\text{PPh}_3)$ [267]. The behavior of $\text{HIr(CO)(PPh}_3)_3$ as a hydrogenation catalyst has been compared with that of other iridium catalysts such as Ir(CO)ClL_2, $\text{IrCl(PPh}_3)_3$ and $\text{H}_3\text{Ir(PPh}_3)_3$ [1716].

3.4 Other Catalysts with Carbonyl and Related π-Acid Ligands

Binary carbonyl compounds of rhodium and iridium have been used as catalysts for the hydrogenation of pentynes [1062], butadiene [821], aldehydes [692] and α,β-unsaturated carbonyl and nitrile compounds [959]. In general, these reactions are slow, even at elevated temperature and pressure. In some cases, co-catalysis of the water—gas shift reaction is the source of hydrogen for the reduction. The reduction of aldehydes, including unsaturated aldehydes, to alcohols has been achieved in this way with $\text{Rh}_6(\text{CO})_{16}$ [325] or $\{\text{Rh}_6(\text{CO})_{16} +$ $N,N,N'N'$-tetramethyl-1,3-propanediamine$\}$ [915] as catalyst; an example is shown in Equation (3.11).

$$\text{Me}_2\text{C}{=}\text{CH(CH}_2)_2\text{C(Me)}{=}\text{CHCHO} \xrightarrow{\;30\,^\circ\text{C, 48 h}\;}$$

$$\text{Me}_2\text{C}{=}\text{CH(CH}_2)_2\text{C(Me)}{=}\text{CHCH}_2\text{OH (94\%)}$$

$$+ \text{Me}_2\text{C}{=}\text{CH(CH}_2)_2\text{CHMeCH}_2\text{CH}_2\text{OH (6\%)} \quad (3.11)$$

The reduction of nitrobenzene to aniline has also been achieved with CO and H_2O in the presence of $\text{Rh}_6(\text{CO})_{16}$ [829]. If CO and D_2O are used for the reduction, deuterium is incorporated in the aniline [1502]. The activity of the catalyst in this system is increased by the addition of γ-aminopyridine [912]. Reductive amination of 3-20 to give 3-21 has been accomplished with $\{\text{CO} + \text{H}_2\}$ and $\text{Rh}_6(\text{CO})_{16}$ as the catalyst [1124].

(3-20)   **(3-21)**

The substituted carbonyl clusters $Rh_6(CO)_{10}L_6$ (e.g. L = PPh_3 or $P(OMe)_3$) are good catalysts for cyclohexene reduction [1465]. Infrared spectroscopic results establish that the major part of the cluster frame is retained throughout the catalytic reaction. There is a report that $Rh_2(CO)_4(PPh_3)_2 \cdot 2C_6H_6$ can be used to catalyse the hydrogenation of 1,3-dienes to alkenes [1418], and that $Rh_2(CO)_2(Ph_2PCH_2PPh_2)_2$ is active for the hydrogenation of acetylene [1013]. The related cationic complex $[Rh_2(\mu\text{-}H)(\mu\text{-}CO)(CO)_2(Ph_2PCH_2PPh_2)_2]^+$ in 1-propanol or bis(2-ethoxyethyl) ether solution catalyses the hydrogenation of alkenes with $\{CO + H_2O\}$ as the source of [H] [1013].

The Ir—carborane complex $2\text{-}[Ir(CO)(PhCN)(PPh_3)]\text{-}7\text{-}C_6H_5\text{-}1,7\text{-}(\sigma\text{-}C_2B_{10}\text{-}H_{10})$ catalyses the homogeneous hydrogenation of 1-alkenes and alkynes [1218]. Hydrogenation of *trans*-1,3-pentadiene occurs specifically at the terminal C=C bond with $[Rh(CO)_2Cl]_2$ as the catalyst [1063]; this reaction occurs under mild conditions (60—80 °C, 1 atm pressure H_2).

Some chelate complexes $Rh(CO)_2(L-L')$ and $Rh(CO)(L-L')(PPh_3)$ have been used in hydrogenation reactions. $Rh(CO)_2(acac)$ is an effective catalyst for the hydrogenation of 4-vinyl-1-cyclohexene [268], and other unsaturated substrates [145], but the triphenylphosphine derivative is not very active. Species formed by adding $NaBH_4$ to $Rh(CO)_2(L-L')$ (L—L' = acac or 8-hydroxyquinolate) or $Rh(CO)(acac)(PPh_3)$ [1579] are catalysts for the selective hydrogenation of 1,5,9-cyclodecatriene to cyclodecane.

The nitrosyl complexes $Rh(NO)L_3$ (L = PR_3 or AsR_3) assist the hydrogenation of terminal and cyclic alkenes, and the activity of the catalyst decreases in the order L = $P(p\text{-}MeOC_6H_4)_3 \simeq P(p\text{-}MeC_6H_4)_3 > PPh_3 > P(p\text{-}FC_6H_4)_3 \simeq AsPh_3 > PMePh_2$ [341, 433, 449]. The effectiveness of these catalysts is increased by the addition of excess L. The related nitrosyl complex $[Rh(NO)(PPh_3)(p\text{-}benzoquinone)]$ catalyses the hydrogenation of 1-hexene, cyclohexene, styrene and 1,3-cyclohexadiene, but the catalytic activity is low [298]. With the nitrosyl-catecholato complexes $Ir(NO)(1,2\text{-}O_2C_6H_4)(PPh_3)$ and $Ir(NO)(1,2\text{-}O_2\text{-}C_6Br_4)(PPh_3)$, the homogeneous hydrogenation of cyclic alkenes has been achieved [596]. The second complex is much less active due to the electron-withdrawing effect of the bromides.

Binuclear rhodium complexes of the type $[Rh_2(\mu\text{-}SR)_2\{P(OR')_3\}_4]$ (e.g.

R = R′ = Me, Ph; R = Bu^t, R′ = CH$_2$Ph) are very active hydrogenation (and hydroformylation) catalysts [900, 901, 902]. Turnover rates as high as 15 min^{-1} are observed at 20 °C and 1 atm pressure H$_2$ in the hydrogenation of 1-hexene or cyclohexene.

A few other rhodium catalysts with π-acid ligands have been used in hydrogenation reactions. The bipyridyl cations [Rh(bipy)$_2$]$^+$ and [Rh(bipy)S$_2$]$^+$ are suitable catalysts for the selective hydrogenation of unsaturated ketones (e.g. cyclohexanones) in the presence of alkenes [1167]. An unusual catalyst has been formed by the co-condensation of rhodium metal vapor with toluene at liquid nitrogen temperature. When this system is allowed to warm, [Rh(MeC$_6$H$_5$)$_x$] is obtained, and this catalyses the hydrogenation of alkenes and arenes under mild conditions [1866].

3.5 Catalysts Derived from Alkene and Diene Complexes of Rh^I and Ir^I

FORMATION OF CATALYSTS FROM [M(ALKENE)$_2$Cl]$_2$;

HYDROGENATION OF ALKENES AND RELATED SUBSTRATES

Alkene or cycloalkene complexes of rhodium(I) are the source of a number of hydrogenation catalysts. Catalysts have been obtained by the addition of a tertiary phosphine (L) to a solution of [Rh(C$_2$H$_4$)$_2$Cl]$_2$ or [Rh(C$_8$H$_{14}$)$_2$Cl]$_2$, and it seems likely that the active catalyst in these systems is RhClL$_2$ [1110]:

$$\tfrac{1}{2}[\text{Rh(alkene)}_2\text{Cl}]_2 + 3\,\text{L} \longrightarrow \text{RhClL}_3 \rightleftharpoons \text{RhClL}_2 + \text{L}$$

These catalysts have been used for the hydrogenation of alkenes, polyenes and alkynes [333, 850, 938, 1658]. In the hydrogenation of vegetable oils, the rate and selectivity have been correlated to the phosphine/metal ratio in the catalyst system [538]. Catalysts for alkene hydrogenation have been formed by the addition of ligands other than tertiary phosphines; these include acetonitrile [1132], ethylenediamine [1429], 2-aminopyridine [1429, 1430] and dialkyl sulfates [1428]. The catalyst formed with 2-aminopyridine is claimed to be 10-fold more active in the hydrogenation of cyclohexene than RhCl(PPh$_3$)$_3$ [1969]. The reduction of unsaturated carboxylic acids is achieved when L = DMA (containing some Cl$^-$) or Et$_2$S [858].

Related catalysts obtained from [Ir(C$_8$H$_{14}$)$_2$Cl]$_2$ can be used for the hydrogenation of alkenes [307, 711, 1843]. However, there is some isomerization of the alkene in these reactions. The stereoselective reduction of 4-t-butylcyclohexanone (3-22 in Equation (3.12)) to 4-t-butylcyclohexanol has been achieved with a catalyst formed from [Ir(C$_8$H$_{14}$)$_2$Cl]$_2$ and (MeO)$_2$POH [141].

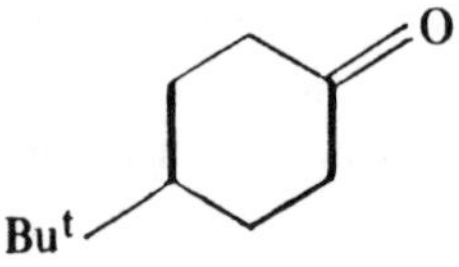

(3-22)

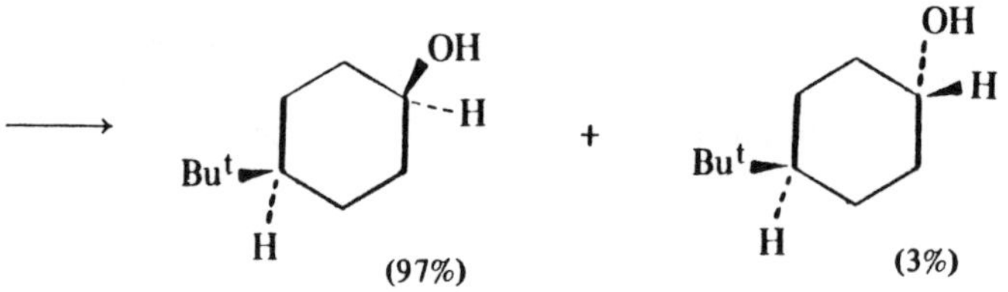

(97%)   (3%)   (3.12)

FORMATION OF CATALYSTS FROM [M(diene)Cl]$_2$

Hydrogenation of alkenes and related substrates

The formation of catalysts from various [Rh(diene)Cl]$_2$ complexes (e.g. diene = COD, NBD) has been studied in considerable detail [239, 1542, 1543, 1570]. The addition of a tertiary phosphine, arsine or phosphite (L) to the diene complex gives [Rh(diene)L$_a$]$^+$ (a = 2 or 3), and treatment with H$_2$ of solutions of this cationic species in polar solvents (e.g., S = acetone or tetrahydrofuran) results in displacement of the diene (Equation (3.13)).

$$[Rh(diene)Cl]_2 \xrightarrow{\text{L}} [Rh(diene)L_a]^+ \xrightarrow[\text{S}]{\text{H}_2} [H_2RhL_aS_b]^+$$
$$\underset{\xrightarrow{\hspace{1cm}}}{\overset{-H^+}{\rightleftharpoons}} \quad HRhL_aS_c \qquad (3.13)$$

The equilibrium between the cationic and neutral hydrido species is sensitive to the nature of L and S, and can be shifted by the addition of acid or base. The neutral complex is a powerful hydrogenation catalyst, but there is concommitant isomerization of the alkene. The cationic species is a moderately active hydrogenation catalyst but a poor isomerization catalyst.

The {[H$_2$RhL$_a$S$_b$]$^+$ $\rightleftharpoons$ HRhL$_a$S$_c$} system (L = PPh$_3$, S = EtOH, dioxan or DMF) has been used for the hydrogenation of alkenes [1570]. When these catalysts are used in primary alcohols, H-transfer from the alcohol to the alkene can occur alongside the main hydrogenation reaction [991]. This leads to the formation of aldehydes which are in turn decarbonylated by the catalyst to form Rh(CO)Cl(PR$_3$)$_2$. In this way, the catalyst is deactivated.

Catalysts can also be formed by the treatment of diene—rhodium complexes with a chelating diphosphine or diarsine (L—L). These catalysts behave differently to those formed from unidentate PR_3 ligands, and this can be understood in terms of stereochemical arguments. When $[Rh(NBD)(L-L)]^+$ (e.g. L—L = $Ph_2PCH_2CH_2PPh_2$) is treated with H_2, the norbornadiene is removed as norbornane [652, 1619], and the solvated cation 3-23 is formed. This cation does not react further with H_2, because *cis* addition of the hydrogen would form a species 3-24 in which H- is *trans* to P-; this is an unfavorable arrangement. The complex cations $[Rh(PR_3)_2S_2]^+$ (3-25) have greater flexibility, and *cis* addition of H_2 gives the dihydrido species 3-26 with a favorable *cis*-H,P arrangement. This effect changes the dominant kinetic pathway for the hydrogenation reactions with the two catalyst systems.

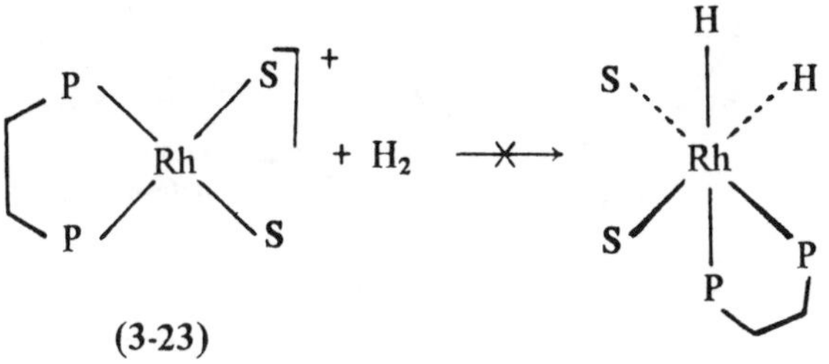

(3-23)   + H₂ —✕→   (3-24)

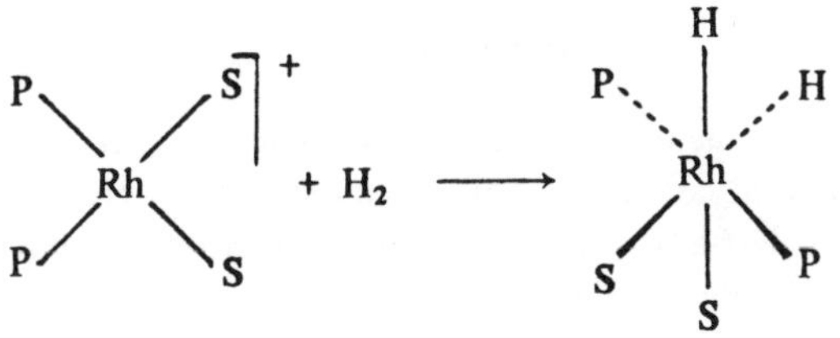

(3-25)   + H₂ ⟶   (3-26)

The solvated $[Rh(PR_3)_2S_2]^+$ system is similar to $RhCl(PPh_3)_3$ and follows a *hydride* pathway. In contrast, the solvated $[Rh(Ph_2PCH_2CH_2PPh_2)S_2]^+$ system seems to follow an *alkene* pathway as shown in Scheme 3.4. At low temperatures, the reductive-elimination of alkane from 3-28 becomes slower than oxidative-addition of H_2 to form 3-27. Under these conditions, there is accumulation of 3-28. The complex 3-29 has actually been intercepted and characterized by multinuclear NMR spectroscopy [304]; this is the first actual observation of an hydrido-alkyl—metal intermediate in a catalytic cycle.

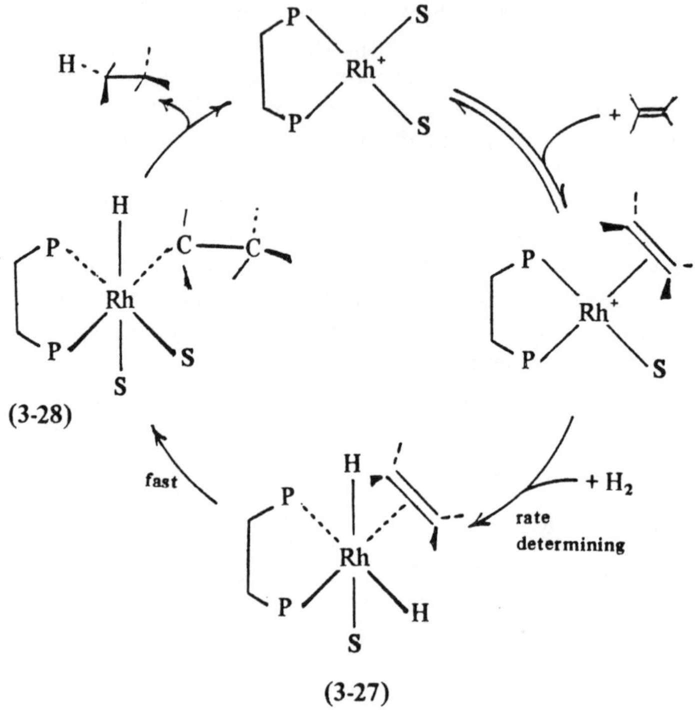

Scheme 3.4

A range of alkenes, including 1-hexene, styrene and acrylic acid can be hydrogenated under mild conditions with catalysts such as $[Rh(Ph_2PCH_2CH_2-PPh_2)S_2]^+$ in methanol [652]. By incorporating bidentate ligands of the type $o\text{-}C_6H_4\{CON(CH_2CH_2PPh_2)_2\}SO_3^-$ into the catalyst, the hydrogenation of alkenes can also be achieved in aqueous media [1274].

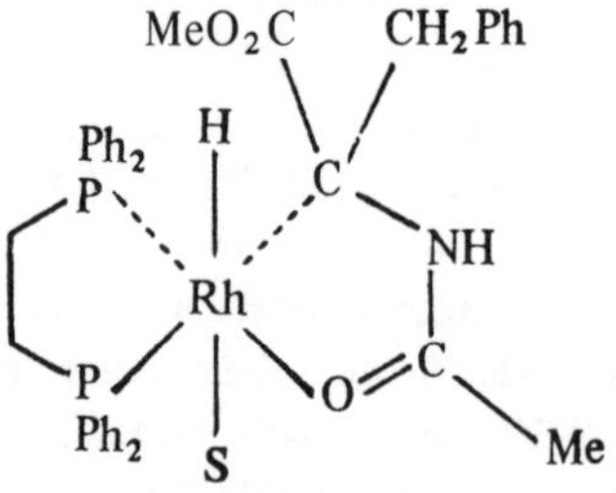

(3-29)

The catalyst derived from $\{[Rh(NBD)(PPh_3)_2]^+ + H_2\}$ achieves the rapid, selective hydrogenation of dienes such as NBD, 1,3-butadienes and 1,3-cyclohexadienes to monoenes [1545]. The selectivity of this reaction is attributed to the exceedingly rapid attack of H_2 on the $[Rh(diene)L_2]^+$ complex, which is the only species in solution in any significant concentration. The H_2 adds both 1,2 and 1,4 to conjugated dienes. The performance of this catalyst system is improved by the addition of H^+ which does not affect the rate of diene hydrogenation, but does slow the isomerization and hydrogenation of the monoalkene product.

There has been a detailed study [1844] of steric effects in the hydrogenation of some octadecadienes with $[RhL_2S_2]^+$ as the catalyst. In the absence of H_2, methyl-(Z,Z)-9,12-octadecadienoate is conjugated rapidly by catalysts with L = PEt_2Ph and PBu^i_3, but slowly when L = PMe_3. Consequently, hydrogenation of the conjugated intermediate rather than the 9,12-diene is achieved with $[Rh(PEt_2Ph)_2S_2]^+$, but with $[Rh(PMe_3)_2S_2]^+$ there is direct hydrogenation of one C=C bond. With these catalysts, hydrogenation of methyl-(E,E)-10,12-octadecadienoate proceeds almost exclusively via 1,4-addition of H_2 to the cisoid conformation, but a 1,2-addition mechanism is almost equally favored in the hydrogenation of methyl-(E,Z)-10,12-octadecadienoate. In the latter case, the trans-C=C bond is preferentially hydrogenated when the catalysts incorporate PMe_3 or PEt_2Ph, but with the bulkier phosphine PBu^i_3 the cis-C=C bond is reduced faster than the trans.

The bisdiene complex $[Rh(NBD)_2]^+$ also catalyses the hydrogenation of dienes such as norbornadiene, but there is extensive dimerization of the substrate [615, 1542]. The polycylic compound 3-30 is the major product of the hydrogenation reaction [1492].

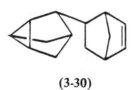

(3-30)

There is a high degree of stereoselection in the hydrogenation of allylic and homoallylic alcohols such as 3-phenyl-3-buten-2-ol with catalysts derived from $[Rh(NBD)(L—L)][BF_4]$ $(L—L = Ph_2P(CH_2)_aPPh_2, a = 4$ or 5). An example is shown in Equation (3.14). Up to 97% stereoselection in formation of (R,S)-threo-3-phenyl-butan-2-ol is achieved [244].

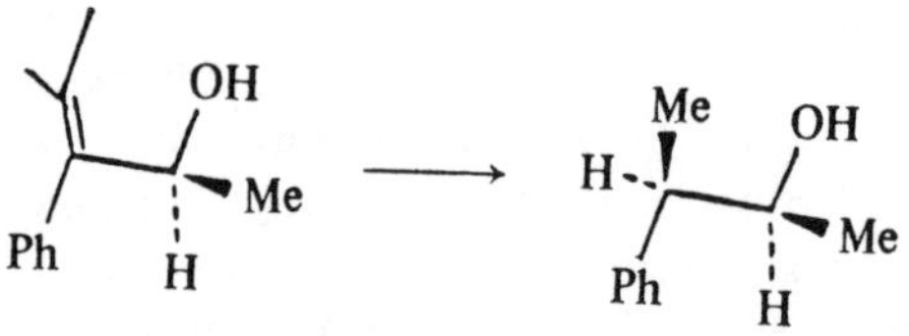

$$\text{(structure: alkene with OH, Me, Ph, H)} \longrightarrow \text{(structure: H, Me, OH, Ph, H)} \qquad (3.14)$$

The reduction of alkynes has been accomplished with the systems $\{[H_2\text{-}RhL_aS_b]^+ \rightleftharpoons HRhL_aS_c\}$ (L = PPh_3, S = EtOH; L = PMe_2Ph, S = acetone) [1544]. For example, 2-hexyne gives 99% *cis*-2-hexene with negligible isomerization.

The activity of $\{[Rh(1,5\text{-hexadiene})Cl]_2 + PR_3\}$ for the hydrogenation of styrene is increased by the addition of triethylamine [1249]*. Apparently, the amine in these systems is involved in the reduction of $Cl-Rh^I$ to $H-Rh^I$ species. Some related hydrogenation catalysts have been formed from $[Rh(COD)L_2]^+$ (e.g. L = MeCN [616, 617] or 2-ethylpyridine [1834]), $[Rh(COD)(PPh_3)py]^+$ or $Rh(COD)(O_2CPh)(PPh_3)$ in benzene containing $PhCO_2H$ and Et_3N [360]. The former systems are effective for the hydrogenation (and isomerization) of 1-hexene. The latter systems are highly selective for the hydrogenation of 1-alkynes to 1-alkenes.

Hydrogenation catalysts have been formed from a similar range of diene—iridium complexes by treatment with PR_3 and H_2 in polar solvents (see Equation (3.15)).

$$[Ir(COD)Cl]_2 \xrightarrow{\ PR_3\ } [Ir(COD)(PR_3)_2]^+$$
$$\xrightarrow{\ H_2/S\ } [H_2Ir(PR_3)_2S_2]^+ \qquad (3.15)$$

These iridium species are effective hydrogenation catalysts [1542, 1570, 1640], but they are less active than the corresponding rhodium compounds for the hydrogenation of alkenes. The activity of the Ir catalysts is affected by changes in the phosphine ligand, with PCy_3 showing the highest activity and selectivity [200, 406]. The hydrogenation of cyclic alkenes and dienes has been achieved with $[Ir(COD)(PMePh_2)_2]^+$ as the catalyst source [365]. However, the catalyst is irreversibly deactivated when alkene has been consumed, or in some cases,

---

* This catalyst system is also very active for the hydrogenation of aromatic nitro compounds to amines; the highest activity is found when the molar ratio of $Rh/PR_3/NEt_3$ = 1:1:2.3 [1025]. The hydrogenolysis of alkyl and aryl halides has also been achieved with this catalyst, and the rates of these reactions are enhanced by the addition of water [1026].

only partly consumed [365]. The *cis*-isomer of $[H_2 Ir(COD)L_2]^+$ hydrogenates coordinated cyclooctadiene at least 40 times faster than does the *cis, trans*-isomer, and this indicates that a coplanar $M(C=C)H$ system may be required for insertion of alkene into the $M—H$ bond [364, 365]. The selective hydrogenation of nonconjugated 1,5- and 1,4-cyclooctadiene to cyclooctene is achieved when the dimeric complex $CODIr(\mu\text{-}Cl)_2 IrH_2 (PPh_3)_2$ is used as a catalyst source [566].

The complex $[Ir(COD)(PCy_3)py] [PF_6]$ is another useful catalyst precursor [361, 362, 365]. It is active when used in dichloromethane at high dilution, but the activity is lost in acetone or other polar solvents. Low concentrations of the catalyst source must be used to inhibit formation of the catalytically inactive dimer $[H_2 Ir_2(\mu\text{-}H)_3(PCy_3)_4] [PF_6]$. The active catalyst is formed in the presence of H_2, presumably via reduction of the cyclooctadiene to cyclooctane. It is particularly effective for the reduction of highly substituted alkenes. For instance, a very high rate of hydrogenation of $Me_2 C=CMe_2$ has been observed. With terminal alkenes such as 1-hexene, the iridium catalyst is about 100 times more active than $RhCl(PPh_3)_3$ under comparable conditions. The catalyst is poisoned by excess PPh_3 but is not affected by O_2. The corresponding Rh complex is less effective by a factor of 10^6.

Further catalysts of this type have been formed from $[M(1,5\text{-hexadiene})\text{-}phen]^+$ [1966]. The iridium is more active than the rhodium system for the hydrogenation of alkenes to alkanes, and of alkynes to alkenes. The corresponding 1,5-COD—rhodium compound does not form an active catalyst.

Reduction of ketones and related substrates

The hydrogenation of ketones such as acetone, butan-2-one and cyclohexanone is catalysed by the $\{H_2 RhL_2 S^+ \rightleftharpoons HRhL_a S_b\}$ system. These reactions require alkyl-substituted phosphines (e.g. PEt_3, which is more basic than PPh_3 [557], or $Pr^i_2 P(CH_2)_3 PPr^i_2$ [1764]) and water or a proton base is required for ketone reduction. The mechanism shown in Scheme 3.5 has been suggested [1541]. The hydrogenation rates are increased when the ketone has electron-withdrawing substituents. In the hydrogenation of unsaturated ketones, saturated ketones and saturated alcohols are formed sequentially [557]. Aldehydes can be reduced with H_2, but catalyst deactivation occurs in the early stages of the reaction due to decarbonylation of the aldehyde and the formation of $Rh(CO)XL_2$ complexes [558].

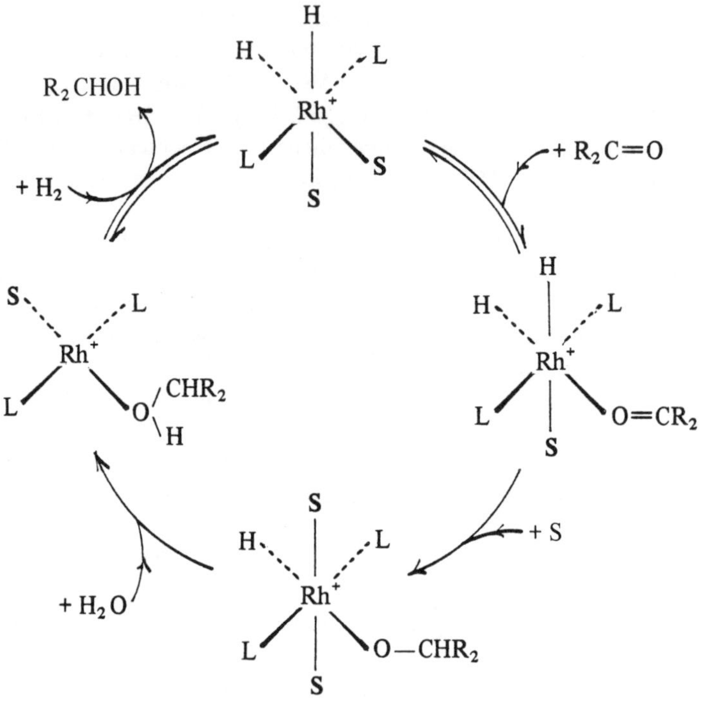

Scheme 3.5

Chelate complexes of the type $[Rh(1,5\text{-hexadiene})(N-N)]^+$ (e.g. $N-N$ = 2,2'-bipy, 1,10-phen) will also catalyse the hydrogenation of ketones when used in alkaline media (e.g. MeOH/NaOH). With excess chelate ligand, the selective reduction of C=O in the presence of C=C can be achieved [1168]. Another catalyst for the hydrogenation of aliphatic and aromatic ketones to alcohols is formed by treating $Rh(COD)Cl(PPh_3)$ with $NaBH_4$ to produce $H_2Rh_2(COD)Cl_2(PPh_3)_2$ [567]. The reduction of Me_2CO to Me_2CHOH has also been achieved with $[Rh(COD)ClL]_2$ (L = benzothiazole) as the catalyst [77].

The hydrogenation of styrene oxide to give $PhCH_2CH_2OH$ and $PhCH_2CHO$ has been achieved with catalysts derived from $[Rh(NBD)L_a]^+$ (e.g. $L_a = (PMe_3)_3$ or $Ph_2PCH_2CH_2PPh_2$) [559, 1208]. The selectivity of formation of these products rather than $PhCH(Me)OH$ or $PhCOMe$ is greatest when L = PEt_3 and least when $L_2 = Ph_2PCH_2CH_2PPh_2$. The reduction of 3,4-epoxy-1-butene to $MeCH=CHCHO$ is achieved in similar manner [556]. In this reaction, the alcohols $MeCH=CHCH_2OH$ and $H_2C=CHCH_2CH_2OH$ are formed as minor products, with the amount of alcohol formed decreasing in the order $L = PEt_3 >$ $Ph_2PCH_2CH_2PPh_2 > PPh_3 \gg PMe_3$.

CATALYSTS DERIVED FROM ALLYL—RHODIUM COMPLEXES

Some hydrogenation catalysts are derived from allyl—rhodium complexes. Thus, the hydrogenation of alkenes has been achieved in the presence of $Rh(\eta^3\text{-}C_3H_5)$- $(PPh_3)_2$ [329], $Rh(\eta^3\text{-}C_3H_5)\{P(OMe)_3\}_3$ [214], $[Rh(\eta^3\text{-}2\text{-}MeC_3H_4)Cl_2]_x$ [1425], $\{[Rh(\eta^3\text{-}C_3H_5)_2Br]_2$ or $[Rh(\eta^3\text{-}C_3H_5)Br_2]_x$ + phosphines, amines or sulfides$\}$ [1369] and $\{[Rh(\eta^3\text{-}2\text{-}MeC_3H_4)Cl_2]_x$ + $Ph_2PCH_2CH_2PPh_2\}$ [1426]. The addition of H_2 to the latter system in ethanol gives $RhCl_2(Ph_2PCH_2CH_2\text{-}PPh_2)$ and trans-$[HRhCl(Ph_2PCH_2CH_2PPh_2)_2]^+$, and the hydrido complex is claimed to be amongst the most active hydrogenation catalysts known. It seems likely that the $Rh(\eta^3\text{-}C_3H_5)L_2$ systems form $[HRhL_2]_x$ under hydrogenation conditions [405, 1618].

3.6. Complexes of M^{III} as the Catalyst Source

A variety of useful hydrogenation catalysts has been derived from complexes of rhodium(III) and iridium(III). These catalyst sources are described in this sub-section.

$\{RhCl_3py_3 + NaBH_4\}$

The addition of 1 equiv of $NaBH_4$ to $RhCl_3py_3$ in DMF gives a pink—brown solution which is an active hydrogenation catalyst for a wide range of unsaturated compounds; these systems are often referred to as *McQuillin catalysts*. A complex of formula $RhCl_2py_2(DMF)BH_4$ has been isolated from the reaction mixture, but in solution under H_2 a transient hydrido—rhodium(I) species probably forms and becomes the active catalyst*.

McQuillin catalysts have been used to hydrogenate a variety of unsaturated substrates. There is very rapid hydrogenation of terminal alkenes under mild conditions [1, 869, 870]. With cycloalkenes, the rate order is cyclohexene > cyclopentene > cycloheptene > cyclooctene [1579]. Studies of the deuteration of unsaturated acids, including maleic and fumaric acids [2], have established that *cis*-hydrogenation products are obtained. *cis*-Hydrogenation also occurs with alkynes such as $PhC{\equiv}CPh$, $MeO_2CC{\equiv}CCO_2Me$ and $HO_2CC{\equiv}CCO_2H$ [2].

---

* Although it is generally thought that McQuillin catalysts are homogeneous, recent evidence based on tests with dibenzo[a,e]cyclooctatetraene establishes that finely divided metal is present (D. R. Anton and R. Crabtree, *Organomet.* 2, 855 (1983)). Thus, this catalyst really behaves as a heterogeneous system.

However, these reactions are not very specific due to further hydrogenation and isomerization of the alkene [1083]. Some polyenes have been reduced. For example, cyclodecene is obtained from 1,5,9-cyclododecatriene [1579], and methyl linoleate gives a monoenoic ester [3]. In the latter case, there is extensive isomerization to form a *trans*-alkene. The reduction of steroid-4-en-3-one compounds has been achieved with McQuillin catalysts [1, 869]; attempted hydrogenations of these steroids in the presence of RhCl(PPh$_3$)$_3$ were not successful. *N*-Heteroaromatics, including pyridine and quinoline, yield saturated heterocycles [1, 870], but isoquinoline is not reduced.

Other unsaturated functional groups have been reduced with {RhCl$_3$py$_3$ + NaBH$_4$} as the catalyst source. Although the ketones acetophenone, benzophenone and benzoin can be reduced [1086], the reactions are exceedingly slow (1–3 d). Amines are obtained from nitro compounds such as nitrocyclohexane and nitrobenzene [1086]. There is reduction of the C=N bond in benzylideneaniline to give benzylaniline [1], and hydrazobenzene is formed from azobenzene [870].

When optically active amines such as PhCH(Me)NHCHO, [PhCH(Me)NHCO]$_2$ or MeCH(OH)C(O)NMe$_2$ are used as solvents, there is induced asymmetry in the hydrogenation product [869, 1156]. For example, methyl-*trans*- or -*cis*-phenylbut-2-enoate gives (+)- or (−)-methyl-3-phenylbutanoate in greater than 50% optical yield.

Some closely related catalyst systems have been used in hydrogenation reactions. Thus, {RhCl$_3$ + *p*-(HO)C$_6$H$_4$CH$_2$C(NH$_2$)HCO$_2$H or *o*-PhNHC$_6$H$_4$CO$_2$H + NaBH$_4$} will catalyse the hydrogenation of maleic acid, diethylmaleate and cyclohexene [96, 97], and the hydrogenation of dienes such as *cis*- and *trans*-1,3-pentadienes to β-alkenes has been achieved with {RhCl(PPh$_3$)$_2$(DMSO) + NaBH$_4$}as the catalyst [546].

(η-C$_5$Me$_5$)M^{III} COMPLEXES

Pentamethylcyclopentadienyl complexes of the type M$_2$(η-C$_5$Me$_5$)$_2$X$_4$ (M = Rh or Ir; X = Cl, Br or I; M = Rh, X = NO$_3$), M$_2$(η-C$_5$Me$_5$)$_2$HX$_3$ (M = Rh, X = Cl or Br; M = Ir, X = Cl, Br or I) and Ir$_2$(η-C$_5$Me$_5$)$_2$H$_2$X$_2$ (X = Cl or Br) can be used as hydrogenation catalysts [594, 1111, 1113, 1501, 1908]. In 2-propanol these complexes are particularly effective in the presence of Et$_3$N. The amine presumably promotes the formation of an Rh—H bond by abstraction of HCl. Substrates that have been hydrogenated with M$_2$(η-C$_5$Me$_5$)$_2$HX$_3$ include the alkenes H$_2$C=CHPr^i, PrMeC=CH$_2$, and *cis*- and *trans*-Pr^{i}CH=CHMe, and the cyclic systems cyclohexene and cyclopentene. The rhodium complex Rh$_2$(η-C$_5$-Me$_5$)$_2$Cl$_4$ catalyses the stereoselective hydrogenation of arenes to cyclohexanes, with the all-*cis*-isomers being obtained. Typical conditions for these reactions are a temperature of 50 °C, an H$_2$ pressure of 50 atm and the incorporation of 15 equiv of Et_3N. The use of these catalysts for the hydrogenation of alkenes and arenes has been reviewed by Maitlis [1112].

The hydroxy complex $[Rh_2(\eta\text{-}C_5Me_5)_2(OH)_3]Cl$ can be used in aqueous solution as an hydrogenation catalyst. With HCHO as the reductant, aldehydes and ketones are converted mainly to alcohols [349]; appropriate conditions are 50 °C and pH > 12.5. Mixtures of $[Rh_2(\eta\text{-}C_5Me_5)_2(OH)_3]^+$ and $[Rh(COD)Cl]_2$ (a 4:1 mixture is the optimum catalyst) have been used in 2-propanol for the hydrogenation of cyclohexene [654].

HYDRIDO—Ir^{III} COMPLEXES; $H_aIrX_bL_c$

The hydrogenation of terminal alkenes can be catalysed by various hydrido—iridium complexes including $H_5Ir(PPh_3)_2$ [1126], $H_3IrL_{2\,or\,3}(L = PPh_3$ or $PEt_2Ph)$ [335, 535, 598, 817], $HIrCl_2L_3$ (L = PPh_3, $AsPh_3$ or $SbPh_3$) [535, 1932] and $H_2Ir(SnCl_3)L_2$ (L = PR_3) [1833]. The $H_3IrL_{2\,or\,3}$ complexes have also been used in the hydrogenation of phenylacetylene to styrene plus ethylbenzene, and of 1,5-cyclooctadiene to cyclooctene [335]. The addition of H_2O_2 to $HIrCl_2(PPh_3)_3$ accelerates the hydrogenation of a variety of substrates including allyl alcohol, diethylfumarate and geraniol [1741]. In contrast to several other iridium hydrogenation catalysts, $H_3Ir(PPh_3)_3$ also achieves the reduction of C=O bonds in ketones and aldehydes [1716].

In acetic acid, $H_3Ir(PPh_3)_3$ forms the acetato complexes $HIr(O_2CMe)_2$-$(PPh_3)_3$ and $H_2Ir(O_2CMe)(PPh_3)_3$, and these species catalyse the hydrogenation of aldehydes and ketones to alcohols [332, 337]. The hydrogenation of alkenes and ketones can also be accomplished in trifluoroacetic acid with $H_3Ir(PPh_3)_3$, $H_5Ir(PPh_3)_2$, $HIrCl_2(PMe_2Ph)_3$ or $IrCl_3(PEt_2Ph)_3$ as the catalyst source [899, 1126]. These reactions are thought to proceed via a carbocation and H^- transfer from the catalyst. The $H_3Ir(PPh_3)_3$ in trifluoroacetic acid system has also been used to form $8\alpha,9\alpha,14\beta$-estrones from dehydroestrones [1850], and in the reduction of imines such as PhCH=NPh [898].

MISCELLANEOUS Rh^{III} COMPLEXES

Other rhodium(III) compounds which have been used to produce catalysts for the hydrogenation of alkenes, dienes and cycloalkenes include: fac-$RhCl_3py_3$ [595], mer-$RhCl_3(PPh_3)_3$ [435, 1344], $RhCl_3(SR_2)_3$ (R = Et or CH_2Ph) [859], $HRh(HDMG)_2(PPh_3)_3$ [1485], $\{RhCl_3$ + N-chelates such as o-phen$\}$ [549], $\{RhCl_3$ + 1,3,5-triphenylbenzene in DMA$\}$ [95] and $H_2Rh(PPh_3)_2$-(O_2CR) (e.g. R = C_5H_{11}, Ph) [1248]. The reduction of aldehydes such as PhCHO and EtCHO has been achieved at 240 atm pressure and 200 °C with a 1:1 H_2/CO mixture in the presence of $RhCl_3 \cdot 3H_2O$ [691]; under these conditions, it seems likely that the active catalyst is a carbonyl—rhodium(I) species. Aromatic nitro compounds are reduced to amines in the presence of catalysts formed from $RhCl_3 \cdot 3H_2O$ and potassium indigo disulfonate [317]. Kinetic results are consistent with a mechanism that involves the formation of $PhNO_2^{\ddagger}$ and then $PhNO$.

3.7 Catalysts Derived from Rh^{II}, Rh_2^{4+}

Rhodium(II) carboxylates such as $Rh_2(O_2CMe)_4$ are effective hydrogenation catalysts when used in the absence of strong, π-coordinating ligands [786, 767]. These catalysts are most effective in solvents such as DMF, DMA, dioxan or THF, and they are selective for the hydrogenation of terminal alkenes. Kinetic evidence indicates that $Rh_2(O_2CR)_4$ is the active species. The initial step in the mechanism is activation of H_2 at only one of the metal centers of the dimer to form $HRh_2(O_2CR)_3$. Subsequent coordination of the alkene is followed by *cis* migration to give an alkyl rhodium intermediate which yields the alkane by hydrogenolysis with H^+.

The Rh_2^{4+} ion is formed when $Rh_2(O_2CMe)_4$ is dissolved in strongly acidic media such as methanolic HBF_4. Although this ion is not a hydrogenation catalyst, it becomes active in the presence of triphenylphosphine at a P/Rh ratio of 2 [1074]. This system will catalyse the hydrogenation of alkenes, alkynes and other unsaturated substances, and it operates in polar media such as methanol. Treatment of Rh_2^{4+} with 3 equiv of triphenylphosphine and an excess of a lithium carboxylate gives $Rh(O_2CR)(PPh_3)_3$, and these compounds will also catalyse the homogeneous hydrogenation of alkenes and alkynes [1192, 1921]. For solutions of $Rh(O_2CMe)(PPh_3)_3$ in methanol with added p-toluene sulfuric acid, the rate of hydrogenation is highly dependent on the acidity of the solution [1645]. This acidified system achieves the rapid, highly selective reduction of cyclic dienes to monoenes. The reduction of $Rh_2(O_2CMe)_4$ with LiH in ethanol provides another catalyst for the hydrogenation of alkenes such as cyclohexene [1970].

3.8 Reductions via H-transfer from Alcoholic Solvents

Some rhodium and iridium compounds promote the transfer of [H] from an alcohol to an unsaturated organic substrate. Perhaps the best known of these are the so-called *Henbest catalysts* which consist of the iridium(IV) salts Na_2IrCl_6 or H_2IrCl_6 dissolved in aqueous 2-propanol containing a phosphite such as $P(OEt)_3$. They are particularly effective for the reduction of ketones to alcohols (Equation (3.15)) [644, 700].

$$\underset{R-C=O}{\overset{Me}{\mid}} + \underset{Me-CHOH}{\overset{Me}{\mid}} \quad \rightleftharpoons \quad \underset{R-CHOH}{\overset{Me}{\mid}} + \underset{Me-C=O}{\overset{Me}{\mid}} \qquad (3.15)$$

For example, cyclohexanones (**3-31**) are reduced to axial alcohols (**3-32**) with a high degree of selectivity. This system has also been applied to steroid ketones, and it is highly specific for the reduction of the 2- and 3-oxo groups [249, 1340]; keto groups in the 11-, 17- and 20-positions are unaffected. This specificity is illustrated by the almost quantitative conversion of **3-33** to **3-34**.

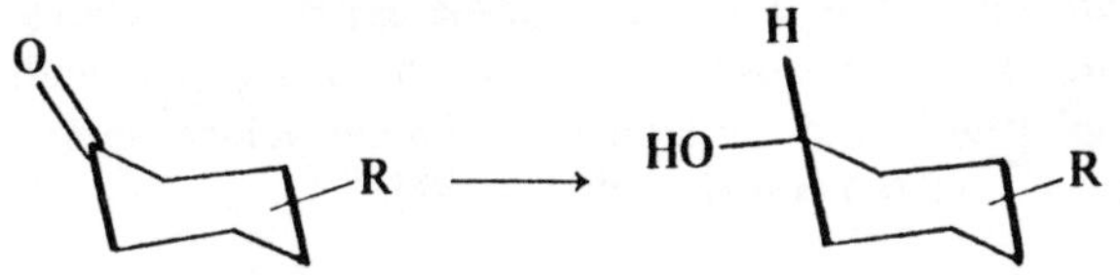

(**3-31**)   (**3-32**)

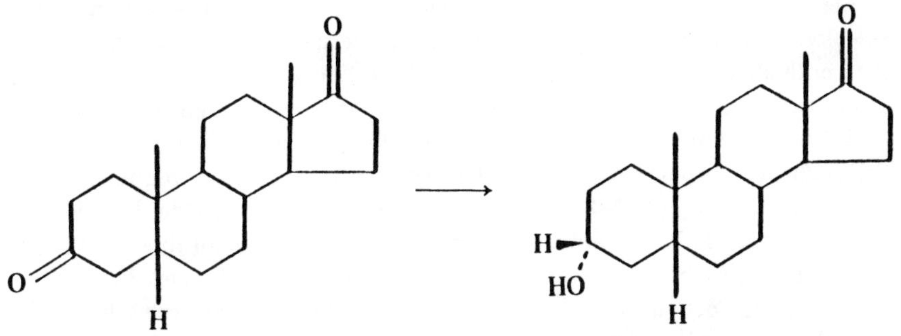

(**3-33**)   (**3-34**)

In some systems, DMSO has been used in place of the phosphite. Thus, the adducts $IrCl_3(DMSO)_3$ and $H[IrCl_4(DMSO)_2]$ will catalyse the H-transfer reaction between cycloalkanones and acyclic alcohols [641, 644]. The reductions of unsaturated ketones such as $PhC(=O)CH=CHPh$ to the saturated ketone [701], and of the unsaturated aldehydes $RCH=CH'CHO$ (R = Ph, R' = H or Me; R = Me, R' = H) to the unsaturated alcohols $RCH=CR'CH_2OH$ [857], have also been achieved with $HIrCl_2(DMSO)_3$ in 2-propanol. The DMSO in these systems stabilizes the iridium against reduction to the metal. Another modification to the Henbest catalyst is the inclusion of $SnCl_2$ in the system [930].

Some chelate complexes of iridium(I), Ir(COD) (L—L)Cl (e.g. L—L = bipy, phen, 4,7-Me$_2$phen), have been used to catalyse H-transfer from alcohols to ketones [1171]. The rate and stereoselectivity of the reactions depend on the concentrations of water and KOH in the systems. The most active of the iridium complexes is the 3,4,7,8-tetramethylphenanthroline derivative. With this catalyst in weakly alkaline media, turnover rates up to 900 min^{-1} have been achieved for H-transfer from 2-propanol to α,β-unsaturated ketones [274, 275, 1170].

Even more active H-transfer catalysts are obtained from $Ir(H_2C=CH_2)_2$ $(3,4,7,8\text{-}Me_4phen)Cl$ [1127].

The hydrido-phosphine complexes $H_3Ir(PPh_3)_2$ and $H_5Ir(PPh_3)_2$ promote H transfer from ethanol to diphenylacetylene giving *trans*-stilbenes [1962]. The active catalyst in these systems is probably $HIr(PPh_3)_2(EtOH)$.

Rhodium(III) halides are much less effective in these types of reactions. The reduction of 3-hexene and of cyclopentadiene has been achieved with $RhCl_3 \cdot 3H_2O$ in an alcohol [992], and the reduction of alkylcyclohexanones has been accomplished in a similar manner with $Rh(NO)Cl_3L_2$ compounds [1571].

The reduction of some dimethylcyclohexanones to the corresponding alcohols has been achieved by H-transfer from 2-propanol in the presence of $RhCl(PPh_3)_3$ [1571, 1573]. The structure of the cyclohexanone affects the rate and stereoselectivity of these reactions; reduction is slower when one methyl is axial than when both are equatorial. These factors are useful in controlling the reduction of steroid ketones, with the 2-, 3- and 17-carbonyl groups being reduced selectively [1340]. The reduction of some substrates by 2-propanol in the presence of $RhCl(PPh_3)_3$ is promoted by KOH. This system has been used to obtain aniline from nitrobenzene [1574], azobenzene [1577], azoxybenzene [1575] and hydrazobenzene [1528]; hydrazobenzene is the most reactive of these substrates. The reduction of benzylideneaniline [1572], 4-alkylpiperidinones [1576] and benzaldimines such as $PhCH=NCH_2Ph$ [629] has been achieved in similar manner. Presumably, hydrido—rhodium(I) species are formed in these systems.

The hydrido complex $HRh(PPh_3)_4$ has been used directly to promote H-transfer from the alcohol $PhCH(CH_3)OH$ to the α,β-unsaturated ketones $RC_6H_4CH=CHCOR'$ (R = H, p-NO, p-Me$_2$N, p-MeO, m-Cl; R' = Me, Ph, Bu^t) [133]. The rates of reduction are increased when the groups R and R' are electron-attracting, and decreased when they are electron-withdrawing. In these reactions, the order of addition of the catalyst and reactants can have an unusual effect on the initial H-transfer rates due to competition between the alcohol and the ketone for the catalyst [132, 134]. To maintain high catalytic activity, the ketone should be added to a solution of the catalyst in the secondary alcohol; the nature of the solvent is also important [412]. With both $RhCl(PPh_3)_3$ and $HRh(PPh_3)_4$, there can be hydrogenolysis of the ligand to form benzene if the catalyst is heated strongly in the H-transfer solvent [1264].

Some other rhodium(I) catalysts have been generated *in situ*. One is formed from $[Rh(COD)Cl]_2$ and R_2NPPh_2, and this promotes the hydrogenation of cyclohexanones by H-transfer from 2-propanol [1738]. Others are obtained from $RhCl_3 \cdot xH_2O$ and triphenyl-phosphine or -phosphite in ROH (R = Me, Et or Pr^i) containing some sodium carbonate. These achieve the hydrogenolysis of organic chlorides such as p-MeC$_6$H$_4$CH$_2$Cl and PhCOCH$_2$Cl to the corresponding hydrocarbons and ethers [625].

The chelate complexes $[Rh(N-N)(1,5\text{-hexadiene})](PF_6)$, $[Rh(N-N)_2-Cl_2]Cl$ (e.g. $N-N$ = bipy, phen, $4,4'\text{-Me}_2$bipy, $4,7\text{-Me}_2$phen) [1169, 1836, 1965] and $[Rh(diene)P_2]^+$ (diene = COD or NBD, P = unidentate phosphine or P_2 = bidentate phosphine) [1650, 1651, 1836] are very active catalysts for H-transfer from 2-propanol to cyclic ketones and other unsaturated substrates. Particularly high rates and selectivities are observed with $[Rh(4,7\text{-Me}_2\text{phen})_2-Cl_2]Cl$ and $[Rh(COD)(Ph_2PCH_2CH_2PPh_2)]^+$.

The cyclooctene complexes $MCl(C_8H_{12})PPh_3$ (M = Rh, Ir) have been used to catalyse the transfer of H from secondary alcohols to alkenes such as 1-pentene and cyclohexene [539]. In addition to Pr^iOH, $PhMeCHOH$ and Ph_2CHOH are suitable H-donors.

H-transfer from solvents other than an alcohol has been observed in a few instances. Thus, $RhCl(PPh_3)_3$ promotes the transfer of H from dioxan to an alkene [1262]. In a related reaction, $\{[Rh(alkene)_2Cl]_2 + L\}$ was used as the catalyst source [1135]. The deuteration of aromatic hydrocarbons such as toluene, pyridine and acetophenone has been accomplished by D-transfer from D_2O. The hydrido complexes $HRh(PPr^i_3)_3$ and $H_2Rh_2(\mu\text{-}N_2)(PPr^i_3)_4$ are suitable catalysts for these reactions which are thought to proceed via oxidative-addition of D_2O to the rhodium(I) complexes [1942]. In some cases, H-transfer from the solvent can lead to hydrogenolysis of the catalyst. This occurs, for example, with pyrrolidine (L) [1263]:

$$RhCl(PPh_3)_3 + L \longrightarrow RhCl(PPh_3)_2L \xrightarrow[6\ h]{100\ ^{\circ}C} \text{benzene (8 mol)}$$

The benzene is formed by stepwise hydrogenolysis of phenyl groups from the triphenylphosphine.

3.9 Supported Hydrogenation Catalysts

A variety of rhodium complexes, including those used for homogeneous hydrogenation reactions, have been attached to insoluble supports. There should be some practical advantages in using these supported catalysts rather than their soluble analogs. In particular, the ease of separation from reactants and products [42, 635, 848, 1407] and the ability to recycle the catalyst [848] should be greatly improved. Moreover, there should be no loss in catalyst activity due to aggregation provided the active sites are isolated from each other on the support [635, 1463]. This isolation of active sites might also enable multistep, sequential catalytic reactions to be carried out [848, 1406]. In fact, the widespread use of supported catalysts for hydrogenation reactions has been hampered by a number of practical and other problems [328]. These will be discussed below.

POLYMERIC SUPPORTS

Functionalization of polymers

Several different polymeric supports have been used for transition metal catalysts [1077]. A common example is a 2% divinylbenzene—styrene copolymer
substituted with phosphine groups which are the sites of attachment of the
metal catalyst. One of the first homogeneous catalysts to be supported on a
polymer of this type was $RhCl(PPh_3)_3$, and alternative procedures for the
formation of such a polymer-attached catalyst are depicted in Schemes 3.6 (a)
[636] and (b) [638].

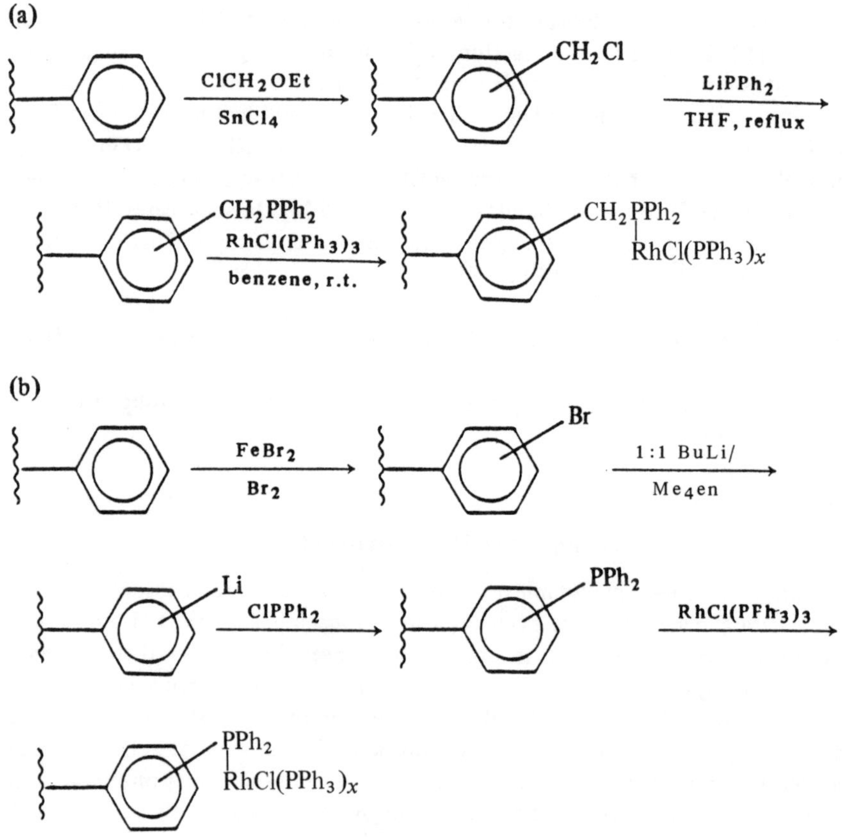

(a)

(b)

Scheme 3.6

In these examples, phosphine groups are introduced into a preformed polymeric resin (actually, some or all of these steps can now be avoided because various chloromethylated and phosphine-substituted polystyrenes are available commercially). It is possible to have the phosphine group further removed from the polymeric backbone. For example, treatment of crosslinked (aminomethyl) polystyrene with $H_2C=CH(CH_2)_8COCl$ and subsequent functionalization with $-PPh_2$ gives supports of type **3-25** [242]. Supports with quaternized phosphines have also been prepared. As an example, $P-SO_3^-[(Me_2N)_3PH^+]$ is obtained from sulfonated styrene–divinylbenzene copolymer by treatment with $(Me_2N)_3P$ [951].

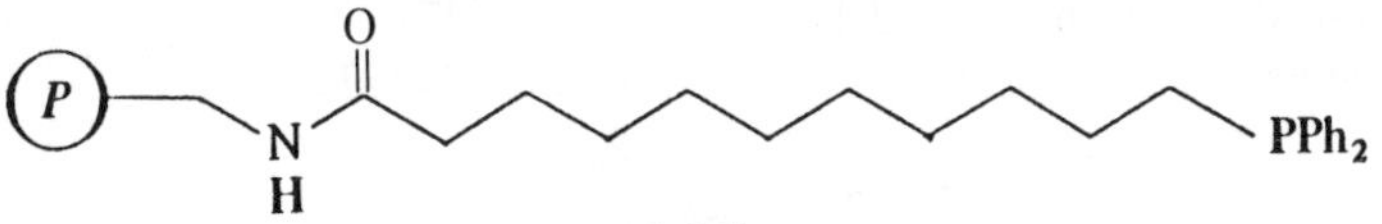

(3-35)

An alternative procedure which is sometimes used to obtain functionalized supports involves the polymerization of prefunctionalized monomeric units. The macro-ligand obtained by polymerizing ethyleneimine provides one example [1380]. Another has been obtained by copolymerizing 4-vinylpyridine with divinylbenzene [286, 288].

In most of the above examples, some divinylbenzene is incorporated into the polymer to achieve a degree of crosslinking, and hence greater insolubility in organic solvents. With 1–2% divinylbenzene crosslinks, gel-type polystyrene polymers are obtained. These are *microreticular* polymers – they must be swollen by solvents before the interior sites are accessible to the catalyst and substrate. This swelling of the gel introduces considerable flexibility to the polymer backbone, and unfortunately this enables widely spaced ligands to bind to the same metal [1463] – thus there is loss of the potential advantages of isolating the ligand sites.

Macroreticular polymers, which have larger diameter pores but are not appreciably swollen by added solvents, can be prepared by alternative procedures. Generally, the copolymerization of styrene and divinylbenzene is achieved in the presence of another organic component, which keeps the reacting monomers in solution but does not cause the crosslinked polymer to swell. It is found that these polymers can be functionalized only at the surface of the polymer bead, and hence they have limited capacity for the binding of metal complexes. Although the rigidity of highly crosslinked macroreticular resins should lead to good isolation of the ligand sites, there are indications that the isolation is time-dependent [117].

Soluble polymer supports can be obtained by attaching appropriate functional groups to noncrosslinked polystyrene [1550]. The relatively high molecular weight of these systems enables them to be separated from the reaction medium by membrane filtration. However, ligand site isolation must be poor.

Attachment of metal complexes; formation of catalysts

Since a considerable range of functional groups can be incorporated into polymers, most of the soluble rhodium compounds that form active hydrogenation catalysts can be converted to supported catalysts. The treatment of phosphine-functionalized polystyrene with $RhCl(PPh_3)_3$ to give catalysts of the type **3-36** has already been mentioned [126, 611, 636, 637]; the kinetics of formation of this and related catalysts have been investigated [713].

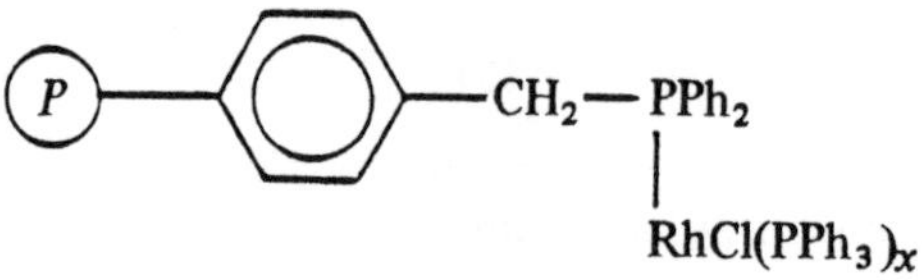

$$\boxed{P} \!-\!\! \bigcirc \!\!-\! CH_2 \!-\! PPh_2$$
$$RhCl(PPh_3)_x$$

(3-36)

These catalysts have been used for the reduction of alkenes such as 1-hexene and cyclohexene. The rate of reduction depends on the extent of swelling of the support and the molecular size of the substrate, with the catalyst showing selectivity for smaller alkenes that can more readily diffuse into the polymer. When the catalyst is suspended in polar solvents, the reduction of nonpolar alkenes in the presence of polar alkenes is achieved. Related catalysts have been formed from $P\!-\!PPh_2$ and $[Rh(C_2H_4)_2Cl]_2$ [149, 642, 1244], and from $\{poly[1,6\text{-}bis(p\text{-toluene sulfonate})\text{-}2,4\text{-hexadiyne}\}\text{-}PPh_2$ and $[Rh(C_8H_{14})_2Cl]_2$ [560, 948]. Treatment of $P\!-\!CH_2C_2B_9H_{11}{}^-$ with $RhCl(PPh_3)_3$ gives an hydrido species $P\!-\!CH_2C_2B_9H_{11}\!-\!Rh(H)(PPh_3)_2$ which is not so effective as $RhCl(PPh_3)_3$ for the hydrogenation of alkenes [309].

Catalysts of the type $P\!-\!M(CO)_aXL_b$ have also been obtained. Some of these catalysts are formed from $M(CO)ClL_2$ (M = Rh, Ir) and $p\text{-}P\!-\!C_6H_4PPh_2$ (P = polystyrene) [340, 848, 849, 1404]. The supported iridium catalyst is selective for the hydrogenation of vinylcyclohexene to ethylcyclohexene, and of 1,5-cyclooctadiene to cyclooctene. It is also useful for selective H-transfer from formic acid to α,β-unsaturated ketones [98]. Other catalysts are formed from $[Rh(CO)_2Cl]_2$ and supports such as the polymeric phosphite $\{-CH_2CH(Me)\text{-}OPPh_2\}_n$ [383, 386], Ph_2P-functionalized poly[1,6-bis(p-toluene sulfonate)-2,4-hexadiyne] [948], crosslinked chloromethylated polystyrene functionalized with pyrrolidine [1206] or polystyrene-coated silica gel functionalized with phosphine groups [73]. Species such as P—Rh(CO)$_2$Cl and P_2—Rh(H)(CO)L (P = functionalized polystyrene [126], or an alternating copolymer of maleic acid and vinyl ether [1841]) are also catalytically active.

Polymers incorporating the functional group 3-37 have been treated with [Rh(CO)$_2$Cl]$_2$, and species of the type Rh(CO)Cl(PPh$_2$-)$_2$ are obtained [1775]. Infrared studies indicate that the product has the *trans* structure 3-38 rather than the *cis* arrangement 3-39 that would result if the bisphosphine acted as a chelate ligand [1521].

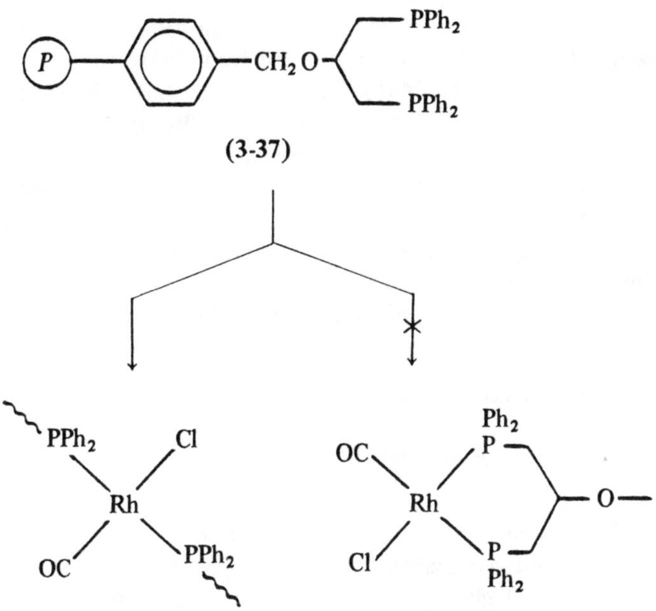

(3-37)

(3-38)   (3-39)

Reactions between polystyrene-bound pyridine or 2,2′-bipyridine ligands and [Rh(CO)$_2$Cl]$_2$ or [Rh(CO)Cl(PR$_3$)]$_2$ result in cleavage of the dimeric complexes. The resulting supported complexes can be used in the catalysis of hydrogenation reactions, but the active species seem to be metal aggregates [463]. Polystyrene-anchored pentane-2,4-dionato complexes of transition metals, e.g. P—acacRh(CO)$_2$, can also be used to hydrogenate cyclohexene [153]. Again, it seems likely that the active catalyst is composed of metal crystallites which are formed on the polymer under the reaction conditions.

Treatment of phosphine-functionalized polystyrene with binary carbonyl clusters such as Rh$_4$(CO)$_{12}$ [340], Ir$_4$(CO)$_{12}$ [1079, 1350, 1454] or Rh$_6$(CO)$_{16}$

[340, 872, 873] also produces hydrogenation catalysts. Their spectroscopic properties indicate that these catalysts are species such as $Ir_4(CO)_{11}PPh_2—P$ and $Rh_6(CO)_{13}\{PPh_2—P\}_3$. The supported Ir_4 clusters are stable during hydrogenation at temperatures below 90 °C. Kinetic results [1078] show that the rate of hydrogenation of ethylene and cyclohexene decreases with an increase in the number of phosphine substituents on the Ir_4 cluster. At temperatures above 120 °C, there is aggregation of the clusters to give crystallites. With $Rh_4(CO)_{12}$ supported on a polymeric amine resin (Amberlyst A-21) as the catalyst, the hydrogenation of α,β-unsaturated carbonyl and nitrile compounds has been achieved [957].

Rhodium diene species of the type $[P—Rh(diene)]^+$ (P = phosphine-functionalized polystyrene) have been obtained from $[Rh(diene)Cl]_2$ (diene = NBD [242, 1724] or COD [48, 1244]); they are active catalysts for the hydrogenation of alkenes. The catalysts obtained from $P—PPh_2$ or $P—PCy_2$ and $[Rh(1,5-COD)Cl]_2$ are also active in the hydrogenation of 1,5-cyclooctadiene [1244]; this contrasts with the behavior of the corresponding homogeneous catalyst.

Some polymer-bound cyclopentadienyl—Rh and —Ir compounds have been formed. The general approach is to functionalize a styrene—divinylbenzene copolymer with $—C_5R_4$ groups and then attach $—M(CO)_2$ or $—M(CO)L$ units [197, 310]; Equation (3.16) provides an example.

$$P—C_6H_4CH_2—(C_5H_4^-)Li^+ + [Rh(CO)_2Cl]_2 \longrightarrow$$

$$P—C_6H_4CH_2—(C_5H_4)—Rh(CO)_2 \qquad\qquad (3.16)$$

This catalyst has been used in the hydrogenation of alkenes, aldehydes and ketones, but it is not very stable. Treatment of polymer-attached cyclopentadiene with $RhCl_3 \cdot 3H_2O$ gives polymer-supported $(\eta\text{-}C_5H_5)RhCl_2$ species [1809]; in the presence of excess Et_3N, this is a good hydrogenation catalyst for alkenes and arenes.

The bidentate ligand anthranilic acid has been anchored to chloromethylated polystyrene beads (Amberlite resin, XAD-4). A particularly effective hydrogenation catalyst is obtained by treating this support with a solution of rhodium trichloride and subsequently with $NaBH_4$ [747]. This supported rhodium(I) catalyst is said to have exceptional activity, long term stability and considerable insensitivity to poisoning. It has been used in the reduction of alkenes, aromatic compounds, carbonyls, nitriles, nitro compounds and vegetable oils [541, 746, 747, 748, 749]. It is one of the more reactive catalysts for the reduction of arenes; the conversion of benzene to cyclohexane, for example, is achieved at 20 °C and 1 atm pressure H_2.

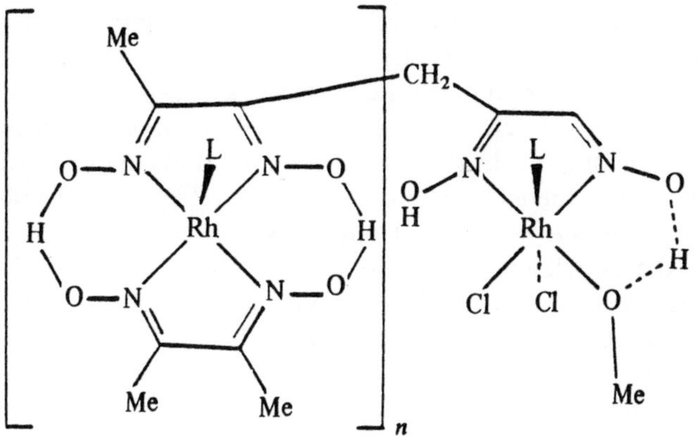

(3-40)

Polymers of 2,3,6,7-octanetetraone tetraoxime can incorporate rhodium (e.g. **3-40**; L = H_2O, MeOH, n = 3,4). This system shows some specificity for the hydrogenation of terminal double bonds under mild conditions [1948]. The activity towards internal C=C bonds is increased markedly when there is conjugation with aromatic rings.

A styrene—divinylbenzene copolymer with attached iminodiacetic acid groups will also incorporate rhodium upon treatment with $RhCl_3 \cdot 3H_2O$. This catalyst is selective for the hydrogenation of C=C bonds in alkenes which contain carbonyl (e.g. 5-hexen-2-one) or aromatic groups (e.g. styrene) [1253].

Properties and characterization of catalysts

The attachment of a metal complex to a polymer support can alter the catalytic properties of the species. Thus, there are reports of activation [635], deactivation [637] and changes in selectivity [636, 637, 1179, 1813] compared to the analogous soluble catalysts. The following examples illustrate the types of changes that have been observed.

With $RhCl(PPh_3)_a$ supported on styrene—divinylbenzene as the catalyst [824], the rates for reduction of cyclohexene, 1-hexene and styrene are found to be lower than those for the homogeneous catalyst. However, the ratio between the rates for 1-hexene and cyclohexene is higher than that in the homogeneous phase. This increase in selectivity is attributed to steric hindrance around the active sites on the resin. In another study [1813], the hydrogenation rates were shown to increase when $RhCl(PPh_3)_3$ was anchored on a rigid copolymer, but decrease when anchoring occurred on the inner phosphine groups of the copolymer.

Extreme caution must be exercised in such comparative studies, because of problems with the detailed characterization of the supported catalyst. Although the relative amounts of support material, ligand sites and attached metal complex can generally be deduced from elemental analyses, more detailed information about the structure and distribution of the active sites is more difficult to obtain [117]. Moreover, there can be problems with leaching and redistribution of the catalyst. The concentration of rhodium complex *in solution* over a polymer-bound hydroformylation catalyst has been measured [1041]; it decreases with an increase in temperature or H_2 pressure and with a decrease in CO pressure. A further problem relates to the poor mechanical and thermal stability of polystyrene and related resins. This can lead to fragmentation of the polymer beads and consequent diminished activity.

It has been established that the radial distribution and the percentage substitution of phosphines in polymer supports can be controlled over a wide range by changes in the preparation time and temperature [638]. An electron microprobe analysis of polystyrene-attached catalysts prepared from $P-PPh_2$ and $[Rh(C_8H_{14})_2Cl]_2$ or $RhCl(PPh_3)_3$ has shown that the distribution of rhodium in the polymer bead can also be controlled by varying the conditions of the attachment reaction [639]. When the attached catalyst is prepared by ligand substitution with a deficiency of complex, the rhodium distribution is confined to the first 180 μ of polymer bead with no metal in the center of the bead. With an excess of the complex, a rhodium distribution similar to that of the phosphine groups is obtained.

With $P-PPh_2 \cdot RhCl(PPh_3)_a$ type catalysts, the ratio of the rates of reduction of 1-hexene/cyclohexene varies from 1.2 to 0.7 depending on the mode of preparation of the catalyst. The changes in selectivity toward primary and secondary alkenes is attributed, at least in part, to the presence of residual, nonpolymeric ligands remaining after catalyst preparation [640]. The stability and activity toward hydrogenation of hexene of the catalyst prepared from $P-PPh_2$ and $[Rh(C_2H_4)_2Cl]_2$ has been investigated as a function of the texture of the polystyrene–divinylbenzene support [642]. The intrinsic catalytic activity depends on the rhodium content, but more significantly on the localization of the rhodium sites. Changes in activity were observed after ageing under H_2, and these were related to the formation of metallic particles which are active for the hydrogenation of benzene.

The effects of changing the degree of crosslinking by divinylbenzene in polystyrene-bound $RhBr(PPh_3)_3$ catalysts has been examined by EXAFS (extended x-ray absorption fine structure analysis) [1463]. With 2% divinylbenzene in the copolymer, the nearest neighbors to rhodium are two P atoms (at 2.16(1) and 2.32(1) Å) and two bromine atoms (at 2.49(1) Å). This indicates that the Rh atoms form part of a dimeric structure. With 20% divinylbenzene, the rhodium is close to three P atoms (one at 2.14(1), two at 2.26(1) Å) and one bromine atom (at 2.50(1) Å), and this indicates a 4-coordinate Rh monomer. Optimum catalytic activity is probably achieved with some degree of crosslinking between 2 and 20%.

Despite all of these difficulties, there have been some studies of the kinetics of hydrogenation reactions with polymer-supported catalysts. One of the best systems for kinetic measurements is $RhCl(PPh_3)_3$ on linear polystyrene [408]. The mechanism of alkene hydrogenation with this catalyst is essentially the same as with the homogeneous catalyst.

A major effect of the support on reaction rates relates to interactions between the solvent and the polymer. The diffusion of reactants and products within the {support + catalyst} medium is often rate-determining, and hence solvents like benzene, which have good penetrating and swelling properties, are vastly superior to those such as ethanol which cause collapse of the polymer.

HIGH SURFACE AREA OXIDE SUPPORTS

The use of supports such as silica or alumina overcomes some of the difficulties experienced with polymeric supports. These oxide supports consist of thermally stable, rigid matrices which are not subject to solvent swelling. Transport occurs by diffusion through the matrix pores.

Some rhodium complexes have been attached to silica supports. Generally, the silica is activated by binding functional groups to reactive Si—OH surface sites [44, 1349] and then adding the rhodium complex (see Scheme 3.7(a)). The silation method is generally preferred, and it can be achieved with X = Cl, OR, O_2CR or related anions and L = NH_2, PPh_2, SH or related groups; the phosphino-silane $(EtO)_3SiCH_2CH_2PPh_2$ is a typical example.

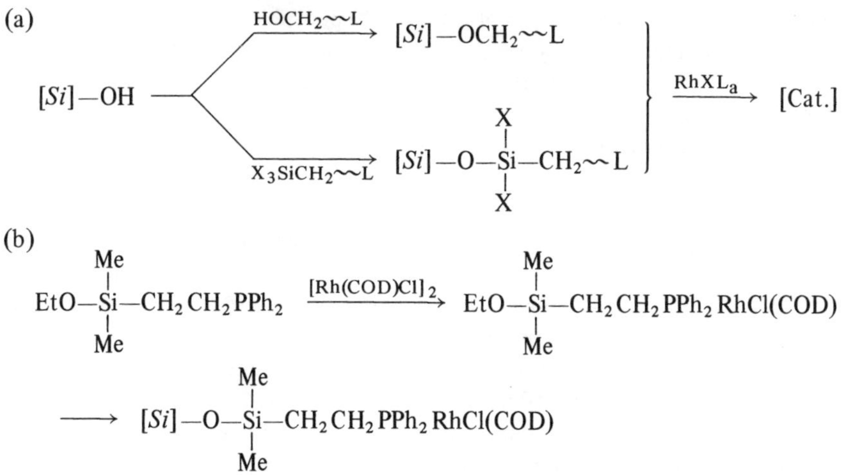

(a)

$$[Si]-OH \xrightarrow{\begin{array}{c} \nearrow \\ \searrow \end{array}} \begin{cases} \xrightarrow{HOCH_2\sim L} [Si]-OCH_2\sim L \\ \\ \xrightarrow{X_3SiCH_2\sim L} [Si]-O-\underset{\underset{X}{|}}{\overset{\overset{X}{|}}{Si}}-CH_2\sim L \end{cases} \xrightarrow{RhXL_a} [Cat.]$$

(b)

$$EtO-\underset{\underset{Me}{|}}{\overset{\overset{Me}{|}}{Si}}-CH_2CH_2PPh_2 \xrightarrow{[Rh(COD)Cl]_2} EtO-\underset{\underset{Me}{|}}{\overset{\overset{Me}{|}}{Si}}-CH_2CH_2PPh_2 RhCl(COD)$$

$$\longrightarrow [Si]-O-\underset{\underset{Me}{|}}{\overset{\overset{Me}{|}}{Si}}-CH_2CH_2PPh_2 RhCl(COD)$$

Scheme 3.7

In an alternative approach, silation of the silica surface is achieved with a silating reagent that already incorporates the rhodium complex (see example in Scheme 3.7(b)). With this method, a better defined surface species is obtained because there is effective control of the L/M ratio in the preparative steps.

In principle, catalysts of these types can be used with almost any solvent. In some instances, the catalyst can be fixed in a gas chromatographic reactor, and this provides a convenient system for the study of hydrogenation reactions in the gas phase [1441].

Some silica-supported catalysts of the type $[Si]-ORhXL_2$ (X = halide, L = phosphine) have been used for the hydrogenation of alkenes and dienes [43, 45]. One method of fixing the rhodium complex to the support is to treat the silica with a solution of RM (R = alkyl or aryl; M = Li, Na, K or MgX) before adding the rhodium compound (see Equation (3.17)) [1358].

$$[Si]-OH + RM \longrightarrow [Si]-OM + RH$$

$$\downarrow RhXL_3$$

$$[Si]-O-RhL_2 + MX \qquad (3.17)$$

Solid-phase catalysts of this type are reported [1813] to be more active than soluble catalysts such as $RhCl(PPh_3)_3$.

A catalyst obtained from diphosphinated silica (*DPSi*) and $[Rh(C_2H_4)_2Cl]_2$ is effective for the reduction of alkenes at 25 °C [1039]. Kinetic data and ESCA spectroscopic results indicate that the active catalyst is mainly (*DPSi*)-Rh(alkene)Cl. A related catalyst prepared from phosphinated silica and $[Rh(C_2-H_4)_2Cl]_2$ undergoes deactivation due to dimerization of the catalytically active species [392]. Another supported catalyst obtained from $[Rh(C_8H_{14})_2Cl]_2$ on phosphinated silica is found to be more active than the analogous homogeneous catalyst obtained from $\{[Rh(C_8H_{14})_2Cl]_2 + (EtO)_3SiCH_2CH_2CH_2PPh_2\}$ [1003].

Other systems have been formed by absorption of solutions of [Rh(1,5-hexadiene)Cl]_2 by 'Aerosil', and the hydrogenation rate with these catalysts depends on the distribution between reaction center and silica support, and also upon the temperature, solvent and alkene structure [761]. Catalysts formed from $[Rh(NBD)Cl]_2$ on phosphinated silica are effective for the hydrogenation of arenes under mild conditions [1324]. When the same catalyst source was supported on phosphinated polystyrene, it was not active in this application. Arenes, including polycyclic compounds, are also hydrogenated with catalysts obtained from allyl–rhodium complexes and silica [1891]. The active catalysts in these systems are hydrido species* such as $[Si]$ —ORh(allyl)H [1892]. The cationic species $[Rh(COD)_2]^+$ has also been adsorbed onto silica, and this catalyses the hydrogenation (at 130 °C) of 1-hexene containing BuSH [49].

Metal carbonyls, including $[Rh(CO)_2Cl]_2$, $M_4(CO)_{12}$ and $Rh_6(CO)_{16}$, have been adsorbed onto oxide surfaces. The supports include γ-Al_2O_3 [1062, 1063], silica [614], phosphine- or phosphite-functionalized silica [116, 165, 1725, 1806], and alumina or silica with attached amine or carboxylic groups [319, 656].

The species formed when $[Rh(CO)_2Cl]_2$ and $Rh_4(CO)_{12}$ interact with γ-Al_2O_3 give infrared spectra with identical CO stretching frequencies (at 2080 and 1997 cm^{-1}); the two systems catalyse the complete hydrogenation of pentynes to pentanes [1062]. The $[Rh(CO)_2Cl]_2/Al_2O_3$ catalyst, in particular, is much more active than the homogeneous catalyst. This catalyst also displays good specificity toward the hydrogenation of the terminal C=C bond of *trans*-1,3-pentadiene [1063].

Treatment of a silica surface with β-diphenylphosphinoethyltriethoxysilane and subsequently with $Ir_4(CO)_{12}$ at 90 °C under CO pressure produces a catalyst which is active for ethylene hydrogenation. The reaction kinetics have been determined for this system. At 27.5 °C, the orders are 0.7 and 1.2 for C_2H_4 and H_2, respectively [1725].

There is decarbonylation of the cluster when $Rh_6(CO)_{16}$ is adsorbed on silica, and particles of varying size have been characterized by electron microscopy [614]. Those corresponding to Rh_6 clusters show little or no catalytic activity, but larger aggregates (> 1.4 nm diameter) are active catalysts for the reduction of benzene and cyclohexene. The species formed when $Rh_6(CO)_{16}$ is added to phosphinated silica at 25 °C seem to retain the Rh_6 cluster frame [977] – thus carbonylation produces $[Si]$ —O—$(PPh_2)_n$—$Rh_6(CO)_{16-n}$ ($n \simeq 3$), which is thermally stable [165]. At 80 °C, the interaction between $Rh_6(CO)_{16}$ and the phosphinated silica produces a mixture of Rh^0 and Rh^I species which can be interconverted under mild conditions by oxidation and reduction. These species undergo structural changes under H_2 [116]. Decarbonylation occurs initially, and finally there is formation of a range of metal aggregates. Somewhere between these two stages, the system shows its highest activity in the hydrogenation of terminal alkenes such as 1-hexene. A spectroscopic study of the catalyst formed from $Rh_6(CO)_{16}$ on silica surfaces modified with ligands such as $(MeO)_3Si$-

* Infrared data substantiate this claim (N. Kitajima and J. Schwartz, *J. Am. Chem. Soc.* **105**, 2221 (1984)). However, other workers find that this catalyst system can contain finely divided metal (H. C. Foley, S. J. De Canio, K. D. Tan, K. J. Chan, J. S. Onuferko, C. Dybrowski, and B. C. Gates, *J. Am. Chem. Soc.* **104**, 3074 (1983)). This again emphasizes the need to be cautious in classifying particular catalysts as homogeneous rather than heterogeneous.

$(CH_2)_3 NH(CH_2)_2 NH_2$ reveals that the Rh_6 octahedron has fragmented to give $Rh(CO)_2$ species [966, 978]. The mononuclear sites are active in gas-phase hydrogenation reactions with alkenes such as ethylene and propene.

Related catalysts have been formed from phosphine-substituted rhodium carbonyl clusters such as $Rh_6(CO)_{13}(PPh_3)_3$, $Rh_6(CO)_{10}(PPh_3)_6$, $Rh_4(CO)_{10}$-$(PPh_3)_2$ and $Rh_4(CO)_{10}\{(Ph_2PCH_2CH_2)_2Si(OEt)_2\}$ [1622]. Treatment of phosphinated silicas with $Ir_4(CO)_{11}(PPh_3)$ or $Ir_4(CO)_{10}(PPh_3)_2$ gives di- and tri-substituted Ir_4 clusters which have been characterized by infrared spectroscopy [293]. Other silica-attached Ir_4 clusters have been generated from $Ir(CO)_2ClL$ (L = p-toluidine) by reduction with zinc [293].

Some rhodium compounds have been intercalated into the swelling layered silicate 'hectorite'. In these systems, the spatial requirements of the substrate into the solvent-swollen interlayers has important effects on the rates of hydrogenation. One catalyst is formed by treatment of dimeric rhodium acetate ions on the interlamellar cation-exchange sites with PPh_3 [1392, 1393]. This hydrogenates 1-hexene without isomerization, and the selectivity is attributed to a preference for the formation of $[H_2Rh(PPh_3)_n]^+$ rather than $HRh(PPh_3)_a$ in the intercalated state. This catalyst also assists the reduction of alkynes selectively to cis-alkenes. Another system has $[Rh(NBD)(Ph_2PCH_2CH_2PPh_2)]^+$ intercalated in the hectorite, and this has been used in the hydrogenation of dienes such as 1,3-butadiene, isoprene and 2,3-dimethyl-1,3-butadiene [1460]. Relative to the homogeneous catalyst the yields of 1,2-addition products are higher, but the rates of reaction are lower.

CATALYSTS ATTACHED TO SILICONE POLYMERS

Functionalized silicones have been used to support rhodium catalysts. The best characterized systems have the idealized formulas $[O_{3/2}Si(CH_2)_nPPh_2]_3$-$RhCl \cdot (O_{3/2}SiMe)_m$, $[O_{3/2}Si(CH_2)_2PPh_2]_4Rh_2Cl_2 \cdot (O_{3/2}SiMe)_m$ and $[O_{3/2}$-$Si(CH_2)_2PPh_2]_3RhCl \cdot (O_{3/2}Si(CH_2)_2PPh_2)_{0.7}$. The first of these catalyst is more active than the other two for the hydrogenation of styrene [259]. A related catalyst has been obtained from $[Rh(CO)_2Cl]_2$ and modified polyphenylsiloxane by prolonged treatment with H_2 [118].

3.10 Asymmetric Homogeneous Hydrogenation with Chiral Rhodium Complexes as Catalysts

REQUIREMENTS FOR ASYMMETRIC SYNTHESIS; AN OVERVIEW

One of the most significant developments in stereoselective synthesis is the comparatively recent discovery [764, 969] of soluble complexes which will catalyse the asymmetric reduction of unsaturated, prochiral substrates. By adding some symmetry principles to the known mechanism of hydrogenation reactions, it is possible to define the conditions necessary to achieve asymmetric hydrogenation [984]. The requirements are that (i) the substrate molecule has prochiral faces, (ii) the catalyst incorporates a chiral site and (iii) the substrate and H_2 are held in close proximity to each other within the diastereomeric transition state which produces the new chiral center. A product of high optical purity* will be obtained only when the prochiral substrate is bound preferentially in one conformation to the vacant site offered by the catalyst. A simplified representation of the formation of a chiral alkane from an alkene with prochiral faces is shown in Scheme 3.8. This scheme does not fully identify the diastereomeric transition states which lead to enantiomeric products (3-41) but they are discussed in more detail later (p. 104). It has been estimated that an energy difference of only $5-10$ kJ mol^{-1} in these transition states is probably sufficient to lead to a completely stereoselective process. Hence, careful chiral reagent—substrate matching [209] is needed to achieve high optical yields of the desired product. The choice of solvent is also important, and benzene is often used [1788] because it is the most nonpolar, aprotic solvent. However, better rates are often achieved in aqueous ethanol or 2-propanol.

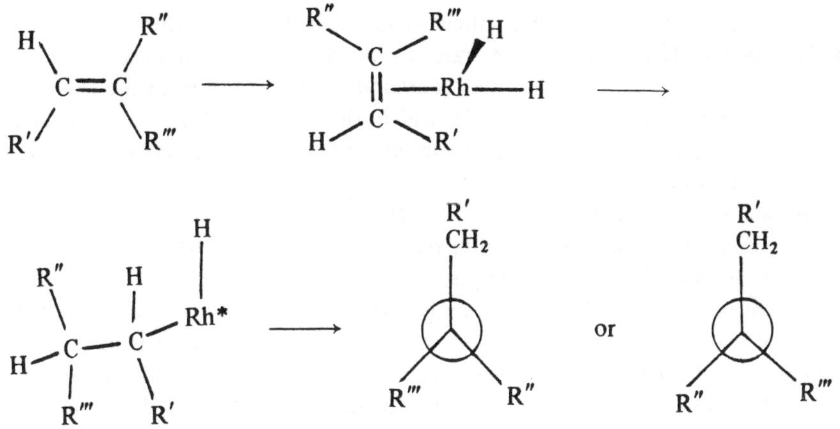

(3-41)

Scheme 3.8

* % optical purity = ($\alpha_{sample}/\alpha_{reference}$) $\times$ 100, where α is the specific rotation at a given wavelength. An alternative expression which is frequently used is % enantiomeric excess (ee) = ([R] $-$ [S]/[R] + [S]) $\times$ 100, where R is the major isomer.

Substituted styrenes were used in the earliest studies of asymmetric hydrog-enation reactions; the products were of relatively low optical purity. More recently, hydrogenation reactions of impressively high stereoselectivity have been achieved with α,β-unsaturated carboxylic acids, particularly α-acetamidoacrylic acids and esters of the type **3-42**. It seems likely that these substrates bind to the catalyst by both the amidooxygen and the alkene, as depicted in **3-43**. Rigidity of this type within the transition state would lead to high optical purity of the product. This mode of attachment to rhodium has been verified by crystal structure analysis [305].

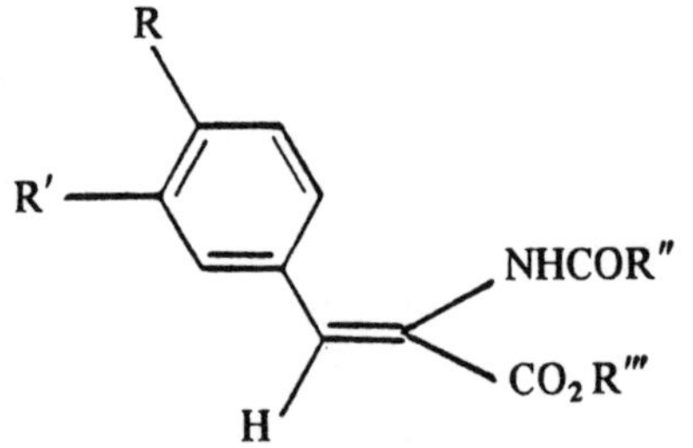

(3-42) 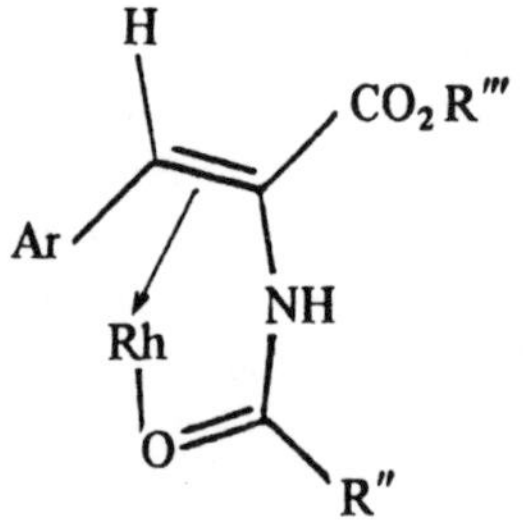   (3-43)

It is possible to increase the optical yields in the hydrogenation of $PhCH=C$ $(NHCOMe)CO_2H$ and related substrates by the addition of amines such as Et_3N [235, 245, 696, 1029, 1336]. This is attributed to ionization of the substrate carboxyl group to give an enamide which then coordinates to the catalytic rhodium complex. Generation of the anion does, however, slow this conversion rate considerably.

Inspection of recent reviews [177, 210, 252, 281, 855, 968, 1228, 1295, 1840] on asymmetric hydrogenation reactions reveals that the most prominent catalysts are soluble *rhodium* complexes which incorporate a chiral center. The formation and use of these catalysts generally follows the procedures already outlined in the preceding sub-section; most often the precatalyst is of the form $RhClL_a$ or $[Rh(diene)L-L]^+$. The asymmetric center can be at the donor atom of a phosphine ligand L, or it may be further removed from the metal at a carbon atom within the ligand L; most of the chiral phosphines that have been bound to rhodium hydrogenation catalysts have been tabulated [1125], see also below p. 90 ff.

The complex $RhClL_3$ (L = o-anisylcyclohexylmethylphosphine; **3-44**) is an example in which P is chiral. This has a particularly effective chiral influence in the homogeneous hydrogenation of α-acetamidoacrylic acids, and acylphenyl-alanines with from 85–90% optical purity have been prepared with this catalyst

[972]. Other tertiary phosphines of this type have been used, and the most effective catalysts all seem to have ligands with three very dissimilar groups. If the ligand has two substantially similar groups as in phenyl-α-naphthyl-4-biphenylphosphine, it induces little or no asymmetry in the hydrogenation products.

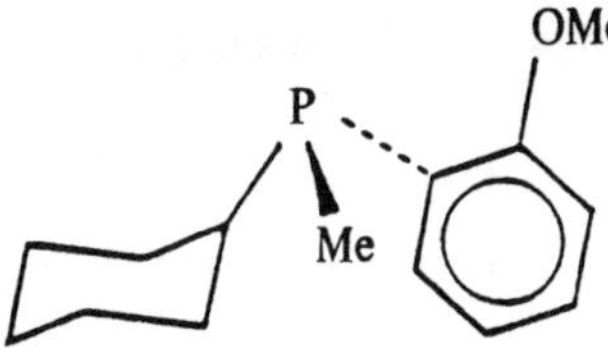

(3-44) 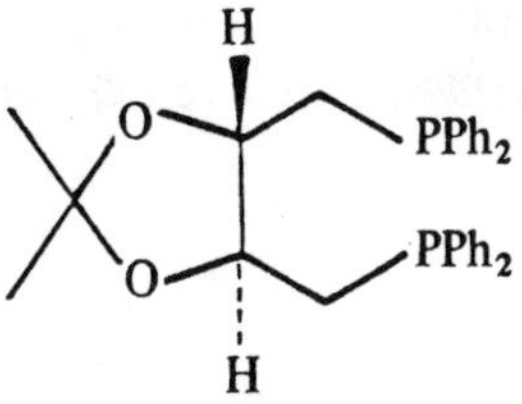   (3-45)

Several ligands that are chiral at the carbon atoms of groups attached to phosphorus have been incorporated in asymmetric hydrogenation catalysts. The ligands $PPh\{CH_2CH(Me)Et\}_2$ [969], $P\{CH_2CH(Me)Et\}_3$ [697], PPh_2-$(CH_2CONHCHRR')$ (e.g. $R = Pr^i$ and $R' = CO_2Bu^t$) [886] and PPh_2(neo-menthyl) [24] are of this type, but perhaps the best known example is the bidentate ligand 2,3-O-isopropylidene-2,3-dihydroxy-1,4-bis(diphenylphosphino) butane (3-45; commonly referred to by the acronym '*DIOP*'). In the asymmetric hydrogenation of α-acetamido-α,β-unsaturated acids to acetamido acids with catalysts incorporating (+)- and (−)-DIOP, members of the L- and D-series respectively have been obtained in very high optical yields (> 90%).

It is possible to have asymmetric centers at both P and C in some ligands. Menthylmethylphenylphosphine provides an example, and this can have the conformations 3-46 to 3-48 [523, 970, 1839]. A complex containing 3-47 attached to rhodium has been used to convert (*E*)-α-acetylamino-6-methylindole-3-acrylic acid to citronellic acid in 65—70% ee [1839].

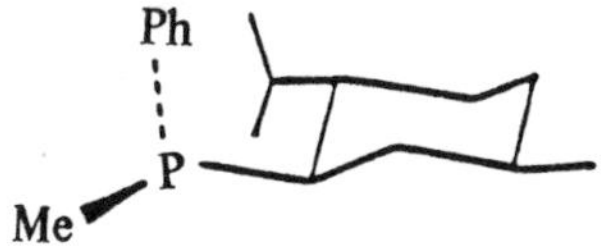

(*S*)-(3-46) 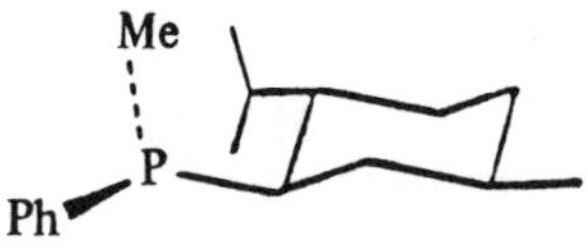   (*R*)-(3-47)   PMePh

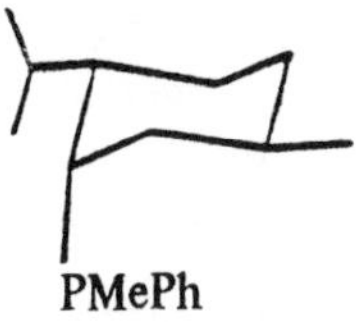

(3-48)

There have been attempts to achieve asymmetric syntheses with catalysts that incorporate chiral ligands other than phosphines. One example is a catalyst obtained from $RhCl(PPh_3)_3$ and Schiff-base ligands such as benzylidene-(*S*)-

alanine [999]. A catalyst has also been obtained from the chiral carboxylate [Rh(COD)(O$_2$CR*)] (O$_2$CR* = L(+)-mandelate) and nonchiral phosphines such as PMe$_3$ [1250]. These are not particularly effective asymmetric catalysts. With the carboxylato complex, for example, the highest optical yield of hydrogenation product was only 13%, and this was attributed to the presence of only one chiral center in the active catalyst. Another species, [Rhpy$_2$(L)Cl-(BH$_4$)]$^+$ where L is a chiral amide such as PhCH(Me)NHCHO, has been used [1086] to reduce (E)-methyl-β-methylcinnamate (Equation (3.18)).

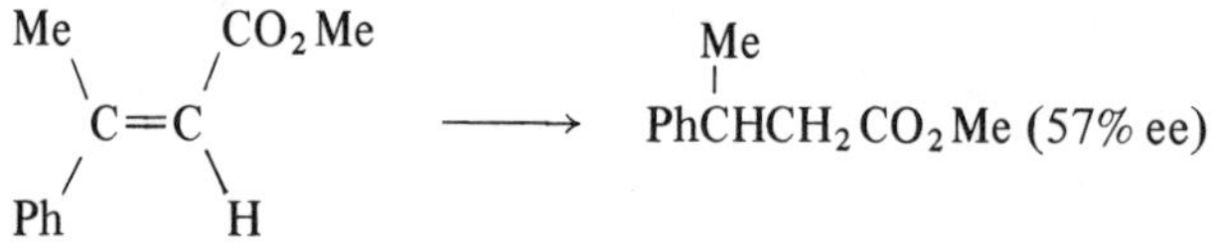

$$\begin{array}{c} \text{Me} \qquad \text{CO}_2\text{Me} \\ \diagdown \qquad \diagup \\ \text{C} = \text{C} \\ \diagup \qquad \diagdown \\ \text{Ph} \qquad \text{H} \end{array} \qquad \longrightarrow \qquad \begin{array}{c} \text{Me} \\ | \\ \text{PhCHCH}_2\text{CO}_2\text{Me (57\% ee)} \end{array} \qquad (3.18)$$

CATALYST TYPES

Chiral tertiary phosphines

Hydrogenation catalysts of the type RhCl(PR^{1}R^{2}R^3) are generally formed from [Rh(C$_2$H$_4$)$_2$Cl]$_2$ or [Rh(1,5-hexadiene)Cl]$_2$ and the chiral phosphine. Asymmetric hydrogenations have been achieved with catalysts containing the following tertiary phosphines:

PMePhR: R = Et [1761], Pr (R- and S-conformers) [697, 762, 763, 764, 969, 970, 972, 1761], Pr^i(S-) [763, 969, 970, 972], Bu (R- and S-) [763], Bu^t (R-) [764], 2-Me−Bu [697, 969], CH$_2$Ph (R- and S-; this ligand is given the acronym '*BMPP*') [697, 970, 1761], Cy [970], o-MeOC$_6$H$_4$ ('*PAMP*') [970, 972, 973, 975] , m-MeOC$_6$H$_4$ [970, 972] and 3-cholesteryl [970]. PMeCy(o-MeOC$_6$H$_4$) ('*CAMP*') [972].
PMePr^i(Ar): Ar = o-MeOC$_6$H$_4$ [760, 972], p-MeOC$_6$H$_4$ [760], o-NH$_2$C$_6$H$_4$, o-Me$_2$NC$_6$H$_4$, p-Me$_2$NC$_6$H$_4$ [760].
PPr^{i}Ph(o-MeOC$_6$H$_4$) [972].
PPh(naphth)(Ar), Ar = o- or p-MeC$_6$H$_4$ ('*NPTP*') [1945].
PPh$_2$(neomenthyl) [1356, 1357].

These catalysts have been used to reduce substrates such as H$_2$C=CR'Ph, R'CH=C(NHCOR'')CO$_2$H, PhCH=CHCH$_2$CO$_2$H and silylenol ethers to give optically active products.

In some asymmetric reductions, the rhodium(III) compounds RhCl$_3$-(PR^{1}R^{2}R^3)$_3$ have been used as the catalyst source. With PPh(4-biphenyl)(1-naphthyl) as the phosphine, the hydrogenation of α,β-unsaturated carboxylic acids such as atropic and cinnamic acids has been achieved [658]. In benzene/ethanol with added Et_3N, $RhCl_3(PMePr^iPh)_3$ and $RhCl_3\{PPh(CH_2CHMeEt)_2\}_3$ assist the reduction of $H_2C{=}C(R)CO_2H$ to $CH_3CH(R)CO_2H$ [969]. The base presumably abstracts HCl from the catalyst source.

Complexes of the type $[Rh(diene)L_2]^+$ (e.g. diene = NBD or COD, L = $PMePh(NEt_2)$) [759], $PPh_2(CHMeEt)$ [1518] or **3-45** [1839]) are effective precatalysts for asymmetric hydrogenations. For example, $[Rh(COD)\{PPh_2{-}(neomenthyl)\}_2]^+$ catalyses the enantioselective hydrogenation of alkyl- and alkenyl-substituted acrylic acids to chiral dihydrogeranic acids [1837, 1838, 1839] (an example is given in Equation (3.19)).

$$Me_2C{=}CHCH_2CH_2CMe{=}CHCO_2H \; (E/Z \simeq 9) \longrightarrow$$

$$(3R){\text -}(\pm){\text -}Me_2C{=}CHCH_2CH_2CHMeCH_2CO_2H \qquad (3.19)$$

These products are useful intermediates in the preparation of vitamin E, citronellal and menthol. The complex $[Rh(NBD)\{PMePh(CH_2Ph)\}_2]^+$ has been used to catalyse the asymmetric hydrogenation of Schiff bases [1853].

The ligand DIOP

Several catalysts with DIOP (**3-45**) as the chiral ligand have been prepared. The neutral complexes of formula RhCl(DIOP)S are typically formed from $[Rh(C_2H_4)_2Cl]_2$ or $[Rh(C_8H_{14})_2Cl]_2$ and 2 equiv of DIOP [891, 892]. Unsaturated substrates that have been reduced with these catalysts include atropic acid and amino acids such as substituted cinnamic acids [234, 312, 430, 431, 891, 892, 893, 961, 1415, 1561, 1610, 1864]. Very high optical yields are often obtained. For example, the preparation of N-acetyl-D-phenylalanine from α-acetylaminocinnamic acid has been achieved with 100% optical purity [1863]. The reduction of (E/Z)-α-acetamidocinnamic acids has been studied using D_2 [431], and it seems likely that some $E-Z$ isomerization occurs prior to reduction.

The related catalyst $HRh(DIOP)_2$ has been used in some asymmetric hydrogenation reactions [385], but it is less effective than RhCl(DIOP)S; hydrogenation occurs more slowly and the optical yields are lower. This compound is prepared by treatment of $RhCl_3 \cdot 3H_2O$ with (+)-DIOP and subsequently with $NaBH_4$ in ethanol; the preparation is described in *Inorganic Syntheses* [513]. The crystal and molecular structures of this catalyst have been determined [112]. The geometry is distorted trigonal-bipyramidal; H occupies an axial position and the two (+)-DIOP ligands are bidentate with each displaying (S,S)-chirality. Multinuclear NMR spectroscopic studies establish that the structure is fluxional in solution. Kinetic and spectroscopic results indicate that one DIOP ligand becomes unidentate during the hydrogenation of an alkene.

Cations of the type [Rh(diene)(DIOP)]$^+$ (diene = COD or NBD) are obtained from [Rh(diene)X]$_2$ and 2 equiv or DIOP [603]; excess DIOP yields [Rh(DIOP)$_2$]$^+$. In some instances, rapid and complete transfer of the DIOP has been achieved via a cuprous halide adduct [1791, 1792, 1793]:

$$CuCl + DIOP \longrightarrow DIOP \cdot CuCl$$

$$DIOP \cdot CuCl + [Rh(diene)Cl]_2 \longrightarrow [Rh(diene)DIOP]^+$$

Corresponding complexes of iridium have been prepared from [Ir(COD)Cl]$_2$ and DIOP [1522], and the molecular structure of Ir(COD){(+)-DIOP}Cl has been determined by crystallographic analysis [251]. The metal atom displays trigonal-bipyramidal geometry (3-49) and the DIOP acts as an apical–equatorial bidentate ligand.

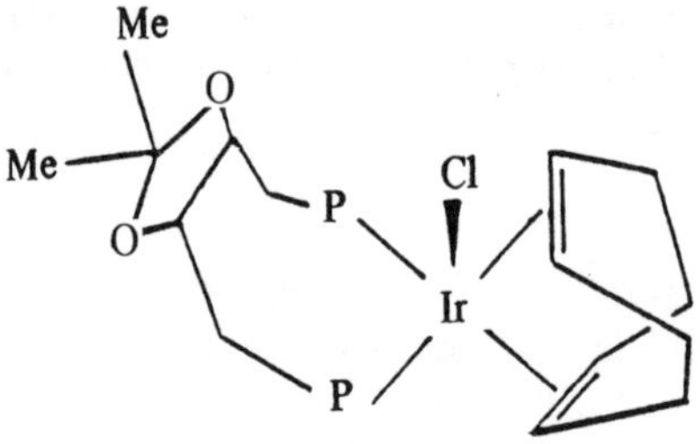

(3-49)

In a kinetic study [603] of asymmetric hydrogenation, it was shown that [Rh(COD)(DIOP)]$^+$ is not air-stable over a period of time. However, reproducible kinetics are obtained after storage under argon. Under hydrogenation conditions, it seems likely [238] that [Rh(COD)(DIOP)]$^+$ and related cations are converted to:

$$[Rh(DIOP)S_2]^+ \rightleftharpoons [H_2Rh(DIOP)S_2]^+$$

The cationic systems are claimed [603] to be more reactive than the neutral RhCl(DIOP) complexes, but both give similar optical yields.

Prochiral substrates that have been reduced in the presence of the cationic Rh–DIOP catalysts include amino acids and their esters [295, 601, 603, 1075, 1320, 1616, 1787], Schiff bases [1853], imidazolinones [522, 1455] and bisdehydrodipeptides [1416, 1472].

Other chelating ligands

Numerous chelating ligands other than DIOP have been incorporated into rhodium complexes to give asymmetric hydrogenation catalysts. Most of these ligands are bisphosphines with chirality directly on one or both P atoms and/or the side-chains. Again, the major applications of these catalysts have been to assist the asymmetric hydrogenations of unsaturated carboxylic acids, esters and ketones, and with many of them very high enantioselectivity ($> 95\%$ ee) is achieved.

The following examples illustrate the wide range of ligands that has been used. The information is organized according to: ligand structure (trivial name); catalyst type (A = neutral $RhCl(L—L)S_a$ or related complex; B = cationic $[Rh(L—L)S_2]^+$ or related complex); notes describing any special features.

Example 1

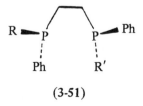

(3-50)

R = H, R' = Me(PROPHOS); A [1292], B [551].
R = H, R' = Ph (PDPE); A [1292], B [1945].
R = H, R' = Cy; B [1472, 1473, 1474].
R = H, R' = CH_2Ph, Pr^i, CH_2OCH_2Ph or CH_2OBu^t; B [56, 147, 148, 1030].
R = R' = Me (CHIRAPHOS); B [954, 985].

In the hydrogenation of $H_2C=C(CO_2Me)CO_2Et$, the catalyst with PROPHOS is capable of breeding its own chirality. The high efficiency of the catalyst with CHIRAPHOS is attributed to the tendency for methyl groups to occupy unhindered equatorial positions, thus fixing the chelate ring in a rigid conformation; hydrogenation of α-acetylaminoacetic acid $\rightarrow 99\%$ ee, but the reaction is relatively slow — it is faster when R = H.

Example 2

(3-51)

R = Ph, R' = neomenthyl; B [955].
R = R' = *o*- or *p*-tolyl or naphthyl [1945].
R = R' = *o*-MeOC$_6$H$_4$ (DIPAMP); B [101, 102, 326, 974, 976, 982, 1506].

DIPAMP is prepared from L-menthol; it is similar in behavior to CHIRAPHOS. The enantioselectivity is much higher than with PMePh(*o*-MeOC$_6$H$_4$) (PAMP), e.g. in the hydrogenation of α-acetylaminocinnamic acid, PAMP → 55% ee, DIPAMP → 95% ee. With geminal-substituted vinyl acetates, efficient hydrogenation in high optical yield is achieved only if there are electron-withdrawing groups *gem* to the acetoxy function [982].

Example 3

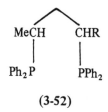

(3-52)

R = H (*S*-CHAIRPHOS), or R = Me (*S,S*-SKEWPHOS); B [1101].

CHAIRPHOS and SKEWPHOS adopt achiral and chiral conformations, respectively; hence stereoselectivity is only achieved with SKEWPHOS–Rh complexes. Related ligands have (a) Ph$_2$PO in place of Ph$_2$P, R = Me [110], (b) *N*-CHRCO$_2$H, R = L-Me or L-CHMe$_2$ in place of the central methylene [942].

Example 4

$$
\begin{array}{c}
\text{R'}\diagdown\qquad\diagup\!\!\cdot\text{R'} \\
\diagup\qquad\diagdown \\
\text{RN}\qquad\quad\text{NR} \\
|\qquad\qquad| \\
\text{Ph}_2\text{P}\qquad\ \ \text{PPh}_2
\end{array}
$$

(3-53)

R = H, R' = CHMePh (PNNP); A [520, 521], B [519, 1152].
R = H, R' = Me; B [32, 1334, 1336].
R = H or Me, R' = Ph; B [518].
R = H, R' = CH$_2$CH(Me)CH$_2$ $\overline{\text{CHCMe}_2\text{CH}_2\text{CH}_2}$ or related group; A [135].

With PNNP, the hydrogenation of α-acetylcinnamic acid → 84% ee. It is necessary to use inert media such as THF because P—N linkages are easily solvolysed. Higher optical yields and/or conversion rates are achieved by addition of Et_3N – this ionizes the substrate carboxyl group which then coordinates to the catalytic Rh complex. Replacement of 2 × NR by 2 × CH_2 reduces enantioselectivity [518, 893]; hence amido-N is probably involved in the bonding to Rh.

Example 5

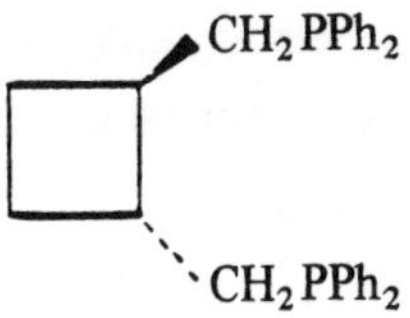

(3-54)

–; B [237, 600, 1791]. A catalyst can also be obtained from ligand + $Rh_6(CO)_{16}$ [1470].

With this and related ligands incorporating the PCH_2CCCH_2P unit, the optical purities of the *N*-acylamino acid and ester products are related to the flexibility of the chelate ring, but there is no simple correlation with the PCH_2CCCH_2P torsional angle [605]. Related ligands include the substituted cyclopentanes 1,3-$(Ph_2PCH_2)_2$ —$\overline{CH(CH_2)_2CHCH_2}$ [135] and 1,2-(Ph_2PO) —$\overline{CH(CH_2)_3CH}$ (BDPCP) [1757].

Example 6

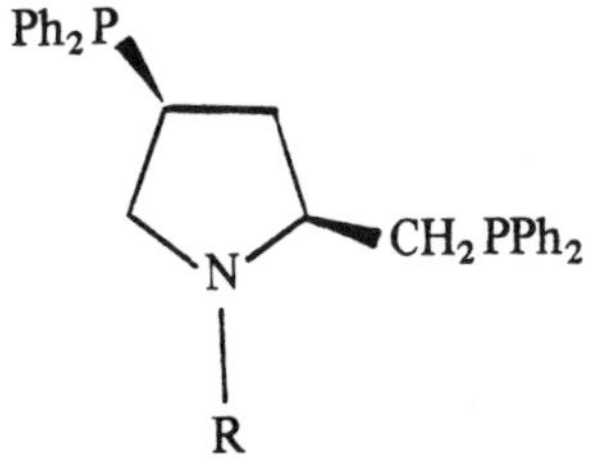

(3-55)

Large variety of R, e.g. CO_2Bu^t (BPPM), $COBu^t$ (PPPM); A [4, 5, 6, 11, 12, 1304], B [7, 9, 10, 15, 16, 17, 430, 961, 1301, 1303, 1304, 1311, 1322, 1506].

A general review is available on the use of this ligand [13]. It is derived from natural (2S,4R)-hydroxyproline. It can be highly efficient with generally fast rates. Changes observed by ^{31}P NMR spectroscopy are an useful aid in mechanistic studies [1299]. Chiral recognition by the Rh catalyst is very sensitive to the stereochemistry of the prochiral alkene [1300] – unwanted D-amino acids are often produced. It has been used in the asymmetric synthesis of dipeptides with diastereomeric purities in excess of 99% [1319]; it is also useful in the asymmetric hydrogenation of α-keto esters [1302]. The catalyst with R = $CO_2C_{27}H_{25}$ is soluble in cyclohexane – it catalyses the conversion of pyruvates to optically active lactates [15]. The related diphosphinite ligand, 2,3-$(Ph_2PO)_2$—$\overline{N(Ph)COCHCHCO}$, can be prepared from natural tartaric acid [217].

Example 7

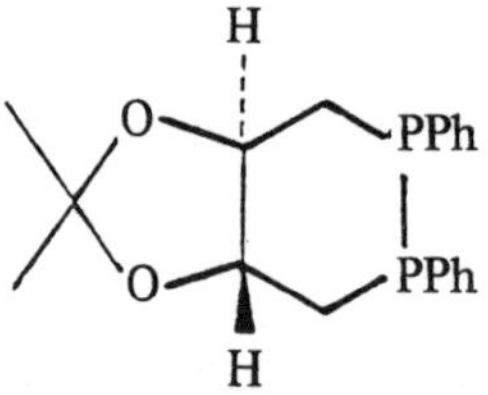

(3-56)

CycloDIOP; A, B [1967].

In the reduction of (Z)-PhCH=C(NHAc)CO_2H, the type A catalyst gives (R)-AcNHCHPhCO$_2$H (79% yield, 71% ee), but the type B catalyst gives a racemic amino-acid product.

Example 8

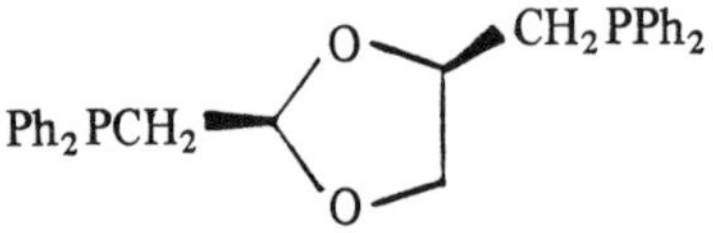

(3-57)

DIOXOP; B [429, 430, 1029, 1031, 1614, 1617].

Derived from 1,6-D-anhydroglucose. The structure of [H$_2$Rh(DIOXOP)(MeOH)]$^+$ has been determined by ^{31}P NMR spectroscopy [236]. High optical yields (*ca.* 80%) are obtained in the hydrogenation of aminoacrylates – the chirality of the product is reversed by the addition of base (e.g. Et_3N). Low optical yields are obtained with esters.

Example 9

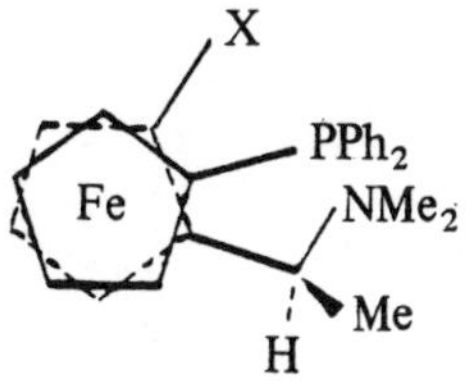

X

PPh₂
NMe₂
Me
H

(3-58)

X = H (PPFA) or PPh_2 (BPPFA); B [382, 387, 388, 677, 680].

Parent member of a range of ferrocene-derived ligands. Preparation [679]; general review [678]. The hydrogenation of α-acetylaminocinnamic acid → 93% ee. The chirality can be varied on the side-chain and on the 1,2-substituted cyclopentadienyl ring [1936]. High enantioselectivity is achieved in the conversion of enolphosphinates to secondary alcohols when the side-chain is a substituted alcohol; however, the reaction rates are slow [676]. The crystal structure of [Rh(NBD)(PPFA)] [PF_6] has been determined [382].

Example 10

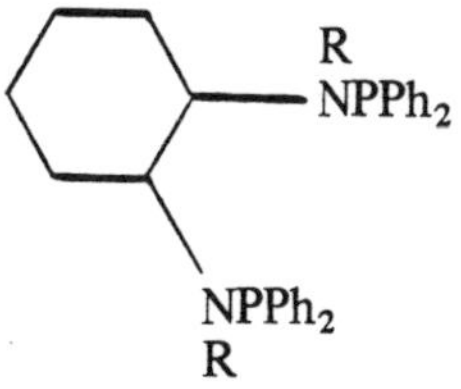

R
NPPh₂

NPPh₂
R

(3-59)

R = H or Me; A [1333, 1334], B [655, 929, 1334, 1335, 1336].

With R = H, the (*R,R*)-ligand always gives the *N*-acyl-(*R*)-amino acid but the (*S,S*)-ligand gives the *N*-acyl-(*S*)-amino acid. With R = Me, the (*R,R*)-ligand gives the *N*-acyl-(*S*)-amino acid. Related ligands are those with NR replaced by CH_2 [604, 961] or O (BDPCH) [682, 1756, 1757], and with heterocyclic N [1247, 1341].

Example 11

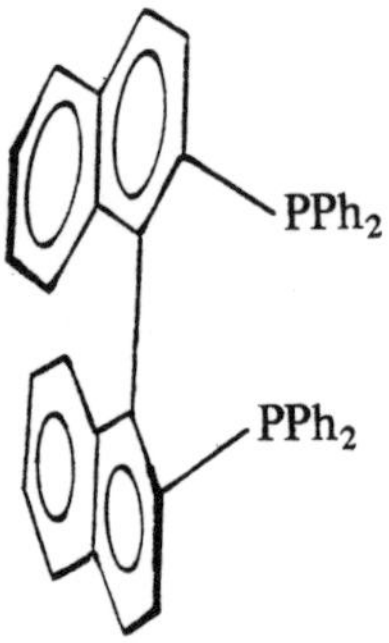

(3-60)

(BINAP); B [31, 1204].

Both the (*R*)-(+)- and (*S*)-(−)-conformers of the ligand are available. The molecular structure of [Rh(NBD){(+)-BINAP}] [ClO$_4$] has been determined. Hydrogenation of (*E*)- and (*Z*)-α-benzamidocinnamic acid → > 95% ee. A related ligand has OPPh$_2$ in place of PPh$_2$ [634].

Example 12

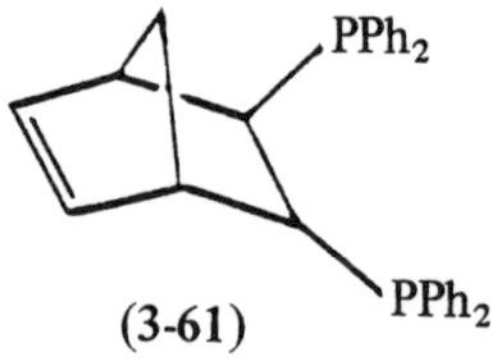

(3-61)

NORPHOS; B [253, 254, 255].

The absolute configuration of (−)-NORPHOS has been shown to be 2*R*,3*R* from X-ray results [255]. The absolute configuration and molecular structure of [Rh(NBD)(NORPHOS)] [ClO$_4$] have also been determined from X-ray data [1028]. The ligand is reduced to the corresponding bicycloheptane under hydrogenation conditions. Hydrogenation of *N*-acylamidocinnamic acid → 96% ee. (+)- and (−)-MeNORPHOS can also be used to form asymmetric hydrogenation catalysts [256]. Other related ligands have 2 × CH$_2$PPh$_2$ on bicycloheptene [1009] or bicycloheptane [703, 704] rings.

Example 13

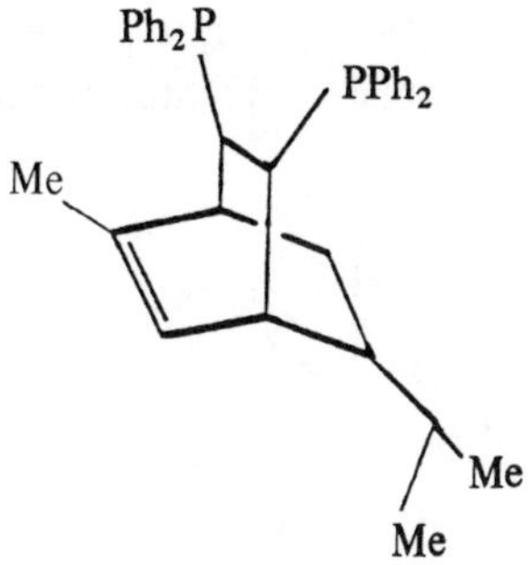

(3-62)

PHELLANPHOS; B [1061].

Improved synthesis of this and related ligand [1519]. The hydrogenation of *N*-acylamidocinnamic acid → 95% ee.

Example 14

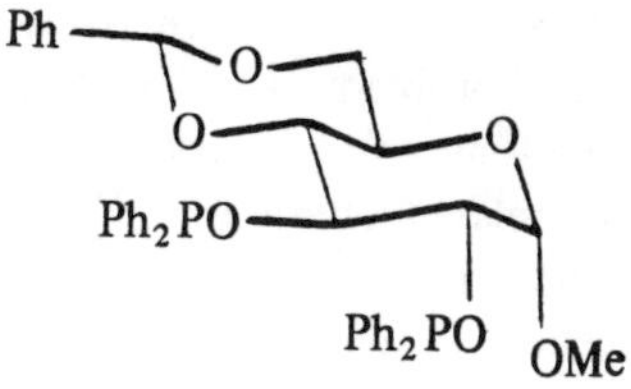

and related sugar derivatives

(3-63)

−; A [1255, 1931], B [847, 1261, 1563, 1564, 1615].

Small changes in stereochemistry lead to significant changes in the optical yield. The related ligand derived from α-L-idofuranose gives a catalyst which is effective in the asymmetric hydrogenation of atropic acid and α-methylcinnamic acid [879].

POLYMER SUPPORTS WITH CHIRAL LIGANDS

In some cases, optically active functional groups may be attached to polymeric materials, and these can be used to form supported catalysts. For example, a catalyst has been prepared from Rh(NBD)acac and polystyrene functionalized with —PR'R'' groups [1723]. The nature of the substituents R' and R'' affects the catalytic behavior of this system. The related catalyst [P—PPh(menthyl)-Rh(NBD)$^+$] [1391], and catalysts formed from P—P(menthyl)$_2$ [1004, 1005], have also been used.

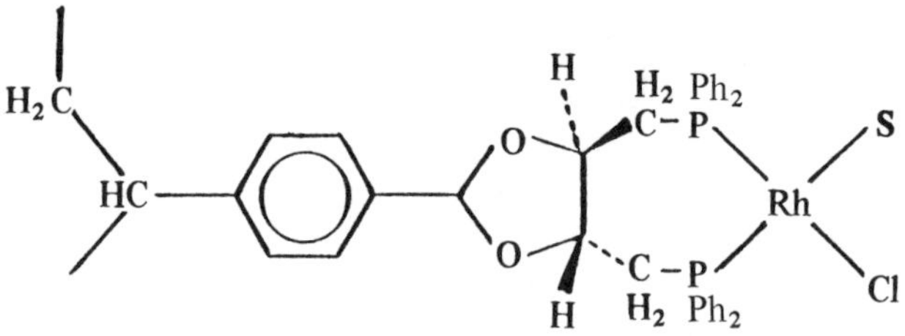

(3-64)

Other supported catalysts have been derived from chelating phosphines. The Rh–DIOP system (3-64) provides one example. Although it is effective for the hydrogenation of nonpolar alkenes in benzene, it is less efficient than the soluble Rh–DIOP analogs [474]. It has also been used for the synthesis of (R)-amino acids and hydratropic acid [1137, 1286, 1743]. However, these polar substrates generally require ethanol as a co-solvent, and this causes coiling of the polymer with loss of activity. This coiling seems not to occur with a noncrosslinked polystyrene support [1286, 1287]. Alternatively, a different copolymer can be used to overcome the problem. Thus, a rhodium catalyst formed from 3-65 is effective in benzene/ethanol for the asymmetric hydrogenation of α-acetamidocinnamic acid and related compounds [553]. Both the reaction rates and the degree of asymmetric induction are similar to those obtained with the corresponding homogeneous catalysts.

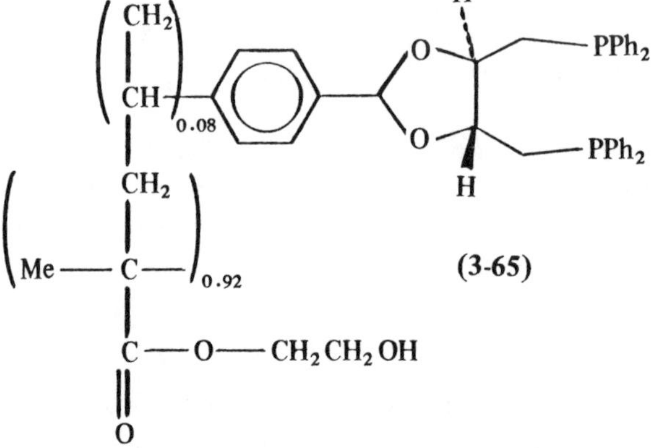

(3-65)

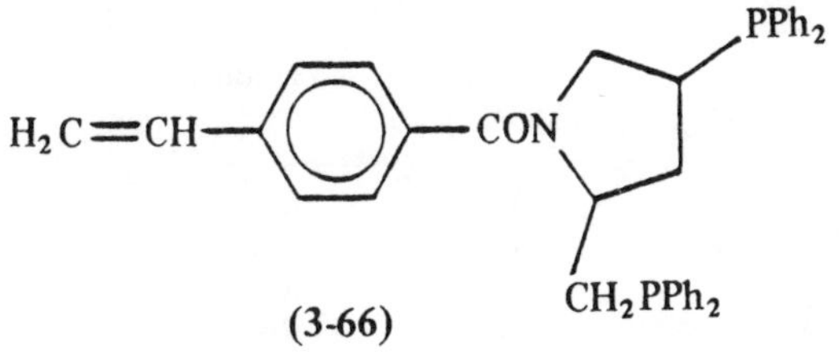

(3-66)

A high optical yield has been reported [14] in the hydrogenation of PhCH=C(NHAc)CO$_2$H with a rhodium catalyst supported on a copolymer of 2-hydroxyethylmethacrylate with the pyrrolidine derivative **3-66**. Several related catalysts have been formed with other pyrrolidine [8, 107, 108, 553] and dioxolan [107] derivatives. One example is the copolymer of hydroxyethylmethacrylate and 2-p-styryl-4,5-bis(tosyloxymethyl)-1,3-dioxolan [1663]; it has been used in the hydrogenation of α-N-acylaminoacrylic acid in ethanol, and the optical yields and absolute configurations of products were the same as those obtained with the homogeneous analogs. There was no loss of activity when the catalyst was removed by filtration and re-used.

Rhodium complexes with chiral aminophosphine ligands can also be supported on oxide supports such as 'hectorite' or 'bentonite' [1151]. The optical yields obtained in the reduction of α-acetamidoacrylic acids are similar to those reported for the soluble catalysts.

FACTORS AFFECTING ENANTIOSELECTIVITY

In the asymmetric hydrogenation of a given substrate, the enantioselectivity of the product is affected by changing the chiral ligand within the catalyst. This is illustrated by the examples shown in Table 3.I.

TABLE 3.I

Effects of ligands in catalysts on enantioselectivity[a]

(a)   *Substrate = α-acetamidocinnamic acid; catalyst = RhClL$_a$S [972]*

L	% ee
PMePrPh	28
PMePh(o-MeOC$_6$H$_4$)	55
PMeCy(o-MeOC$_6$H$_4$)	85

(b)   *Substrate* = H_2C=$C(Ph)Et$; *catalyst* = $[Rh(L-L)S_2]^+$

L—L	% ee	Product configuration	Reference
DIOP	24	R	[682]
(+)-BDPCH	33	R	[1757]
(+)-BDPCP	60	R	[682]
S,S-CHIRAPHOS	99	R(D)	[550]

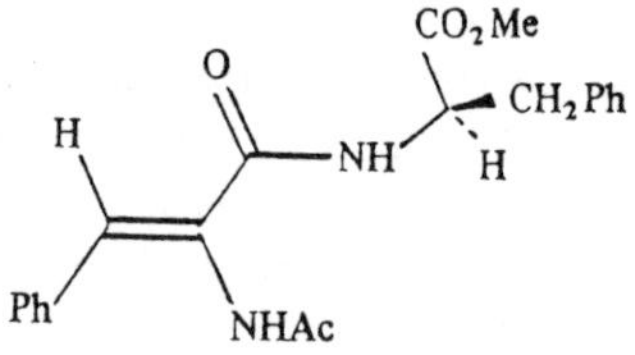

(3-67)

(c)   *Substrate = 3-67* [1174]

Catalyst	% ee	Product configuration
RhCl$\{Ph_2P(CH_2)_3PPh_2\}$	20	S
RhCl$\{$(+)-DIOP$\}$	90	S
RhCl$\{$(−)-DIOP$\}$	80	R
RhCl(BPPM)	> 90	R
[Rh(COD)$\{$(+)-DIOP$\}$]$^+$	88	S
[Rh(COD)(BPPM)]$^+$	> 90	R
[Rh(NBD)(DIPAMP)]$^+$	> 90	S

a DIOP = **3-45**; BDPCH, see **3-60**; BDPCP, see **3-55**; CHIRAPHOS = **3-51**; BPPM = **3-56**; DIPAMP = **3-52**.

Interesting inversions of stereoselectivity are sometimes induced by small changes in the chiral ligand. This occurs, for example, in the hydrogenation of α-acetamidoacrylic acids when the phenyl groups in DIOP are replaced by o-MeC$_6$H$_4$ [243], and when the R groups in ligands of type **3-60** are changed from H to Me [929, 1337]; for example, **3-60** (R = H) → (R)-amino acid and **3-60** (R = Me) → (S)-amino acid. This change is attributed to the chiral helical conformation of the phenyl groups attached to the phosphorus atom in the ligand. If the helicity is left-handed, the (R)-amino acid is obtained; with right-handed helicity, formation of the (S)-amino acid is preferred.

In the hydrogenation of alkenes with particular catalysts, there is a considerable change in the enantioselectivity as the substrate is varied. This is illustrated by the examples given in Table 3.II.

TABLE 3.II
Effects of substrate substituents on enantioselectivity

(a)   Substrate = H_2C=$CPh(R)$; catalyst = $[Rh(DIOP)S_2]^+$ [682, 894]

Substrate, R =	% ee	Product configuration
CO_2Et	63	S
CO_2Me	7	R
Et	24	R

(b)   Substrate = 3-68; catalyst = $[Rh(DIPAMP)S_2]^+$ [326]

3-68, R =	X =	% ee
H	OH	35
Me	OMe	88
H	OMe	55
Me	OH	88
H	NHR	90

(c)   Substrate = 3-69; catalyst = $[Rh(DIOP)S_2]^+$ [602]

3-69, R =	% ee
Me	69
Pr^i	15
Bu^t	0

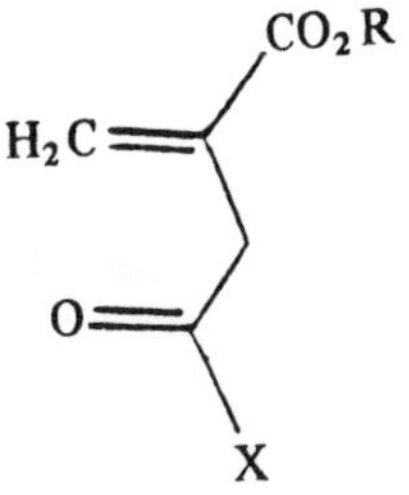

(3-68)

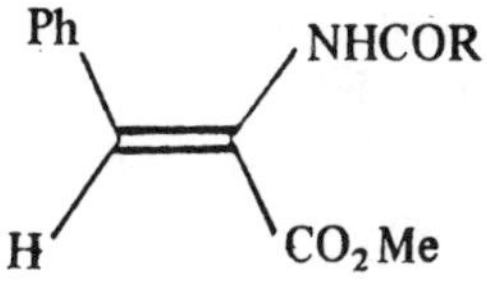

(3-69)

With acetamidocinnamic acid esters, hydrogenation of the E- and Z-conformers can be compared. The data in Table 3.III relate to the reactions with the catalyst $[Rh(DIPAMP)S_2]^+$ [306]. The faster and more enantioselective synthesis with the Z-conformer is generally ascribed to steric hindrance. The conformation of the alkene is not so crucial with the 2-amido-3-alkylacrylates [1554], particularly with catalysts incorporating DIPAMP.

The enantioselectivity can also be affected by the reaction conditions. The examples in Table 3.IV illustrate the effects of temperature, pressure and presence or absence of base (MeCHPhNH$_2$). Similar effects are observed when the catalyst is formed from chiral pyrrolidinodiphosphines [1312].

TABLE 3.III
Effects of substrate conformation on rates and enantioselectivity

Conformation of substrate	Relative rate	% ee
E-	1	23
Z-	15	96

TABLE 3.IV
Effects of reaction conditions on enantioselectivity[a]

Temp.(°C)/pressure (atm)	Base	% ee	Product configuration
0/1	yes	95	S
0/1	no	6	R
100/1	yes	75	S
100/1	no	68	S
25/1	no	13	S
25/100	no	30	R

[a] Substrate = PhCH=C(NHAc)CO$_2$H; catalyst = [Rh(DIOXOP)S$_2$]$^+$ [1614] (DIOXOP = 3-58).

KEY INTERMEDIATES

As mentioned earlier, one of the requirements for effective asymmetric catalysis is recognition by the substrate of the chiral 'hole' presented by the ligand within the catalyst. The synthesis of a large number of chiral ligands, combined with trials of their effectiveness in asymmetric catalysis, has provided an interesting but largely empirical approach to this problem. Attempts to characterize the diastereomeric transition state that leads to the enantiomeric products are more objective but more difficult approaches. Various avenues have been followed, but multinuclear NMR spectroscopy [235, 304] has been one of the most rewarding.

In the overall reaction pathway, it is generally agreed [240, 246, 431, 983] that the rate-limiting process is the addition of H$_2$ to an Rh^I–substrate complex (see Scheme 3.9), thus, this is the step that will determine the enantioselectivity of the product. In particular systems, detailed information about the stereochemistry of the catalyst-substrate adduct has been obtained from changes in the ^{31}P NMR spectra. Identification of the species [Rh{(S,S)-Ph$_2$PCH(Me)-CH(Me)PPh$_2$}{(Z)-PhCH=C(NHAc)CO$_2$Et}] [306] is an example. The structure corresponds to the diastereomer in which the [C$_\alpha$-re, C$_\beta$-si] face of the alkene is bound to rhodium. This is unexpected, because the hydrogenation product ((R)-PhCH$_2$CH(NHAc)CO$_2$Et, > 95% ee) is derived from the alternative diastereomer. Although the latter is present in concentrations too small to detect, its rate of reaction with H$_2$ is very much faster than that of the major diastereomer. It is this difference in reaction rates, rather than the preferred conformation of the transition state, that determines the enantioselectivity in the overall asymmetric hydrogenation reaction (J. Halpern, *Inorg. Chim. Acta* **50**, 11 (1981); cf. references [234, 1299, 1300, 1312]). This is an important principle, that has not always been appreciated.

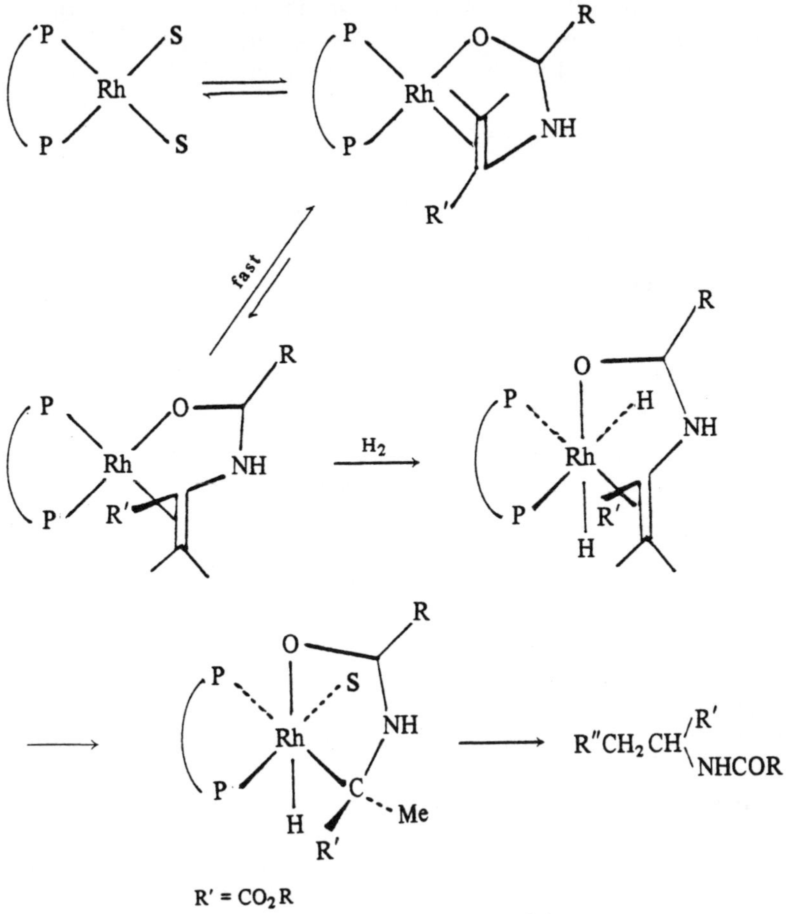

R' = CO$_2$R

Scheme 3.9

Other studies have established that the type of intermediate formed is affected by changes in the substituents on the substrate (e.g. X in $H_2C=CHC(O)X$ and R in $H_2C=CRCO_2H$ [247]) and by structural changes in the chiral ligand [245, 247]. It is interesting that 5- and 7-ring chelate bisphosphine—Rh complexes can exhibit markedly different catalytic properties [605]. For instance, the former are much more effective in the reduction of enamides and other substrates that are potentially tridentate. The chelate angle P—Rh—P may also play a critical role in asymmetric catalysis.

When bis(diarylphosphines) are complexed with rhodium, there is an array of four aryl groups around the metal center. It seems likely that a preferred conformation of these groups is the origin of the asymmetric bias. The ligand conformation in $[Rh(COD)(DIPAMP)]^+$ has been established from X-ray studies and is shown in 3-70 — it presents an alternating 'face—edge' arrangement of the ring planes. It has been suggested [968, 1865] that linear prochiral substrates would lie more easily along the face-exposed aryls and this would result in steric control.

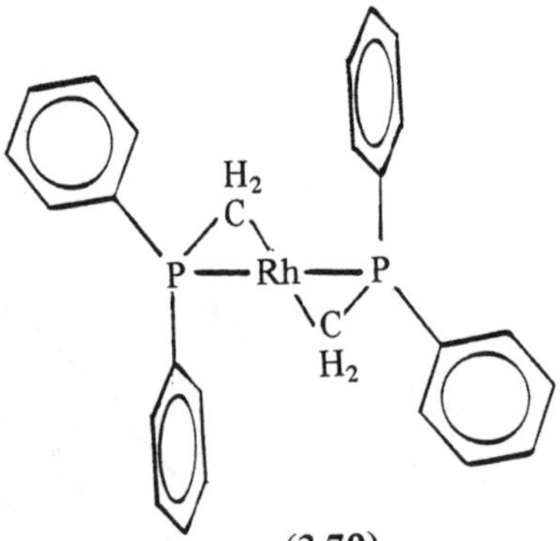

(3-70)

These effects are less easy to apply to seven-membered chelates. However, Glaser [599] has studied space-filling models of Rh{(+)-DIOP} systems to determine which conformers show the least steric strain. Although this approach does not take into account polar factors or alkene stereochemistry, it successfully predicts the configuration of products from numerous asymmetric hydrogenation (and hydrosilation) reactions.

There is much to be learned about the subtle interplay between substrate and ligand in these systems, but given the short period of investigation the present level of understanding is impressive.

A COMMERCIAL APPLICATION; THE SYNTHESIS OF L-DOPA

To date, there has been one commercial application of the asymmetric hydrogenation reaction. A synthesis of L-dopa (3,4-dihydroxyphenylalanine), which is used for treating Parkinson's disease, has been developed by Monsanto. A substituted cinnamic acid (3-71) is the prochiral alkene, and the catalyst is formed from a diene–rhodium(I) complex by treatment with the bidentate phosphine DIPAMP (see 3-52) in methanol [971, 1865]. Yields of the product are affected markedly by the pH of the solution and by the purity of the H_2. Up to 90% optical purity is achieved in the synthesis.

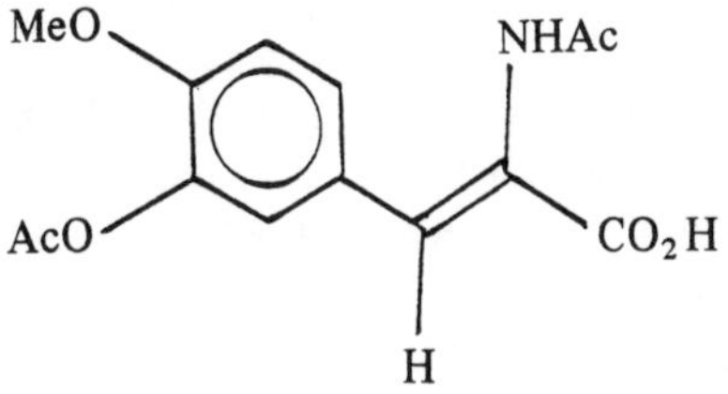

(3-71)

THE ASYMMETRIC REDUCTION OF KETONES

Complexes of the type $[Rh(diene)L_2]^+$ (e.g. diene = NBD or COD, L = PMePh-(CH_2Ph) or $PMeCy(o-MeOC_6H_4)$) [201, 1638, 1639, 1759, 1787] have been used to prepare optically active alcohols by the asymmetric reduction of ketones. For example, methylphenylketone gives 1-phenylethanol enriched in the R-(+)-enantiomer, and ethylmethylketone yields 2-butanol enriched in the R-(−)-enantiomer [201]. The optical yields are highest in carboxylic acid solvents [1641], and they depend on the P/Rh ratio [1787].

Similar asymmetric reductions of ketones have been achieved with RhCl-(L—L) and $[Rh(diene)(L—L)]^+$ (L—L = DIOP [1075, 1292, 1302, 1788, 1789] or BPPM [1302]). The data in Table 3.V relate to the hydrogenation of $MeCOCO_2Pr$ at 20 °C and 20 atm pressure H_2 for 24 h with these catalysts. In the reduction of acetophenone with $[Rh(NBD)DIOP]^+$ as the catalyst source, the enantioselectivity is reversed by the addition of achiral tri-n-alkylphosphines [1787, 1789]. Silylenol ethers are also reduced asymmetrically with RhCl(DIOP) as the catalyst [1761].

Catalytic activity in ketone reductions is increased significantly if Et_3N is added to $[Rh(NBD)L_2]^+$ (L = PPh_2(menthyl), PPh_2(neomenthyl) or DIOP) [696]. In these systems, the optical yield depends on the amount of base added. When water is added, there is no acceleration of the reaction rate [1302].

TABLE 3.V

Effects of catalyst and solvent on asymmetric reduction of MeCOCO$_2$Pr

Catalyst	Solvent	Conversion	% ee	Product configuration
Rh{(−)-DIOP}ClS	MeOH	57	25	R
	THF	100	42	R
	C$_6$H$_6$	100	35	R
Rh(BPPM)ClS	C$_6$H$_6$	98	76	R
[Rh(COD){(−)-DIOP}]$^+$	C$_6$H$_6$	99	16	S
[Rh(COD)(BPPM)]$^+$	C$_6$H$_6$	99	37	R

Hence, the mechanism for these reactions differs from that which operates in the reduction of ketones with [H$_2$Rh(PPh$_3$)$_2$S$_2$]$^+$ as the catalyst.

The neutral complex RhCl(DIOP) has been anchored to supports such as 'Amberlite' [1288]. In the asymmetric hydrogenation of prochiral ketones, the catalytic efficiency of this supported catalyst is similar to that of the homogeneous system.

Various iridium compounds catalyse the transfer of [H] from 2-propanol to prochiral ketones such as acetophenone. With catalysts formed from [Ir-(COD)Cl]$_2$ and bidentate phosphines, the optical yield depends on the type of phosphine [1649] (see Table 3.VI). Related iridium complexes containing optically active Schiff bases (e.g. (−)-2-pyridinalphenylethylimine) also catalyse H-transfer from 2-propanol to PhCOR (e.g. = Me, Pr^i) [1963, 1964].

TABLE 3.VI

Effects of ligand in {[Ir(COD)Cl]$_2$ + L−L} in the asymmetric reduction of acetophenone by H-transfer from Pr^{i}OH

Chiral ligand	Optical yield (%)
NMDPP	18
PROPHOS	30
CHIRAPHOS	28
DIOP	14

3.11 Hydrosilation Reactions

HOMOGENEOUS CATALYSTS

Many systems which catalyse the homogeneous hydrogenation of C−C multiple bonds lack activity towards the reduction of other unsaturated bonds. For example, $RhClL_3$ complexes are inactive toward the hydrogenation of carbonyl compounds; moreover, they may achieve decarbonylation of the organic substrate. Hydrosilation of the carbonyl group, followed by hydrolysis (e.g. with aqueous methanolic KOH), provides a means of overcoming these problems (see Equation (3.20)).

$$\text{$>$C$=$O} + HSiR_3 \xrightarrow{\text{[cat.]}} \begin{array}{c} H \\ | \\ -\overset{|}{\underset{|}{C}}-OSiR_3 \\ | \\ H \end{array}$$

$$\xrightarrow{\text{hydrolysis}} \begin{array}{c} H \\ | \\ -\overset{|}{\underset{|}{C}}-OH \\ | \end{array} \tag{3.20}$$

Generally, both parts of the reaction are rapid at ordinary temperatures. Moreover, the reaction is completely regiospecific because of the high affinity of Si for O.

There have been numerous reports of the hydrosilation of alkenes and alkynes to produce alkylsilanes. In the reactions with alkenes, the silicon generally attaches to the less crowded end of the C=C bond. Terminal alkenes undergo hydrosilation much more readily than internal alkenes. In those cases where internal alkenes appear to react, it seems likely that isomerization to a terminal alkene precedes the hydrosilation. This is illustrated by the following example:

$$cis\text{-}CH_3CH_2CH{=}CHCH_3 + R_3SiH \xrightarrow{\text{[cat.]}}$$

$$CH_3(CH_2)_4SiR_3$$

A steric influence is demonstrated by the observation that trans-2-pentene will not participate in the hydrosilation reaction [301].

The relative importance of hydrosilation and isomerization depends to some extent on the nature of the silane [303]. The purity of the reagents used can be important also. Thus, with phosphine-rhodium catalysts, the hydrosilation of alkenes is completely inhibited if all reagents are carefully purified and traces of oxygen are excluded [442]; under these conditions, isomerization is rapid. A dramatic promotion of the hydrosilation reaction is achieved by photolysis of the mixture in the presence of air [499]. It is sometimes difficult to achieve selectivity in the alkyne addition reactions. An example is shown in Equation (3.21).

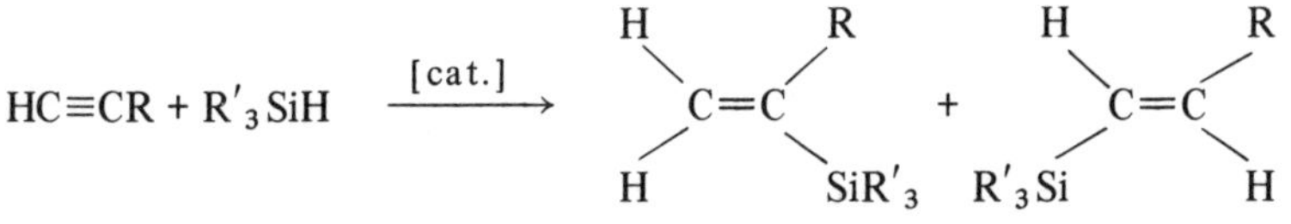

$$HC{\equiv}CR + R'_3SiH \xrightarrow{\text{[cat.]}} \overset{H}{\underset{H}{>}}C{=}C\overset{R}{\underset{SiR'_3}{<}} + \overset{H}{\underset{R'_3Si}{>}}C{=}C\overset{R}{\underset{H}{<}} \tag{3.21}$$

Many metal complexes have been used as hydrosilation catalysts [665, 1088, 1089, 1321]. Although platinum complexes such as H_2PtCl_6 have probably been the most thoroughly studied [1643], the rhodium complexes $RhClL_3$ are also excellent hydrosilation catalysts. They have been used to reduce a variety of ketones and aldehydes [356, 1307, 1310, 1317, 1318, 1566, 1733]. As shown in Equations (3.22) and (3.23), the hydrosilation of alcohols [356, 1298, 1308, 1309] and amines [993] has also been achieved.

$$Pr^iOH + Et_3SiH \longrightarrow Et_3SiOPr^i \tag{3.22}$$

$$Et_2NH + Ph_3SiH \longrightarrow Ph_3SiNEt_2 \tag{3.23}$$

The complexes $RhClL_3$ have been widely used in the hydrosilation of alkenes [301, 322, 336, 442, 673, 674, 818, 986, 987, 988, 1020, 1313, 1467, 1736, 1737, 1894], dienes [1296, 1314, 1466] and alkynes [336, 442, 443, 818, 995, 996, 1315, 1448, 1449, 1895, 1896]. The reaction between 1-hexene and Et_3SiH in the presence of $RhCl(PPh_3)_3$ is a specific example. With excess alkene as the solvent, the yield of (1-hexyl)$SiEt_3$ is 80% after 2 h at 60 °C [301, 674]. Some isomerization of the alkene is observed. The type of solvent used can affect the extent of hydrosilation of phenylalkenes with $RhCl(PPh_3)_3$ as the catalyst; thus, the conversion of styrene is very high in hydrocarbons but decreases substantially in polar solvents [1021].

A variety of products is obtained by the hydrosilation of alkenyl sulfides [1879, 1883, 1884, 1885] and dialkenyl sulfides [1880]; Equations (3.24) and (3.25) provide two examples.

$$EtSHC{=}CH_2 + Et_3SiH \longrightarrow EtSCH_2CH_2SiEt_3 \text{(major)} +$$

$$EtSCH(Me)SiEt_3 \text{(minor)} \tag{3.24}$$

$$S(HC{=}CH_2)_2 + Et_2SiH_2 \longrightarrow Et_2HSiCH(Me)SCH{=}CH_2 +$$

$$Et_2HSiCH_2CH_2SCH{=}CH_2 \tag{3.25}$$

In some instances, however, $C-S$ bond cleavage to give a thiosilane such as $(EtO)_3SiS(CH_2)_nCH{=}CH_2$ predominates. Cyclization to give heterocyclic systems containing Si and S (see Equation (3.26)) also occurs in some cases [1880, 1881].

$$Et_2HSiCH(Me)SCH{=}CH_2 \longrightarrow$$

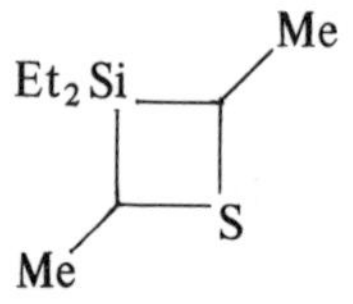

(3.26)

Several trends have been observed in hydrosilation reactions with $RhClL_3$ compounds as catalysts. In general, aryl ketones react more slowly than alkyl ketones, and di- or tri-hydrosilanes react more rapidly than $HSiR_3$. For the addition of $HSiPr_{3-n}(OEt)_n$ to alkenes, the rate decreases from $HSi(OEt)_3$ to $HSiPr_3$ [1737]; the rate of addition of silanes to 1,3-butadiene decreases in the order $HSiMe_3 > HSi(OEt)_3 > HSiCl_3$ [1466]. For the addition of $HSiEt_3$ to 1-hexene with $RhCl(PPh_3)_3$ as the catalyst, the rate increases in the order $L = PCy_3 \ll PMePh_2 < PCy_2Ph \simeq PPh_3 < PCyPh_2 \simeq PEtPh_2$. With ethylene as the substrate, the relative rates of hydrosilation with $HSiMe_2(C_6H_4X)$ and different catalysts have been estimated [1737]; the order is H_2PtCl_6 (relative rate $\simeq 4000$) $> RhCl(PPh_3)_3$ ($\simeq 200$) $> Co_2(CO)_8$ ($\simeq 1$).

The stereochemistry of the hydrosilation of ketones such as camphor and menthone is of interest because two products, 3-72 and 3-73, are possible [1318]. With silanes such as H_3SiPh, H_2SiEt_2 or H_2SiPh_2, the predominant product is generally the alcohol formed by H-transfer from the least hindered direction (3-72). Increasing amounts of 3-73 are obtained with more bulky silanes such as $HSiEt_3$. Similar selectivity effects are observed in the hydrosilation of 4-*t*-butylcyclohexanone [1566].

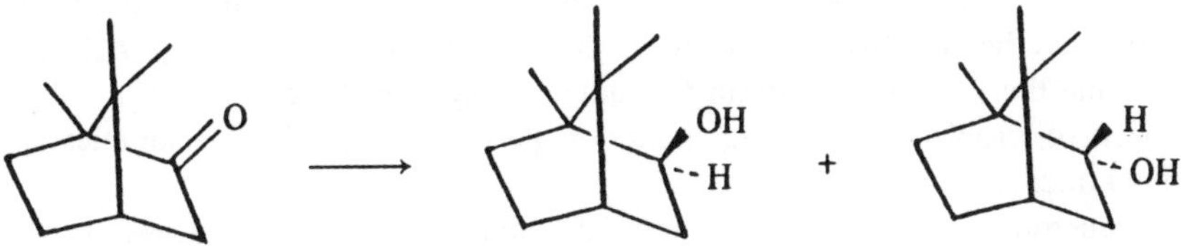

(3-72)   (3-73)

There has also been interest in the stereochemistry of the hydrosilation of alkynes. With $RhCl(PPh_3)_3$ as the catalyst, there is stereoselective *trans* addition of $HSiMe_2Ph$ to $PhC{\equiv}CH$ [1896] and to alkyl acetylenes [1315] to give the *cis* products. Under some conditions, complete isomerization of *cis* to *trans* products occurs during the hydrosilation reaction. The stereoselectivity of these reactions is increased when the $RhCl(PR_3)_3$ catalyst contains bulky R groups [995]. With $Rh(CO)Cl(PPh_3)_2$ as the catalyst, the stereoselective reduction occurs at a faster rate and there is no isomerization of the product.

The mechanism of the hydrosilation reaction with $RhClL_3$ as the catalyst is probably closely related to that of hydrogenation. For the reaction with ketones, a suggested pathway is shown in Scheme 3.10. Some of the initial oxidative-addition products, $HRh(SiR_3)ClL_2$, have been isolated and the following orders of stability to dissociation have been established: $SiMe_3 < SiMe_2Cl < SiMeCl_2 < SiCl_3$; $SiPh_3 < SiEt_3 < Si(OEt)_3 < SiCl_3$; $PCy_3 < PCy_2Ph <$

$PCyPh_2 \leq PPh_2(C_6H_4Me) \simeq PPh_3$; $PMePh_2 < PEtPh_2 < PPr^iPh_2 \simeq PPh_3$; $AsPh_3 < PPh_3$. The dependence of the stability on L reflects a combination of steric and electronic effects [674].

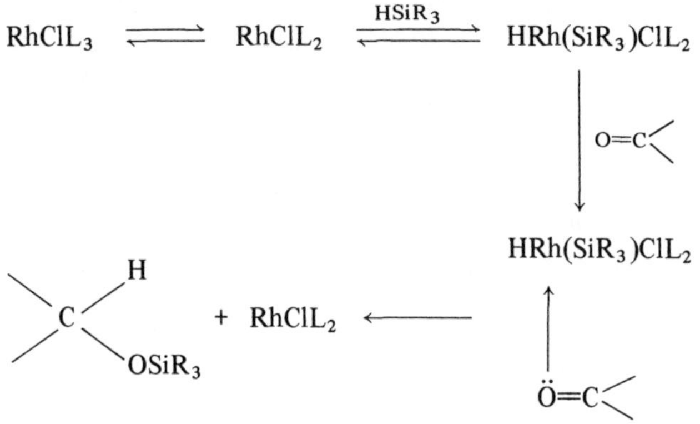

$$RhClL_3 \rightleftharpoons RhClL_2 \xrightarrow{\text{HSiR}_3} HRh(SiR_3)ClL_2$$

$$\downarrow \quad O=C\hspace{-3pt}\diagup$$

$$HRh(SiR_3)ClL_2$$

$$\begin{array}{c} H \\ \diagdown C \diagup \\ \diagup \quad \diagdown OSiR_3 \end{array} + RhClL_2 \longleftarrow \quad \uparrow$$

$$\ddot{O}=C\hspace{-3pt}\diagup$$

Scheme 3.10

The reactions with alkenes may follow a related pathway; alternatively, the alkene may be coordinated prior to the Si—H oxidative-addition step. Attempts to define the precise mechanism for individual systems have been plagued by a variety of problems, including induction periods, irreproducible kinetics and competing reactions.

Numerous other rhodium compounds, and some iridium species, are also effective hydrosilation catalysts. Compounds related to $RhCl(PPh_3)_3$ include $RhX(PPh_3)_3$ (X = Br, I or Me) [996, 1466, 1467], $RhCl(SbPh_3)_3$ [996], $[RhClL_2]_2$ (e.g. L = PPh_3, PCy_3) [219, 768] and $HRh(PPh_3)_a$ [994]. Carbonyl compounds such as $Rh_4(CO)_{12}$ [65, 66, 318, 1104, 1105, 1107, 1108], [Rh-$(CO)_2Cl]_2$ [128, 301, 1882], $Rh(CO)ClL_2$ (L = PPh_3 [301, 996, 1449] or $PEtPh_2$ [359, 428]) and $HRh(CO)(PPh_3)_a$ [301, 1449] are other sources of active catalysts. The $RhX(PPh_3)_3$ compounds are more effective catalysts than the carbonyl—rhodium complexes for the addition of monosilanes to $BuCH=CH_2$, but less effective for additions to $PhCH=CH_2$ [1467]. For the hydrosilation of 1,3-butadiene, the $RhX(PPh_3)_3$ species are more effective in the addition of $HSiMe_3$ and $HSiCl_3$, but less effective than the carbonyl compounds with $HSi(OEt)_3$ [1466].

Catalysts of the type $[RhL_2S_2]^+$ have been used in the hydrosilation of α,β-unsaturated ketones, and either the C=C or the C=O function can be reduced depending on the type of silane used (see Scheme 3.11).

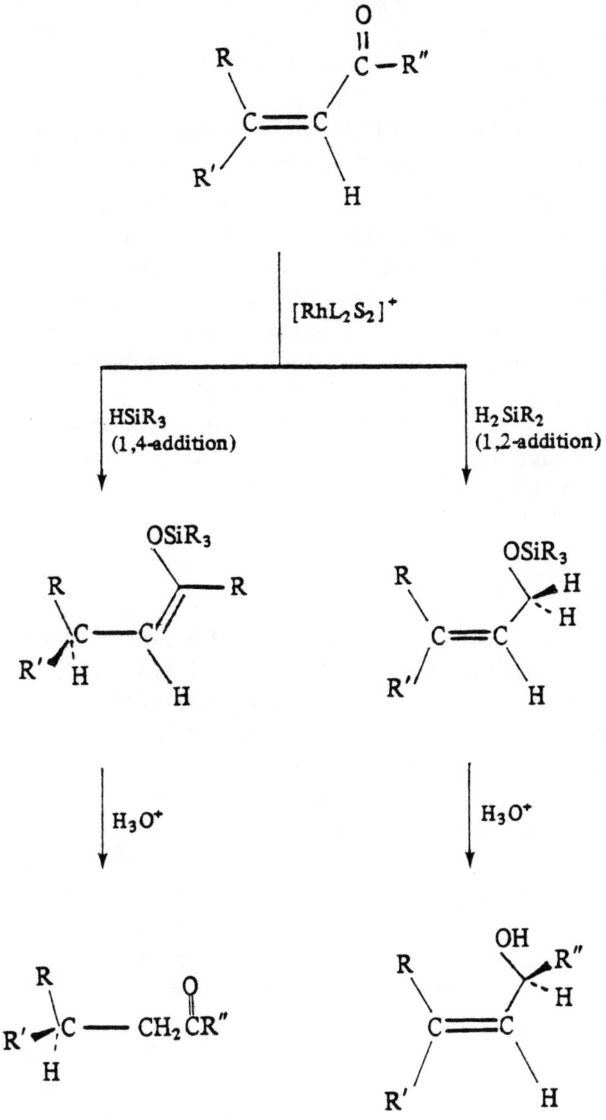
O
R
C—R″
C═C
R′
H
[RhL$_2$S$_2$]$^+$
HSiR$_3$
(1,4-addition)
H$_2$SiR$_2$
(1,2-addition)
OSiR$_3$
R
R
C—C
R′
H
H
OSiR$_3$
R
H
H
C═C
R′
H
H$_3$O$^+$
H$_3$O$^+$
R
O
R′
C — CH$_2$CR″
H
OH
R
R″
H
C═C
R′
H

Scheme 3.11

Other citations of hydrosilation catalysts have included the following rhodium compounds: $Rh(CO)_2(L-L)$ (HL_2 = acetylacetone [67, 314, 1000, 1739] or 8-hydroxyquinoline [314]), $Rh(CO)(PPh_3)(acac)$ [314], $[Rh(C_2H_4)_2-Cl]_2$ [1735], $Rh(\eta-C_3H_5)\{P(OMe)_3\}_3$ and $Rh(\eta-C_3H_5)(acac)$ [214, 314], $\{[Rh(COD)Cl]_2 + Ph_2PCH_2CH_2PPh_2$ and related ligands$\}$ [1246], $(\eta-C_5Me_5)_2-Rh_2Cl_4$ [1183], carbene—Rh complexes such as $HRhCl_2(CHNMe_2)(PPh_3)_2$ and $RhCl(COD)\overline{CNRCH_2CH_2N}R$ [717], carborane complexes including 3,3-$(PPh_3)_2$-4-C_5H_5N-3,1,2-$RhC_2B_9H_{10}$ [1960], 3,3-$(Ph_3P)_2$-3-H-3,1,2-$RhC_2B_9H_{11}$ and 2,2-$(Ph_3P)_2$-2-H-2,1,7-$RhC_2B_9H_{11}$ [1958, 1961], $[Rh(DMG)_2(PPh_3)]_2$ [1359, 1578], MX_3 compounds including $RhCl_3 \cdot 3H_2O$, $Rh(OH)_3$, $Rh(OR)_3$ [302, 587, 1644], $Rh(acac)_3$ and $Rh_2(\mu-OAc)_4$ [353].

Iridium compounds that provide active hydrosilation catalysts include $Ir(CO)XL_2$ [180, 308, 1642], $Ir(N_2)Cl(PPh_3)_2$, $[Ir(C_8H_{14})_2Cl]_2$, $IrClL_2$ [180] and $IrCl_3$ [302, 587, 1644]. With $Ir(CO)XL_2$ as the catalyst, the proposed mechanism [180, 1642] involves oxidative-addition of the silane to give $HIrX(SiR_3)(CO)L_2$ followed by reductive-elimination of $ClSiR_3$. This gives $HIr(CO)L_2$ which adds HCl to form $H_2Ir(CO)ClL_2$.

POLYMER-SUPPORTED CATALYSTS

Phosphine-functionalized polystyrene has been used to support some of the hydrosilation catalysts, including $RhCl(PPh_3)_3$ [129, 280, 835, 1175, 1176, 1464, 1734], $Rh(CO)ClL_2$ [1175, 1176], $HRh(CO)L_3$ [1175, 1176] and $[Rh(C_2H_4)_2Cl]_2$ [1006]. One study of the catalyst prepared from $[Rh(CO)_2-Cl]_2$ and dimethylaminomethylated-styrene—divinylbenzene copolymer with different degrees of crosslinking has shown that the catalytic activity towards hydrosilation increases with increasing surface area and pore radius [446].

Silica has also been used as a support for hydrosilation catalysts. Although $RhCl(PPh_3)_3$ on silica gives a stable catalyst for the hydrosilation of ketones such as 2- and 4-heptanone, the reactivity is low compared to that with the analogous homogeneous catalyst [835]. A lower activity of the silica-supported catalyst is also observed in the reactions of organosilicon hydrides with organosilanols [1177].

Silica functionalized with phosphine [1175, 1176] or amine [1120] groups has also been used to support hydrosilation catalysts such as $RhCl(PPh_3)_3$. When the catalyst is formed from $[Rh(C_2H_4)_2Cl]_2$ and $Ph_2P(CH_2)_n-$ groups linked to silica, the activity decreases significantly as the $Ph_2P(CH_2)_n-$ chain length is increased. This is attributed to the greater activity of the mononuclear complexes formed with $Si-CH_2PPh_2$ compared to the binuclear species obtained with $Si-(CH_2)_{2-6}PPh_2$ [1178]. With catalysts formed from $RhCl(PPh_3)_3$ and $[Ph_2P(CH_2)_nSiO_{3/2}]_x$ (n = 2 or 3), kinetic data for the hydrosilation of alkenes show that the support has essentially no effect on the kinetic parameters and the mechanism [1121].

Another supported catalyst has been obtained by treating silica with Ph_2P-$(CH_2)_2Si(OEt)_3$ and then with $[Ir(C_8H_{14})_2Cl]_2$. This is an effective catalyst for the hydrosilation of primary alcohols with $HSiR_3$ (R = Et, OEt) or Me_3-$SiO[Si(H)(Me)O]_nSiME_3$ [477].

ASYMMETRIC HYDROSILATION

The asymmetric reduction of prochiral ketones by hydrosilation in the presence of chiral rhodium complexes has been achieved [894, 895, 1125]. In some cases, asymmetric reaction occurs at both Si and C centers (see Equation (3.27)).

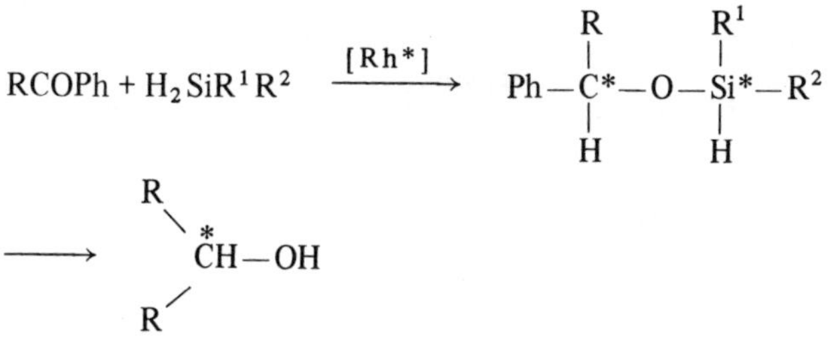

$$RCOPh + H_2SiR^1R^2 \xrightarrow{[Rh^*]} Ph-\overset{\overset{\displaystyle R}{|}}{\underset{\underset{\displaystyle H}{|}}{C^*}}-O-\overset{\overset{\displaystyle R^1}{|}}{\underset{\underset{\displaystyle H}{|}}{Si^*}}-R^2$$

$$\longrightarrow \quad \overset{\displaystyle R}{\underset{\displaystyle R}{\diagdown}}\overset{*}{C}H-OH \diagup \qquad (3.27)$$

As with the asymmetric hydrogenation of alkenes, the chiral catalyst can be of the type $RhCl\{PMe(CH_2Ph)Ph\}_a$ [685, 1306, 1308, 1316, 1935], RhCl(DIOP) [139, 357, 599, 895, 1303, 1385], $[Rh\{PPh_2(neomenthyl)\}_2S_2]^+$ [1222] or $[Rh(DIOP)S_2]^+$ [895, 1320]. Other chiral ligands that have been incorporated in the Rh hydrosilation catalysts include carboranes [1956], Schiff bases [257], glucophosphinites [878], methyl-4,6-O-benzylidene-2,3-bis-O-(diphenylphosphino)-α-D-glucopyranoside [279] and (S)-N,N-dimethyl-1-(O-(diphenylarsino)-phenyl)ethylamine [1377]. In some cases, the chiral catalyst has been anchored to a polymer support; —Rh(DIOP) [474] and —$RhX\{PMePh(CH_2CH_2CH_2-Si(OEt)_3\}$ [278] are typical examples. In general, the optical yields are modest (ee $\leq$ ca. 50%) for simple ketones and α,β-unsaturated ketones [1321, 1385]. One of the better conversions is given by the example in Equation (3.28)

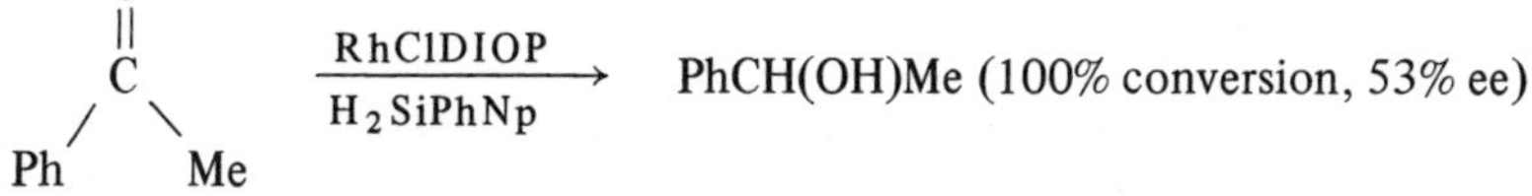

$$\overset{\overset{\displaystyle O}{\|}}{\underset{\underset{\displaystyle Ph \quad Me}{\diagup \diagdown}}{C}} \xrightarrow[H_2SiPhNp]{RhClDIOP} PhCH(OH)Me \text{ (100\% conversion, 53\% ee)} \qquad (3.28)$$

In these reactions, the optical yields seem to be strongly dependent on the nature of the silane [895], but few generalizations can be made as yet with confidence. Higher optical yields are attained with 1,2- and 1,4-dicarbonyl compounds [1305, 1321] (see Equation (3.29) for an example).

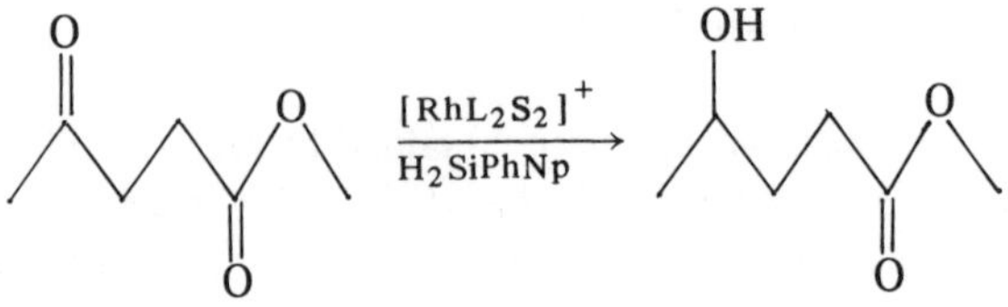

$$(3.29)$$

(L = PMePh(CH$_2$Ph)); 88% conversion, 76% ee of S-conformer)

With these systems, the second carbonyl group probably coordinates to the Rh, thus fixing the orientation of the ketone at the chiral catalytic center. The asymmetric synthesis of amines can also be achieved by hydrosilation (Equation (3.30)) with a chiral complex of rhodium as the catalyst [1042].

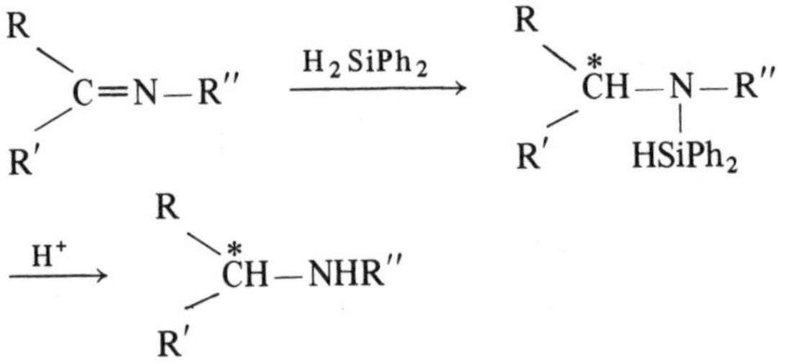

$$(3.30)$$

Kinetic data have been determined [989] for the conversion of PhCOBu^t to PhCH(OSiHPh$_2$)Bu^t with a catalyst formed from [Rh(1,5-COD){(−)-DIOP}]-[ClO$_4$]. The initial induction period is due to displacement of the diene by the silane to give [Rh{(−)-DIOP}(C$_6$H$_6$)$_a$]$^+$. There is then oxidative-addition to form [HRh{(−)-DIOP}(SiHPh$_2$)]$^+$ and subsequent attack by the ketone in the rate-determining step. Further kinetic results indicate that a similar mechanism operates when [Rh(DIOP)$_2$] [ClO$_4$] is used as the catalyst source [990]. Spin-trapping experiments on the asymmetric hydrosilation of RCOMe by H$_2$SiPh(α-naphthyl) in the presence of RhCl(DIOP) are consistent with a mechanism which involves the formation of an [Rh(CMePh){OSiHPh(α-naphthyl)}] intermediate [1386].

HYDROGERMYLATION

The hydrogermylation of unsaturated compounds, including cyclohexene [355] and phenylacetylene [354], has been achieved with RhCl(PPh$_3$)$_3$ as the catalyst. There is retention of configuration in the adducts formed in these reactions.

ADDITION OF AMINES TO CARBONYL COMPOUNDS

Imines are formed from the reaction between primary amines and carbonyl compounds such as cyclohexanone (Equation (3.31)).

$$\text{\textbackslash}C{=}O + NH_2R \longrightarrow \text{\textbackslash}C{=}NR + H_2O \qquad (3.31)$$

A catalyst for this reaction can be formed from $H[Rh(DMG)_2Cl_2]$ + $NaBH_4$ [962].

A similar reaction with aldehydes under H_2 pressure gives tertiary amines (equation 3.32). The catalyst for this reaction can be formed from $RhCl_3$, CO and H_2 under pressure [807]

$$RCH({=}O) + R_2'NH + H_2 \longrightarrow RCH_2NR_2' + H_2O \qquad (3.32)$$

CHAPTER 4

CARBONYLATION AND HYDROFORMYLATION REACTIONS

Many soluble metal compounds catalyse the addition of CO to organic substrates, and these reactions are useful for the formation of oxygenated compounds such as carboxylic acids, acid halides, ketones and isocyanates. Some of the more important reaction types are indicated below:

Alcohols:	ROH	$\longrightarrow$	RCO_2H
Alkyl halides:	RX	$\longrightarrow$	$RCOX$
Amines:	RNH_2	$\longrightarrow$	$RNHCHO$
	$2\ RNH_2$	$\longrightarrow$	$RNHCONHR + H_2$
Nitro compounds:	RNO_2	$\longrightarrow$	$RNCO$

The best known carbonylation reactions are those that involve the formation of aldehydes and alcohols by the simultaneous addition of CO and H_2 to an alkene. This type of reaction is described as '*hydroformylation*', and has been used industrially since the early 1940s. The most recent commercial carbonylation reaction involves the conversion of alcohols to carboxylic acids; this process first came on-line in the early 1970s. The present discussion is concerned with carbonylation reactions carried out in the presence of soluble rhodium and iridium catalysts.

4.1. The Water–Gas Shift Reaction

Hydrogen, CO and mixtures of these two gases ('*synthesis gas*') are important feedstocks in the chemical industry. The industrial production of H_2 is generally achieved through the water–gas shift reaction:

$$H_2O + CO \rightleftharpoons H_2 + CO_2; \quad K_{127} = 1.5 \times 10^3$$

This reaction can also be used to adjust the H_2/CO ratio in synthesis gas. The CO in these systems can be generated from a number of sources, including natural gas, petroleum fractions and coal. Although natural gas is currently the major source, there is likely to be a growing dependence on coal over the next 10–20 years [1906].

The water–gas shift reaction is favored thermodynamically ($\Delta H^\circ = -40$ kJ mol^{-1}, $\Delta G^\circ_{298} = -20$ kJ mol^{-1}), but it is slow. Useful rates of reaction are achieved only when catalysts and high temperatures (e.g. Fe_3O_4 at 300–350 °C; or copper–zinc oxide mixtures at 200–250 °C) are used. At these temperatures, however, there are significant equilibrium concentrations of both reactants and products, because the equilibrium constant exhibits a negative temperature-dependence. Hence, there is a need to find better catalysts.

Recently, there have been several reports of catalysis of the shift reaction by soluble metal complexes, including several compounds of rhodium and iridium. Although these homogeneous catalysts may never be used commercially, they do provide an insight into the role of the metal in the reaction. One of the best understood catalyst systems is a solution of $[Rh(CO)_2I]_2$ in aqueous acetic acid [106, 316, 484]; it converts CO to an equimolar mixture of H_2 and CO_2 at low pressures. Although the rate is slow at 80–90 °C, it is much faster at higher temperatures. This catalyst can be used in the presence of acqueous HI, and a high-pressure infrared study [1613] has established the nature of the predominant rhodium species present in solution at 185 °C and 400 psi. If the HI concentration is low and the water concentration high, the rhodium(I) anion $[Rh(CO)_2I_2]^-$ predominates; at high HI and low H_2O concentrations, most of the metal is present as the rhodium(III) species trans-$[Rh(CO)_2I_4]^-$. In these systems the conversion of CO and H_2O to CO_2 and H_2 is thought to proceed according to Scheme 4.1. The involvement of a high-valent metal in the first step is believed to be important because it will cause the attached CO to be more susceptible to attack by nucleophiles. Kinetic studies [106] have established that the rate-limiting step of the catalysis changes with temperature. The production of CO_2 from the reduction of the Rh^{III} species is rate-determining at higher temperatures (> 80 °C), but at lower temperatures the formation of H_2 from the oxidation of the Rh^I species is the rate-limiting step. The reactions at the higher temperatures are also affected by pH [1613]. Thus, the oxidation of rhodium(I) becomes the rate-controlling step at low acidity.

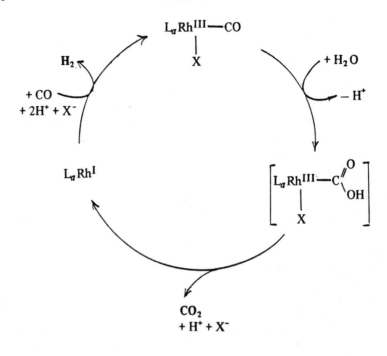

Scheme 4.1

Some other rhodium compounds show catalytic activity in the water–gas shift reaction. Catalysts formed from hydrido–rhodium(I) compounds such as $HRhL_3$ (L = PEt_3, PPr^i_3), $HRh(PBu^t_3)_2$ and $H_2Rh_2(\mu\text{-}N_2)(PCy_3)_4$, are active above *ca.* 50 °C [1943]. Under the conditions of the water–gas shift reaction, these complexes are converted to carbonyl compounds; thus,

$$HRhL_3 \xrightarrow[-L]{CO} \textit{trans-}HRh(CO)L_2 \xrightarrow[-H_2,-L]{CO} Rh_2(CO)_aL_b$$

The treatment of $Rh_2(CO)_3L_3$ (L = PPr^i_3) with water in pyridine results in the evolution of H_2 and the formation of $[Rh(CO)L_2py]^+$. Further reaction between this cationic complex and CO results in the formation of CO_2 together with *trans-*$HRh(CO)L_2$ [1941]. Both $[Rh(CO)L_2py]^+$ and *trans-*$HRh(CO)L_2$ are effective catalysts for the water–gas shift reaction.

Another rhodium carbonyl system that has been used to catalyse the shift reaction is the molecular 'A-frame' complex $[Rh_2(\mu\text{-}H)(\mu\text{-}CO)(CO)_2\text{-}(Ph_2PCH_2PPh_2)_2]^+$ in the presence of 2 mol equiv of LiCl [1012]. Catalysis is achieved under mild conditions (1 atm pressure CO, 90 °C, in PrOH). Various rhodium carbonyl clusters, including $Rh_4(CO)_{12}$ $Rh_6(CO)_{16}$ and

$[Rh_{12}(CO)_{30}]^{2-}$, are also active catalyst sources; they are most effective when ethylenediamine is added as a base and a solvent for the reaction [913]. These are better catalysts than those obtained from water-soluble rhodium compounds such as $RhCl_3 \cdot 3H_2O$ plus an aqueous base (e.g. Na_2CO_3) and a water-miscible alcohol (e.g. MeOH). With this system, catalysis is achieved only at 175–260 °C and 150–200 atm pressure [1271]. Amongst the relatively few iridium compounds that have been used to catalyse the shift reaction are $Ir_4(CO)_{12}$ [917] and the cations $[Ir(COD)L_2]^+$ (e.g. L = $PMePh_2$, L_2 = phen) [931]. One of the most active of these catalysts is $[Ir(COD)(phen-S)]^+$ (phen-S = 4,7-diphenyl-phenanthroline sulfonate).

4.2. The Hydrogenation of CO; Synthesis Gas Chemistry

REACTION PRODUCTS

A large array of hydrocarbons and oxygen-containing compounds can be obtained from the direct reaction between CO and H_2. This type of chemistry originated in Germany in the 1920s, and is referred to as *Fischer–Tropsch* (F–T) synthesis. During World War II, it was used extensively to convert synthesis gas to motor fuels. The production of fuels and chemicals from coal became noncompetitive in the post-war years because of the cheapness of oil and gas. With the current search for new feedstocks [1580], there is a surge of research activity aimed at developing an understanding of the mechanism of F–T chemistry and at finding new catalysts that will improve the economics of the reaction.

F–T chemistry is a composite of many reactions and the available catalysts lack selectivity. The classical heterogeneous catalysts are supported iron and cobalt, and these metals are regarded primarily as olefin formers:

$$CO + 2H_2 \quad \rightarrow \quad (-CH_2-) + H_2O$$

Typical products are straight chain hydrocarbons and oxygenated organic molecules of the types C_nH_{2n+2} and $C_nH_{2n+1}OH$ (n = 4–7). With other catalysts, the products may contain different major components.

Some related chemistry is based on the conversion of synthesis gas to lower molecular weight products such as methane, methanol and ethylene glycol:

$$CO + 3H_2 \quad \rightarrow \quad CH_4 + H_2O$$

$$CO + 2H_2 \quad \rightarrow \quad CH_3OH$$

$$2CO + 3H_2 \quad \rightarrow \quad HOCH_2CH_2OH$$

Above 450°, the thermodynamics allow only the formation of methane; the higher the temperature the greater the tendency to form methane. The reactions to form methanol and other oxygenated products are favorable, and significant quantities of these products are obtained from the gaseous mixture particularly when high pressures are applied. Because of the incredible complexity of the overall system, it is impossible to make predictions based on thermodynamics about product distributions. Moreover, little can be established about the gross kinetics of the system.

In the new look at F–T and synthesis gas chemistry, a variety of targets are apparent. Fuels and petrochemicals are still of interest, but greater importance is attached to the formation of alcohols such as methanol and ethylene glycol. There is hope that cluster compounds will provide selective catalysts for some of these syntheses, and that soluble catalysts will furnish insight into the mechanisms of the reactions [169, 712, 941, 1238, 1484]. Further discussion of these systems will be restricted to the use of catalysts derived from rhodium and iridium compounds.

CATALYSTS

Although the reduction of CO to methane is achieved with 'catalysts' such as $Ir_4(CO)_{12}$, the conversion is exceedingly slow under the conditions used (140 °C, 2 atm pressure, toluene solution) [417, 1778]. The rate of formation of methane is enhanced in the presence of trimethylphosphite, but this is due to a reaction between hydrogen and the methyl groups of the phosphite [1548]. This hydrogenolysis reaction is catalysed by the cluster complex. Some improvement in the catalytic activity of $Ir_4(CO)_{12}$ is observed when it is used in an $NaCl–AlCl_3$ melt at 170–180 °C [1890]. Under flow conditions at 1 atm pressure, the major products are isobutane and propane; the formation of ethane is favored by long reaction times.

Aliphatic hydrocarbons have also been produced with catalysts formed from rhodium compounds. Combinations of $Rh_4(CO)_{12}$, $[Rh(CO)_2Cl]_2$, $Na_2[Rh_{12}(CO)_{30}]$ or $RhCl(PPh_3)_3$ with $AlBr_3$ and a reductant such as Mg or Al have been used [1044, 1046, 1047]; the most effective catalyst is $\{Rh_4(CO)_{12} + AlBr_3 + Al\}$. In these systems, CO conversion is increased dramatically if the CO/H_2 ratio is decreased from 3:1 to 1:4 [1045]; the distribution (vol. %) of hydrocarbons in the products is then CH_4 (47), C_2H_6 (8), C_3H_8 (14), isobutane (30) and n-butane (1).

The hydrogenation of CO bound to thermally stable supported catalysts has been studied. Methane is formed when metal carbonyls supported on alumina are thermally decomposed in a stream of hydrogen [221, 774]. Polynuclear carbonyls including $Rh_4(CO)_{12}$, $Ir_4(CO)_{12}$ and $Rh_6(CO)_{16}$ have been used in these investigations, and the greatest amount of methane was produced from $Ir_4(CO)_{12}$ (11.2 CH_4 per $Ir_4(CO)_{12}$). It appears that these reactions are stoichiometric rather than catalytic, and that the stoichiometric reaction is faster than catalytic methanation involving CO(g). Presumably, then, good methanation catalysts will result from the use of supported metals that can (a) form good hydride-transfer species and (b) regenerate M—CO species under mild conditions. Since methane can be obtained from supported $Mo(CO)_6$, it is evident that polynuclear sites are not essential for CO hydrogenation.

The impregnation of Al_2O_3 or SiO_2 supports with potassium salts of carbonylate anions (e.g. $K[Ir(CO)_4]$), followed by thermal decomposition under hydrogen, also provides active catalysts for CO hydrogenation [1161]. This approach yields catalyst sites in which potassium is adjacent to the transition metal, and this increases the promotional effects of the potassium. The hydrogenation of CO with these catalysts produces mainly C_2 to C_5 alkenes. Ethylene and propylene are the major products obtained from CO when the catalyst is prepared by treating SiO_2 with $Rh_6(CO)_{16}$ or $[Rh(CO)_2Cl]_2$ and then $\{CO + H_2O\}$ [841].

A variety of rhodium complexes catalyse the transformation of synthesis gas to mixtures of ethylene glycol, propylene glycol, glycerol* and lower alcohols including methanol. The reaction conditions are generally severe, but selectivity to ethylene glycol ranges up to 70%. The choice of rhodium compound appears not to be vital, and $Rh_4(CO)_{12}$, $Rh_6(CO)_{16}$ and $Rh(CO)_2(acac)$ are all effective catalyst sources; the iridium compound $Ir(CO)_2(acac)$ can also be used [1925]. Generally, the reaction is assisted by promoters such as 2-hydroxypyridine, o-phenanthroline or cesium acetate, and stabilizers such as 1,3-dimethyl-2-imidazolidinone or crown ether and cryptate ligands [138, 297, 454, 606, 667, 919, 920, 921, 922, 923, 924, 925, 926, 927, 1889].

The formation of ethylene glycol (16.1 g), methanol (3.9 g), glycerol (2.7 g) and propylene glycol (1.2 g) from a 1:1 H_2/CO mixture at 220 °C and 20 000 psi for 3 h in the presence of $Rh(CO)_2(acac)$ as catalyst and 2-hydroxypyridine as promoter is typical of these reactions. This reaction could reach commercial scale in the near future; already, the process has been developed to the semiworks scale by Union Carbide. Commercial success may, however, be hindered by the severity of the reaction conditions and the high concentration of catalyst required.

Optimum yields of glycerol are obtained at high pressures (1000—4000 atm) when methanol is present and the stoichiometric ratio of $CO + H_2$ is used; $Rh_6(CO)_{16}$ seems to provide the most active catalyst [415]. Promotion of the formation of propylene glycol is achieved by the addition of aluminum compounds such as $Al(OCH_2CH_2)_3N$ [928].

Cluster anions including $[Rh_6(CO)_{15}C]^{2-}$ [1857, 1858, 1862], $[Rh_9(CO)_{21}P]^{2-}$ [1859, 1862] and $[Rh_{17}(CO)_{32}S_2]^{3-}$ [358, 1536, 1862] have been used as Fischer—Tropsch catalysts. In one instance $[Rh_{12}(CO)_{30}]^{2-}$ has

* Alternative name, glycerin or 1,2,3-propanetriol.

been used in combination with hydride complexes such as $[H_3Mo(PMePh_2)_3]^+$ or $[H_4W(PMePh_2)_4]^+$ [395]. Other catalysts are based on mixtures of ruthenium(III) acetylacetonate with rhodium(III) acetylacetonate [965] or rhodium(II) acetate [1910] in basic solvents.

Some catalysts have been formed by impregnating oxide supports with solutions of metal carbonyl compounds. With these catalysts, the conversion of synthesis gas to lower alcohols seems to be favored, and the nature of the support strongly influences the product distribution. With catalysts prepared from $Rh_6(CO)_{16}$ or $[Rh(CO)_2Cl]_2$ on silica, the major products are ethanol and acetaldehyde if the supported catalyst is pretreated with $\{H_2 + H_2O\}$ [841]. Catalysts can be prepared from rhodium and iridium carbonyl compounds deposited on oxides of zinc and magnesium; subsequent decomposition gives metal crystallites which catalyse the formation of methanol from $CO + H_2$ under mild conditions [791]. With rhodium crystallites on oxides of titanium, zirconium or lanthanum, ethanol is obtained in addition to methanol, methane and other products [792, 796, 801]. These crystallites can be formed by the pyrolysis of carbonyl clusters such as $(\eta\text{-}C_5H_5)_2Rh_2(CO)_3$, $Rh_4(CO)_{12}$, $Rh_6(CO)_{16}$, $(NEt_4)_3[Rh_7(CO)_{16}]$ and $(NBu_4)_2[H_{2-3}Rh_{13}(CO)_{23}]$ on the oxide support. The specific yield of ethanol is greatest with catalysts formed from $Rh_4(CO)_{12}$ [793]. The formation of ethanol from CO and H_2 at atmospheric pressure is achieved with catalysts formed from Rh_4–Rh_{13} carbonyl clusters impregnated onto silica containing Zr and Ti oxides [800]. Spectroscopic results show that there is initial interaction between the clusters and acid sites on the support to form hydrido species such as $[HRh_6(CO)_{15}]^-$ and $[H_3Rh_{13}(CO)_{23}]^{3-}$. Aggregation of the Rh clusters is apparently inhibited by the ZrO_2 and TiO_2, which also serve to enhance CO dissociation and to improve the selectivity in ethanol formation. Ethanol is also obtained, together with AcH and AcOH, when the catalyst is $Mg_3[RhCl_6]_2$ or $Mg[Rh(CO)_2Cl_2]_2$ on SiO_2 or $MgSiO_3$ [1531]. Various additives (CeO_2, CaO, Na_2O) affect the behavior of the catalyst formed from $Rh_6(CO)_{16}$ on ThO_2 [113]. With the undoped catalyst, the hydrogenation of CO at 250 °C gives MeOH (2100 ppm), EtOH (3500) and Me_2CHOH–PrOH (50); under the same conditions, a CeO_2-doped catalyst yields MeOH (4600), EtOH (4900) and Me_2CHOH–PrOH (470).

MECHANISM

The complexity of the catalyst system, the sensitivity to reaction conditions and the diversity of products obtained all combine to make it incredibly difficult to define a comprehensive mechanism for the F–T reaction. Several quite different pathways have been proposed [712, 1484]. In some, the starting point is the dissociative chemisorption of CO to give 'surface-C' species. Subsequent hydrogenation can give methane or 'methylene' intermediates which associate to form higher hydrocarbons. There is no clear perception of the steps involved in $C-C$ bond formation. The formation of CO_2 or H_2O accounts for the oxygen which is liberated by $C-O$ bond scission.

With homogeneously catalysed synthesis gas reactions, a carbonyl–rhodium anion is probably the active catalyst. There has been a high-pressure infrared investigation [1860] of the species formed when solutions of $Rh(CO)_2(acac)$ are treated at 500–1000 atm pressure with CO and H_2. At 50–200 °C, the anions $[Rh_5(CO)_{15}]^-$ and $[Rh(CO)_4]^-$ were the only species detected. Similar results were obtained when clusters such as $Rh_4(CO)_{12}$, $[Rh_{12}(CO)_{30}]^{2-}$ and $[H_3Rh_{13}(CO)_{24}]^{2-}$ were used in place of $Rh(CO)_2(acac)$. At higher temperatures (200–290 °C), which correspond to the optimum catalytic conditions, the observed spectral changes are more ambiguous. They are probably consistent with the formation of high nuclearity species such as $[H_xRh_{13}(CO)_{24}]^{n-}$, $[Rh_{14}(CO)_{25}]^{4-}$ and $[Rh_{15}(CO)_{27}]^{3-}$. The addition of N-methylmorpholine to solutions of $[H_3Rh_{13}(CO)_{24}]^{2-}$ under CO pressure results in the formation of $[H_2Rh_{13}(CO)_{24}]^{3-}$ and subsequent fragmentation into $[Rh_5(CO)_{15}']^-$ and $[Rh(CO)_4]^-$. Hydrogenation of the resulting solutions regenerates the original cluster.

It has been suggested that bound CO on the active catalyst reacts with hydrogen to give hydroxymethyl and related species. The hydroxymethylene growth reaction outlined in Scheme 4.2 shows one means of achieving chain growth to produce methanol and ethylene glycol [1431].

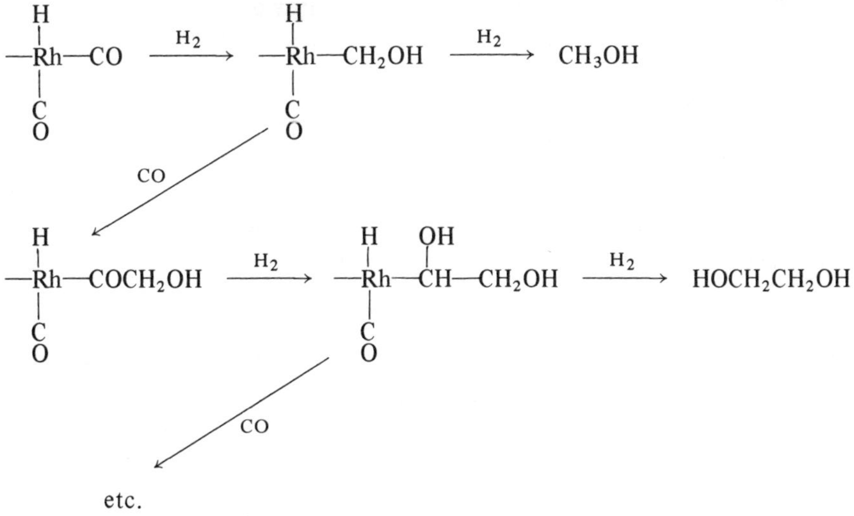

Scheme 4.2

[14]C-tracer studies have established that methanol and ethylene glycol are primary products of the reaction, and that they do not undergo secondary transformation under the reaction conditions [1365]. There is, however, a report [473] of a rhodium-catalysed (e.g. $RhCl_3 \cdot 3H_2O$) homologation of methanol:

$$MeOH + CO + H_2 \quad \rightarrow \quad EtOH$$

This reaction shows high selectivity when the H_2/CO ratio is 40.

Another proposed mechanism for these reactions has HCHO as the key intermediate [495]. The formation of HCHO from $\{CO + H_2\}$ is not thermodynamically favored, but sufficient would be formed for it to act as a transient intermediate. The complexes $Rh(CO)XL_2$ (e.g. $X = Cl$, $L = P(p\text{-}MeC_6H_4)_3$) have been used to hydroformylate HCHO to $HOCH_2CHO$. Moderate yields but high selectivity are achieved with N,N-disubstituted amides such as DMF at 110 °C and 1800 psi pressure [1647]. In other solvents, hydrogenation products predominate. Deuteration studies show that the mechanism of this reaction is analogous to that of alkene hydroformylation [1648].

There are many variations and modifications of these schemes, and it is not yet possible to state the definitive mechanism.

4.3. The Catalytic Reduction of NO by CO

In the control of car exhaust emissions, there is catalytic reduction of NO and oxidation of CO. Various platinum metals adsorbed on ceramic supports are used to achieve these conversions (Equations (4.1) and (4.2)).

$$2\,NO \quad \xrightleftharpoons{Rh} \quad N_2 + O_2 \tag{4.1}$$

$$CO + \tfrac{1}{2}\,O_2 \quad \xrightarrow{Pt/Pd} \quad CO_2 \tag{4.2}$$

There are homogeneous catalytic cycles in which nitric oxide and carbon monoxide are converted simultaneously to nitrous oxide and carbon dioxide:

$$2\,NO + CO \quad \longrightarrow \quad N_2O + CO_2$$

Although the reduction of NO by CO is strongly favored thermodynamically, the reaction does not proceed at appreciable rates in the absence of a catalyst. The dinitrosyl cations $[M(NO)_2(PPh_3)_2]^+$ (M = Co, Rh, Ir) [152, 686, 889] and the dicarbonyl anion $[Rh(CO)_2Cl_2]^-$ [705, 706, 1173] are effective catalysts. Rhodium ions in zeolite Y are also effective [804]. Clearly, the soluble catalysts can never be used as emission control catalysts. They may, however, add considerably to our understanding of the chemistry of these systems.

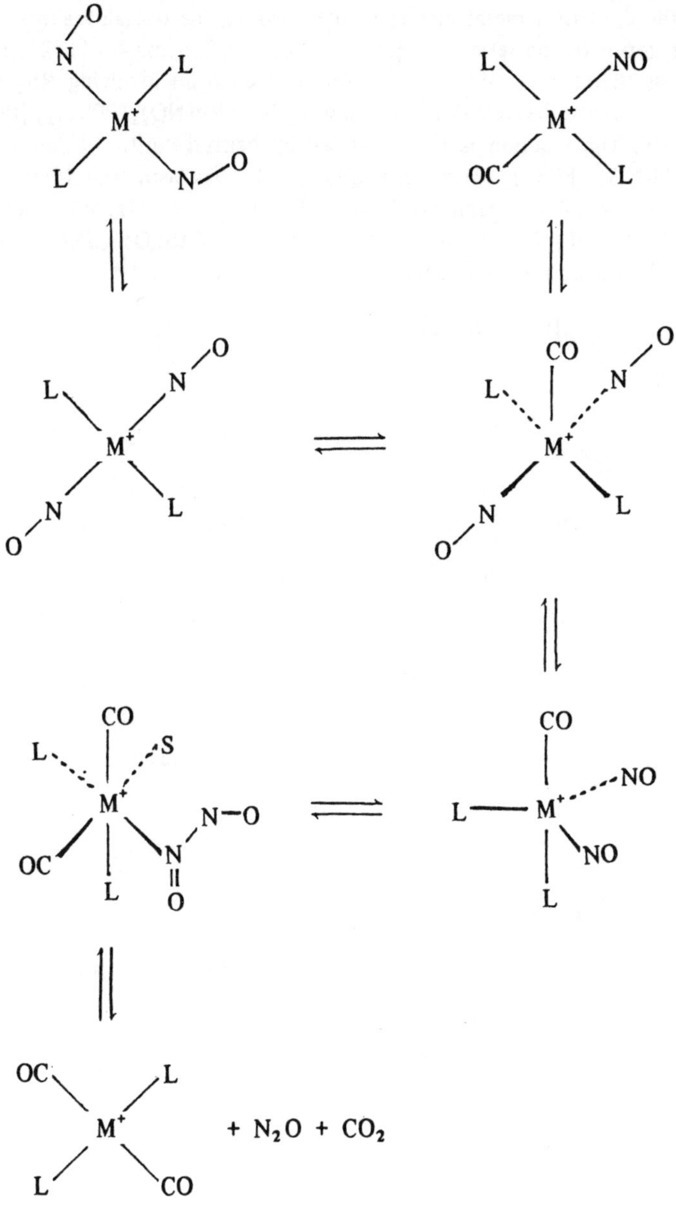

$+ N_2O + CO_2$

Scheme 4.3

With the dinitrosyl metal species as the catalyst, the overall reaction can be written in terms of the series of equilibria shown in Scheme 4.3 [152]. There is kinetic and thermodynamic evidence for a mechanism involving Rh_2 species and solvent when the reaction is catalysed by $[Rh(NO)_2(PPh_3)_2][PF_6]$ in DMF [890]. The reaction is also catalysed by neutral dinitrosyl complexes of the type $M(NO)_2(PPh_3)_2$ (M = Fe, Ru, Os), and it has been established that the relative activity of the catalysts is $Rh > Ir > Co > Ru > Os > Fe$. Since the trend in the M—N—O angle is $Rh(159°) < Ir < Co < Ru,Os < Fe(178°)$, there is apparently a structure–reactivity correlation [889].

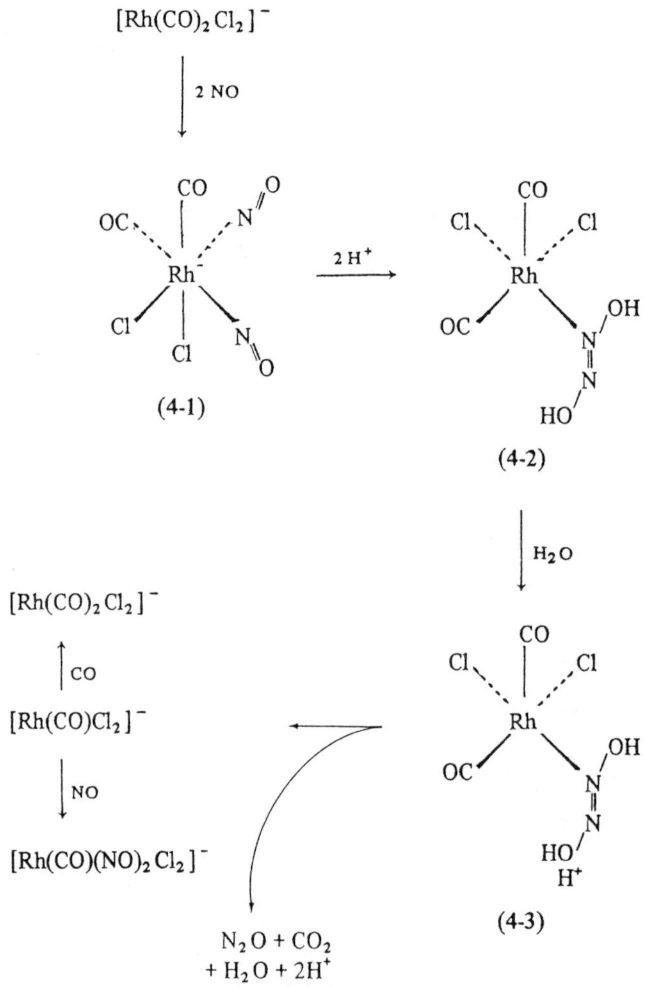

$[Rh(CO)_2Cl_2]^-$

Scheme 4.4

The most detailed investigations of the mechanism of the NO + CO reaction have used $[Rh(CO)_2Cl_2]^-$ as the catalyst precursor [805, 706, 1173], and the proposed reaction pathway is shown in Scheme 4.4. It is suggested that $[Rh(CO)_2Cl_2]^-$ reacts with NO to form a dinitrosyl-carbonyl intermediate which may have the composition $[Rh(CO)(NO)_2Cl_2]^-$. A square-pyramidal geometry would be predicted for this anion, with a bent nitrosyl ligand in an axial position and a linear NO ligand in an equatorial site. The linear $\rightleftarrows$ bent transition for coordinated nitrosyl probably requires relatively little energy, and this complex can become coordinatively unsaturated by conversion of the linear nitrosyl to a bent nitrosyl. It can then coordinatively add CO and this is regarded as the rate-determining step. A rapid H^+-catalysed ligand migration reaction involving the cis-nitrosyls in 4-1 gives a hyponitrite complex (4-2). Water can then hydrogen-bond to the hyponitrite oxygen prior to nucleophilic attack on the coordinated CO. The reductive decarboxylation of 4-3 presumably occurs according to Equation (4.3),

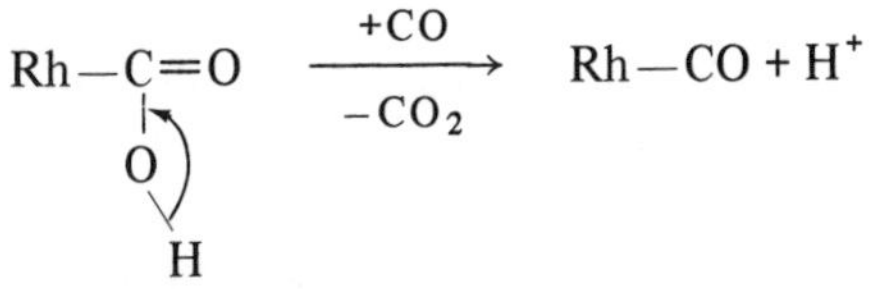

$$Rh-\underset{\underset{H}{\overset{|}{\underset{O}{\overset{\nwarrow}{}}}}}{C}{=}O \quad \xrightarrow[-CO_2]{+CO} \quad Rh-CO + H^+ \tag{4.3}$$

and the generation of N_2O and H_2O from 4-3 is analogous to the $RhCl_3$-catalysed decomposition of hyponitrous acid (see Equation (4.4)).

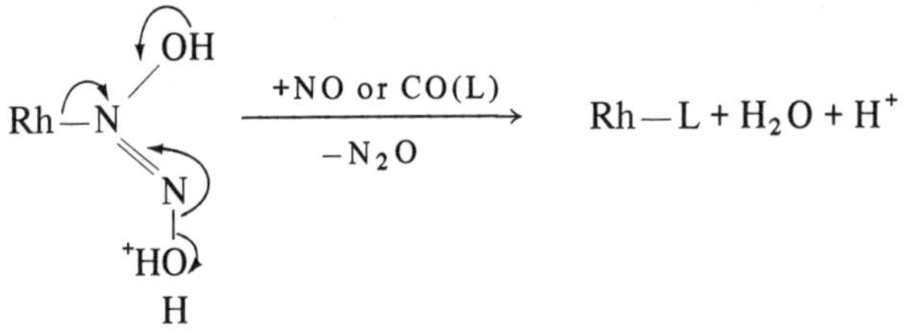

$$Rh{-}\underset{\underset{\underset{H}{\overset{|}{\overset{+}{HO}}}}{\overset{\nwarrow}{N}}}{\overset{\overset{\displaystyle OH}{\diagup}}{N}} \quad \xrightarrow[-N_2O]{+NO\ or\ CO(L)} \quad Rh-L + H_2O + H^+ \tag{4.4}$$

4.4. Carbonylation Reactions

CONVERSION OF ALCOHOLS TO CARBOXYLIC ACIDS AND ESTERS

Several approaches to the synthesis of acetic acid and its derivatives are used commercially. The oxidation of ethylene via acetaldehyde is assisted by Pd catalysts, and Co^{II} compounds catalyse the oxidation of butane or naphtha. The carbonylation of methanol is one of the newer methods. The original process was developed by BASF in the 1960s and used a catalyst system consisting of $Co_2(CO)_8$ plus an iodo compound; severe conditions ($> 200\ °C$, 700 atm pressure) were required, but the selectivity exceeded 90%. A newer process, which was put on-stream by Monsanto in the 1970s, operates under mild conditions (180 °C, 30–40 atm pressure) with a rhodium compound as catalyst and an iodide promoter [486, 1209]. To reduce corrosion, equipment constructed from Mo–Ti alloy has been recommended [1138].

The details of this system have been described in several reviews [1490, 1491, 1555, 1556]. Almost any rhodium compound is a suitable catalyst source, and the compounds used range from rhodium(III) species such as $RhCl_3 \cdot 3H_2O$ to rhodium(I) complexes including $[Rh(CO)_2Cl]_2$, $[Rh(CO)_2X_2]^-$ and $Rh(CO)ClL_2$. The promoter component can be aqueous HI, MeI, I_2 or $CaI_2 \cdot 3H_2O$; the role of these compounds is to convert MeOH to the more electrophilic MeI. Both the catalyst and the promoter are sufficiently expensive to require recycling. The solvents include water, acetic acid, methanol and benzene. The formation of acetic acid is favored in aqueous systems, but esters are obtained in alcoholic solvents:

$$MeOH + CO \xrightarrow{\ ROH\ } MeCO_2R$$

A recent study of solvent effects has shown that ketonic solvents such as acetophenone and benzophenone keep the activity and selectivity of some catalysts high, even at elevated temperatures [1148]. Greater than 99% selectivity in the formation of acetic acid or its methyl ester is achieved. The yields are reduced, however, if hydrogen is present due to hydrogenolysis of some methanol to form methane [468]:

$$ROH + H_2 \longrightarrow RH + H_2O$$

Catalyst systems that have been used for the conversion of methanol to acetic acid are given below:

$RhCl_3 \cdot 3H_2O$, $Rh_2O_3 \cdot 5H_2O$ [1375];

$RhI_3 + MeI$ [1273];

$IrCl_3 \cdot xH_2O + MeI$ [529, 1373, 1375];

Rh [1530, 1827] or Ir [1827] phthalocyanin complexes + a sulfonate or sulfonic acid + an alkyl halide;

$RhCl_3$ and C_6Cl_5SH; this catalyst gives MeOAc and some Me_2O at 175 °C and 50 atm pressure [465].

$[Rh(CO)_2Cl]_2$ or $Rh(CO)Cl(PPh_3)_2$ [230, 491, 1375];

$RhCl(PPh_3)_3$ [230];

$RhCl(Ph_2PCH_2CH_2PPh_2)_2$ and related complexes [119, 120, 121];

$[Rh\{P(OPh)_3\}_4]^+ + Na_3PO_4/RCO_2H + MeI$ [1907].

$[Rh(COD)X]_2$ with X = Cl, I, OMe, OAc, OPh [230];

$[Rh(COD)I]_2 + Na_3PO_4/RCO_2H + MeI$ [1907];

Rh(COD) Xpy [230, 343].

$[Ir(CO)_3Cl]_x + MeI$ [1373];

Ir(CO)Cl(PPh$_3$)$_2$ [1375] or Ir(CO)Cl(Ph$_2$PCH$_2$CH$_2$PPh$_2$) [122];
[Ir(COD)(N—N)](PF$_6$), N—N = α,α'-bipy or o-phen [229];
Ir$_4$(CO)$_{12}$ [1375].

A comparison of reaction rates and product distributions indicates that the various sources of rhodium and iodide may yield the same catalytically active species [1491]. The probable catalyst precursor is the anionic species [Rh(CO)$_2$X$_2$]$^-$, and this can be formed from a rhodium(III) halide in hydroxylic media by reaction with CO [525]. A detailed mechanism for the carbonylation of methanol which uses this complex as the starting point is presented in Scheme 4.5 [526, 527, 528]. The pathway shown is based on spectroscopic studies of the sequential steps and isolation of the key intermediate **4-4**. A kinetic study of the system is consistent with the proposed mechanism, and it shows that the rate is zero order in reactants and first order in catalyst and promoter [729, 730]. The oxidative-addition of MeI to the rhodium(I) complex is the rate-determining step.

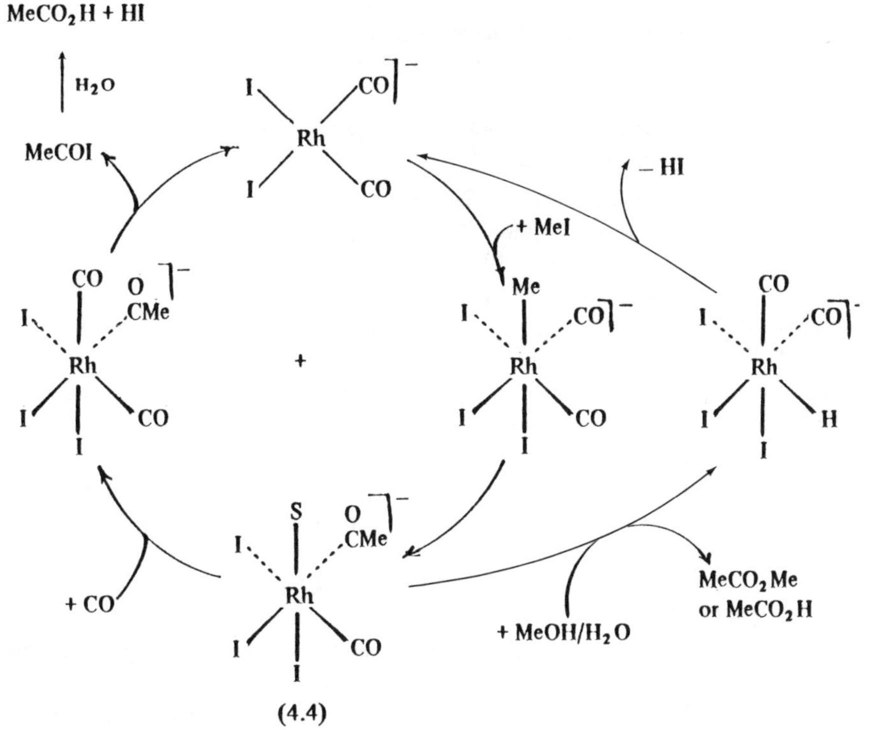

MeCO$_2$H + HI

(4.4)

Scheme 4.5

A comparative study of rhodium and iridium catalysts for the carbonylation of methanol has not revealed any major differences in gross behavior but the iridium compounds are more sensitive to the actual ligands attached [229]. The kinetics of methanol carbonylation have been studied with $IrCl_3 \cdot xH_2O$ as the catalyst precursor and MeI as the promoter [529, 1149]. The observed kinetics indicate that the mechanism is more complex than with the Rh catalysts. Two principal catalytic cycles predominate, and under many conditions a competitive water–gas shift reaction is observed.

A number of supported catalysts have been used for the carbonylation of methanol. They include:

$RhCl_3$ or Rh_2O_3 supported on carbon + MeI or I_2 [555, 802, 1481, 1546];
$RhCl_3$ anchored to $P-C_6Cl_5SH$; P = lightly crosslinked polystyrene [1904];
$RhCl_3$ on oxide supports (e.g. Al_2O_3, SiO_2, TiO_2 [1010] but not MgO [61]);
$RhCl_3$ [1057] or $[Rh(NH_3)_5Cl]Cl_2$ [61, 1559] on Linde 13X zeolite;
$RhCl(PPh_3)_3$ anchored to $P-C_6Cl_5SH$ (P = styrene–divinylbenzene co-polymer – attached C_6Cl_5SH acts as promoter) [1904];
$RhCl(PPh_3)_3$ or $Rh(CO)Cl(PPh_3)_2$ adsorbed on alumina [1011];
$Rh(CO)Cl(PPh_3)_2$ supported on $P-CH_2PPh_2$ (P = styrene–divinylbenzene copolymer) [871];
$[Rh(CO)_2I_2]^-$ supported on anion-exchange resin [464];
Rh and Ir carbonyls on zeolites [585].

ESCA results have established that the zeolite-supported catalysts contain rhodium predominantly in the +3 oxidation state. The zeolite framework facilitates the removal of ligands from the coordination sphere of the rhodium, and it also assists the dissociative adsorption of MeI [61]. There have been kinetic studies of the reactions with $RhCl_3/C$ [1481], $Rh(CO)Cl(PPh_3)_2/$ $P-CH_2PPh_2$ [871], and Rh and Ir carbonyls/zeolites [586]. With the rhodium carbonyl compounds, oxidative-addition of MeI is the rate-determining step. With the iridium catalyst, however, the rate-determining step is oxidative-addition of MeOH.

Alcohols other than methanol can be carbonylated in similar manner. The formation of phenylacetic acid by carbonylation of benzyl alcohol has been studied in detail, and the reaction is found to be sensitive to the reaction conditions [1136]. With a catalyst formed from $RhCl_3 \cdot 3H_2O$ in the presence of benzyl iodide, the optimum conditions are ca. 30 kg cm^{-2} pressure and 140 °C. After 90 min, 100% conversion is achieved, and the selectivity to phenylacetic acid is 94%. Kinetic studies of the reactions with 2-propanol [731] and benzyl alcohol [1136] are consistent with the mechanism depicted for methanol to acetic acid conversion in Scheme 4.5.

The conversion of other higher alcohols and diols to carboxylic acids and esters has been achieved as follows:

ROH, C_2-C_{20} → carboxylic acids and esters; catalyst = $[Rh(CO)_2 X]_2$ with X = Cl or Br, $Rh(CO)Cl(PPh_3)_2$, $RhCl(PPh_3)_3$ [1372], $RhCl(Ph_2PCH_2$-$CH_2PPh_2)_2$ and related complexes [120, 121], $IrCl_3$, $[Ir(CO)_3Cl]_x$ and other Ir compounds + MeI [1371, 1372, 1373].

EtOH + ÈtI → ethyl propionate; catalysts include $RhCl_3$, $[Rh(NH_3)_5 Cl]Cl_2$ and $HRh(CO)(PPh_3)_3$ on zeolites [1558].

PhOH → PhOAc in AcOH; catalyst from $RhCl_3$ + MeI, reaction at 180 °C under CO pressure (30 atm) [836].

Allyl alcohols, C_1-C_{18} → allyl esters; catalyst = $RhCl_3$ or $RhBr_3$ [222].

Glycols, C_3-C_{20} → dicarboxylic acids (e.g. 1,4-butanediol → adipic acid); catalyst = $Rh(CO)Cl(PPh_3)_2$, $RhCl(PPh_3)_3$, $Ir(CO)Cl(PPh_3)_2$ + MeI or HI [1211, 1371].

Glycerol and other alkane polyols → alkanoic acids; catalyst = $RhCl_3 \cdot 3H_2O$ or $[Rh(CO)_2 Cl]_2$ + HI, MeI or I_2 [1252, 1254].

4-Alkoxy-2-butanone, $MeC(O)CH_2 CH_2OR$ → levulinic ester, $MeC(O)CH_2$-CH_2COOR; catalyst = $[Rh(CO)_2 Cl]_2$, $RhCl(PPh_3)_3$ + MeI [1140].

REACTIONS OF ETHERS, ESTERS AND RELATED COMPOUNDS

Esters can be formed by the carbonylation of ethers. For example, ethyl-propionate is obtained when diethyl ether and ethanol are treated with CO in a solvent such as tolylpropionate [837, 838]. The conversion is about 90% at 190 °C and 41 atm pressure with a catalyst derived from $RhCl_3 \cdot 3H_2O$, EtI and Bu_3N. The carbonylation of cyclic oxiranes such as ethylene oxide gives β-lactones; $Rh(CO)Cl(PPh_3)_2$ has been used as the catalyst for this conversion [908]. Further carbonylation of lactones in saturated aliphatic carboxylic acids as the solvent results in ring-opening to produce dicarboxylic acids. For example, γ-valerolactone (4-5) is converted to 2-methylglutaric acid (65%) and adipic acid (20%) by carbonylation in aqueous $Me(CH_2)_4CO_2H$ with $RhCl_3 \cdot 3H_2O$ + MeI as the catalyst source [1504, 1505].

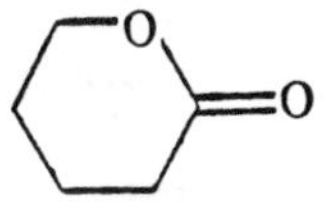

(4-5)

Acetic anhydride is formed by carbonylation of dimethyl ether or methyl acetate. Catalysts for these reactions have been formed from $RhCl_3$ + MeI with the addition of species such as 1-methyl-3-picolinium iodide [489], $\{Cr(CO)_6 + Bu_3P$ or $Ph_3As\}$ [1412, 1423, 1443] or $\{Zr$ or $Hf + Ph_3As\}$ [1444, 1445, 1446].

Treatment of methyl acetate with CO and H_2 in the presence of rhodium catalysts gives ethylidene diacetate as the major product [466, 1014, 1198, 1480]. Typical conditions are 1:2 CO/H_2 at 140 atm pressure and 150 °C with $RhCl_3 \cdot 3H_2O$, 3-picoline and MeI as the catalyst source [1480]. The early stages of this reaction probably proceed in a similar manner to methanol carbonylation to produce an acyl–rhodium intermediate. Cleavage with AcOH and H_2 produces $(MeCO)_2O$ and MeCHO respectively, and the diacetate is formed from subsequent reaction between these products.

In some related reactions with esters, there is co-production of carboxylic acids and esters [1585], for example

$$MeOAc + CO + H_2 \longrightarrow EtOAc + AcOH$$

Catalysts for this process are obtained from $RhCl_3$, $RuCl_3$ and ZnI_2 plus a crown ether (e.g., 15-crown-5).

The conversion of methyl formate to acetic acid is achieved in 99% yield by treatment with CO at 500 psi and 170 °C in the presence of $RhCl(PPh_3)_3$ and MeI [70].

Homologation of acetic acid to propionic acid has been achieved with CO and H_2. Soluble rhodium compounds plus MeI provide suitable catalysts for the reaction [964].

CARBONYLATION OF ORGANIC HALIDES

A discussion of the reaction between CO and alkyl or aryl halides is included in a general review on the transition metal catalysed reactions of organic halides [689]. There are relatively few examples of such reactions catalysed by rhodium or iridium compounds. The conversion of methyl chloride to acetyl chloride has been achieved at 75 atm pressure and 180 °C with a catalyst formed from $RhCl_3 \cdot 3H_2O$, $CrCl_3 \cdot 6H_2O$, Ph_3PO, $MePh_3PI$, MeI and MeCl in heptane [490]. The related formation of $PhCH_2COCl$ from $PhCH_2Cl$ and CO is catalysed by $Rh(CO)Cl(PPh_3)_2$ [1802], and analogous reactions occur with other $ArCH_2Cl$ compounds. Rhodium(III) salts will catalyse the carbonylation of some unsaturated hydrocarbons in the presence of HCl. Thus $H_2C = CHCOCl$ is formed by the carbonylation of $\{HC \equiv CH + HCl\}$ [1527], and RCH_2CH_2COX and $RC(COX)HCH_3$ compounds are formed in a similar manner from $\{RH_2C = CH_2 + HX\}$ [1796].

CARBONYLATION OF NITROSO, NITRO AND AZIDO COMPOUNDS;

THE FORMATION OF ISOCYANATES

Supported rhodium metal is a suitable catalyst for the carbonylation of nitro-benzene, and mechanistic aspects of this heterogeneous reaction have been considered [485]. It is suggested that heterolytic chemisorption occurs to give $PhN-(Rh_3)$ species which are subsequently carbonylated to form phenyl-isocyanate. Soluble complexes of rhodium and iridium are also useful catalysts for the carbonylation of aromatic nitroso compounds to give arylisocyanates [1826]. Detailed theoretical consideration has been given to the suitability of compounds of the type $M(CO)XL_2$ (M = Rh, Ir; X = Cl, CN, OH; L = PH_3, NH_3, CO, et.) [1500].

Phenylisocyanate has been obtained from nitrobenzene and CO with $RhCl(PPh_3)_3$ as the catalyst [1355]. Similar conversions are achieved with $[Rh(CO)_2Cl]_2$ and co-catalysts such as $MoCl_5$, YCl_4 [1115, 1551, 1823, 1824] or pyridine hydrochloride [1116, 1117, 1270]. In some instances, the $[Rh(CO)_2Cl]_2$ is produced in situ from $RhCl_3 \cdot 3H_2O$ and CO. In the presence of $Rh(CNPh)_4Cl_3$ as a catalyst, 2,4-dinitrotoluene can be carbonylated to yield 2,4-toluene diisocyanate [540].

The conversion of $2,4,6-R_3C_6H_2NO$ to $2,4,6-R_3C_6H_2NCO$ occurs readily at 90 °C and 1 atm CO pressure with $M(CO)Cl(PR_3)_2$ compounds as catalyst [1825]. trans-$Ir(CO)Cl(PPh_3)_2$ is the most active and selective catalyst, and the reaction rate decreases in the order $2,6-Me_2C_6H_3NO > 2,4,6-Me_3C_6H_2NO \gg$ $4,2,6-R'Me_2C_6H_2NO$ (R' = MeO or NO_2). These reactions follow first-order kinetics with respect to the nitroso compound.

There is participation of the solvent in some reactions. In benzene with $Rh_4(CO)_{12}$ or $Rh_6(CO)_{16}$ as the catalyst, the formation of arylisocyanates from nitroarenes and CO is followed by reaction with C_6H_6 to give N-aryl-benzamides, RC_6H_4NHBz (R = H, p-MeO, m-Me) [1189].

The carbonylation of nitrobenzene in methanol gives methyl-N-phenyl-carbamate. Various triphenylphosphine complexes of rhodium and iridium have been used as catalysts, and the best yields are obtained with $HRh(CO)$-$(PPh_3)_3$ and $Rh(CO)Cl(PPh_3)_2$ [610]. In other reactions, aromatic nitro compounds are treated with CO (10–500 kg cm^{-2} pressure) and ammonia in alcoholic solvents at 80–260 °C; aromatic urethanes are formed. Various metal compounds, including Rh halides and carbonyls, are used as catalysts in conjunction with a Lewis acid such as $FeCl_2$ [1205].

The carbonylation of azides provides another route to isocyanates. Thus, PhNCO is formed at atmospheric pressure from PhN_3 with $[Rh(Ph_2PCH_2-CH_2PPh_2)_2]Cl$ or $Rh(CO)Cl(PPh_3)_2$ as the catalyst [1040]. In the presence of EtOH and $PhNH_2$, the products of this reaction are $PhNHCO_2Et$ and

PhNHCONHPh, respectively. 1,3-Diarylureas are also obtained by the reductive carbonylation of arylazides with CO and H_2 in the presence of $Rh_6(CO)_{16}$ as catalyst [831]. Complexes of the type Rh(CO)ClL$_2$ have also been used to catalyse the conversion of p-ethoxycarbonylphenyl azide to the corresponding isocyanate. Much of the activity of these catalysts ($L = Ph_2PCH_2CH_2Si(OEt)_3$) is lost when they are fixed on porous silica glass [1552]. 2-Arylazirines (4-6) can also be carbonylated under mild conditions (0–5 °C, 1 atm pressure CO) with [Rh(CO)$_2$Cl]$_2$ as the catalyst. The products can be isolated as the isocyanates, 4-RC$_6$H$_4$C(NCO)=CR$'_2$ (e.g. R = H and R$'$ = H or Me; R = Me, OMe or Br and R$'$ = H), or as the carbamates or ureas [1513].

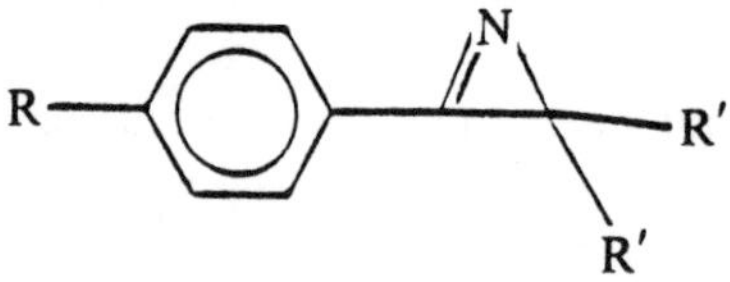

(4-6)

CARBONYLATION OF ORGANIC AMINES

Amides and ureas are formed when primary aliphatic amines are carbonylated in the presence of rhodium and iridium catalysts. For example, the formation of BuNHCHO and BuNHCONHBu from BuNH$_2$ and CO is catalysed by [Rh(CO)$_2$Cl]$_2$ [476] and HRh(CO)(PPh$_3$)$_3$ [1058]. The selectivity of the [Rh(CO)$_2$Cl]$_2$-catalysed reaction is improved by the addition of phosphines. Thus, BuNHCHO is the only product when PMe$_3$ is added to achieve an Rh/P ratio of 1:6 [476]. The carbonylation of secondary amines with RhCl(PPh$_3$)$_3$ as the catalyst also gives amides, R$_2$NCHO [687]. In the presence of hydroxy compounds such as alcohols, the carbonylation of organic amines gives carbamate esters. For examples, PhNHCO$_2$Me is obtained from PhNH$_2$ and CO in methanol in the presence of rhodium carbonyl compounds as catalysts [1231].

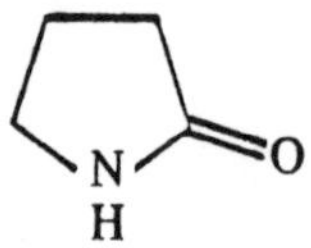

(4-7)

Carbonylation of the unsaturated, primary amine H_2C=CHCH$_2$NH$_2$ gives the γ-butyrolactam 4-7 with Rh(CO)Cl(PPh$_3$)$_2$ as the catalyst [963]; the reaction is achieved at 150 °C and 136 atm pressure. Related γ-lactams are formed when cyclopropylamines are carbonylated in the presence of catalytic amounts of Rh$_6$(CO)$_{16}$ or RhCl(PPh$_3$)$_3$ [828]. With RhCl(PPh$_3$)$_3$ supported on silica—alumina as the catalyst, the carbonylation of allyl amine gives 2-pyrrolidinone [1153].

CARBONYLATION OF CYCLIC, POLYCYCLIC AND UNSATURATED HYDROCARBONS

With $[Rh(CO)_2Cl]_2$ as the catalyst, the carbonylation of cyclopropane yields cyclobutanone and dipropylketone [715]. A polyketone is obtained in the corresponding reaction with methylene—cyclopropane. There is no reaction with phenylcyclopropane.

Several polycyclic dienes have been carbonylated with $[Rh(CO)_2Cl]_2$ as the catalyst. Thus, dicyclopentadiene (4-8) gives the trishomocubanone 4-9 [196], and the tricycloheptane systems 4-10 and 4-12 yield the bicyclic ketones 4-11 and 4-13 [699]. Carbonylation of 4-14 could not be achieved under the conditions used. With $Rh(C_2H_4)_2(acac)$ as the catalyst, the carbonylation of 2,3-homotropilidenes gives bicyclo[3.3.1]-, [3.3.2]- and [4.2.1]-nonadienones [94].

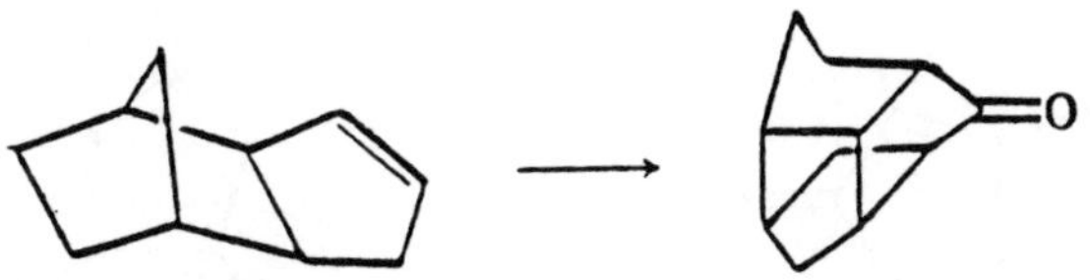

(4-8)   (4-9)

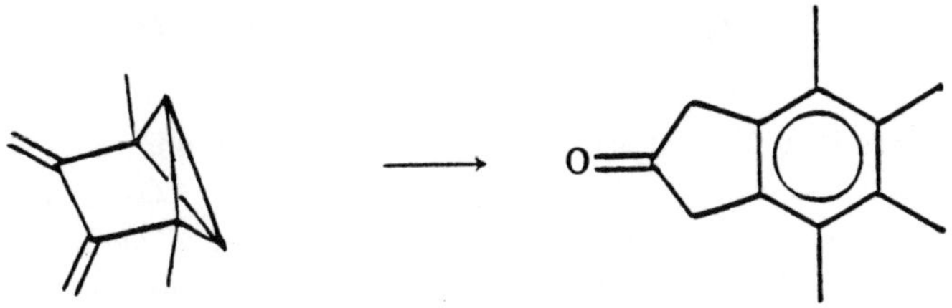

(4-10)   (4-11)

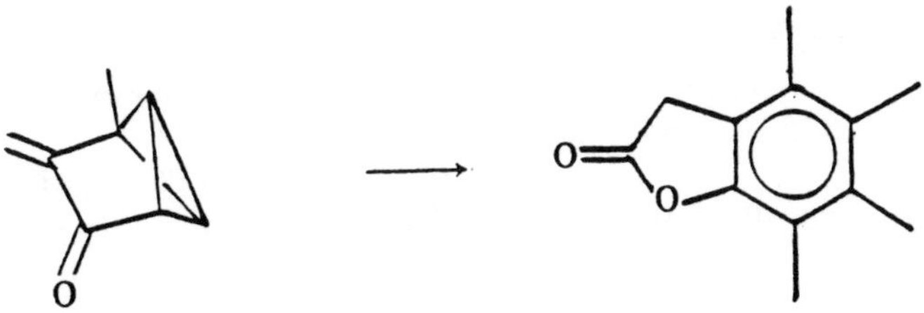

(4-12)   (4-13)

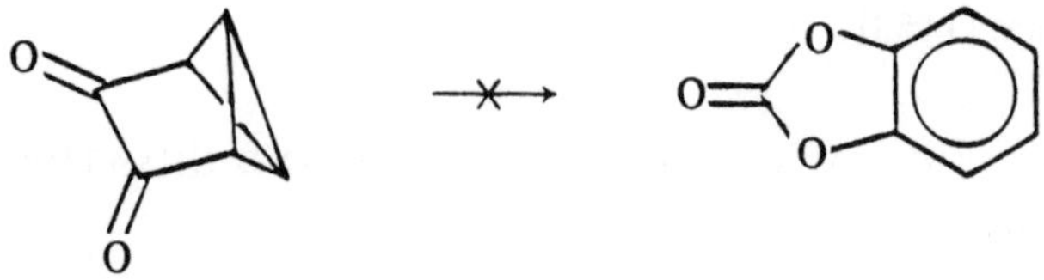

(4-14)

The catalytic carbonylation of some alkynes has also been achieved. Thus, hydroquinone and quinhydrone are obtained from HC≡CH and CO at 850–900 atm pressure and 100–150 °C with $RhCl_3$ or $[Rh(CO)_2Cl]_2$ as catalysts [770]. The formation of 3,6-diphenyl-1-oxa-[3.3.0]bicycloocta-3,6-diene-2,4-dione **(4-15)** from PhC≡CH is achieved under mild conditions in the presence of $[Rh(CO)_2Cl]_2$ [949].

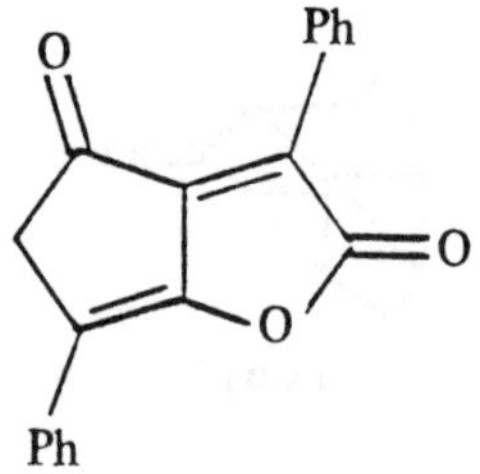

(4-15) 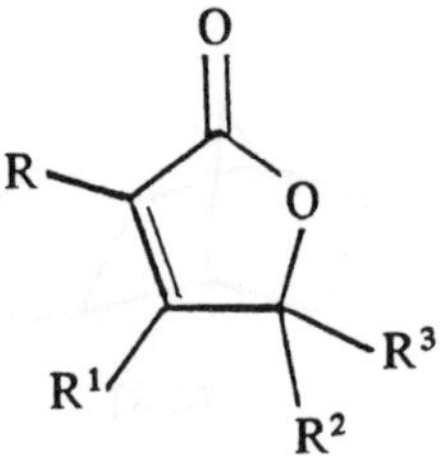   **(4-16)**

Treatment of $Rh(CO)Cl(PPh_3)_2$ and related compounds with Na(Hg) under CO, and subsequently with RHgX, gives $RHg \cdot Rh(CO)_{4-n}L_n$ complexes. These are useful catalysts for the carbonylation of alkenes [826]. Various iridium compounds (e.g. $[HIr(CO)_3I_2]$) have also been used as catalysts in the carbonylation of alkenes [1213].

Furanones are obtained when alkynes are carbonylated in the presence of alkenes and protic solvents with rhodium carbonyl catalysts [753, 1190]. For example, **4-16** (R = R^1 = Ph, R^2 = H, R^3 = Et) is obtained from PhC≡CPh and $H_2C=CH_2$ with $Rh_4(CO)_{12}$ or $Rh_6(CO)_{16}$ as catalysts. In reactions with unsymmetrical alkynes, some regioselectivity is observed. Thus, the corresponding reaction with PhC≡CMe gives **4-16** (R = Me, R^1 = Ph, R^2 = Et, R^3 = H) and **4-16** (R = Ph, R^1 = Me, R^2 = Et, R^3 = H) in 48% and 4% yields, respectively [753].

CARBONYLATION OF ORGANOMERCURY COMPOUNDS;

THE FORMATION OF KETONES

The carbonylation of organomercury compounds at room temperature and 1 atm pressure provides a convenient and high-yield synthesis of ketones. Yields as high as 99% have been obtained from RHgCl or R_2Hg (e.g. R = Ph, p-MeC_6H_4) in $(Me_2N)_3PO$ containing $Bu_4N^+I^-$ and $[Rh(CO)_2Cl]_2$ [261]. In a similar manner, divinylketones are obtained from vinylmercury chloride with $\{ [Rh(CO)_2Cl]_2 + 2 LiCl\}$ as the catalyst [1054].

4.5. Hydroformylation; the Formation of Aldehydes from Alkenes

Aldehydes are large-volume organic chemicals. As shown in Scheme 4.6, they are used to make monohydric alcohols.

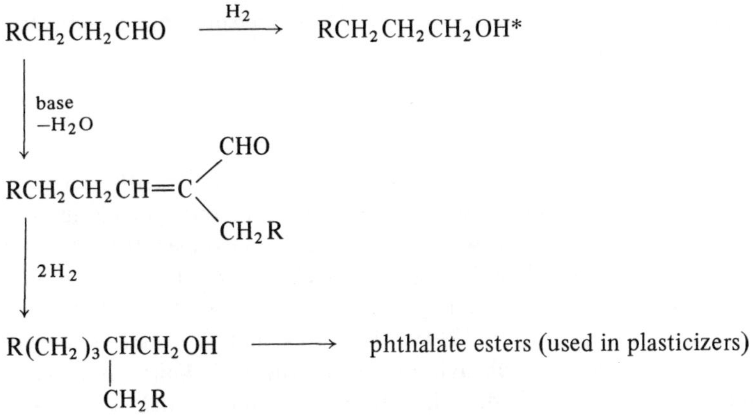

$$RCH_2CH_2CHO \xrightarrow{\quad H_2 \quad} RCH_2CH_2CH_2OH^*$$

$\downarrow$ base $-H_2O$

$$RCH_2CH_2CH{=}C\Big\langle {}^{CHO}_{CH_2R}$$

$\downarrow$ $2H_2$

$$R(CH_2)_3\underset{\underset{CH_2R}{|}}{C}HCH_2OH \longrightarrow$$ phthalate esters (used in plasticizers)

(* R = Me, industrial solvent; R = long-chain alkyl, used in the synthesis of biodegradable detergents)

Scheme 4.6

The 'Oxo' reaction, which uses an alkene and synthesis gas ($CO + H_2$) as raw materials, is the usual process for the synthesis of aldehydes on an industrial scale:

$$RCH{=}CH_2 + CO + H_2 \longrightarrow RCH_2CH_2CHO + R\underset{\underset{CH_3}{|}}{C}HCHO$$

Until recently, all Oxo-process plants used $Co_2(CO)_8$ as the catalyst precursor, with $HCo(CO)_4$ becoming the active catalyst under the reaction conditions (e.g., 120–145 °C, 200–300 atm pressure). A mechanism for this reaction has been developed, principally by Heck and Breslow [1784]. Several disadvantages associated with the cobalt-catalysed reaction (in particular, the formation of significant amounts of the branched aldehyde and of byproducts such as alcohols and aldol condensation products) are overcome when certain rhodium compounds (e.g. $Rh_4(CO)_{12}$) are used as the catalyst precursor. The rhodium catalysts are most effective when used with excess tertiary phosphine, and this system is now used by Union Carbide [537, 1431] in the commercial production of butyraldehyde. Several recent reviews [352, 1027, 1123, 1396, 1432, 1597] discuss the essential chemistry of the rhodium-catalysed hydroformylation reaction.

HYDROFORMYLATION CATALYSTS

Catalysts for hydroformylation reactions have been prepared from a wide variety of rhodium compounds.

$Rh_4(CO)_{12}$ and $Rh_6(CO)_{16}$

With $Rh_4(CO)_{12}$ as the catalyst precursor, reaction is conveniently achieved at ca. 75 °C and ca. 100 atm pressure H_2. The more thermodynamically stable $Rh_6(CO)_{16}$ is a better catalyst source for reactions performed at temperatures above ca. 110 °C. Both carbonyl compounds yield a catalyst which is effective for the hydroformylation of a variety of alkenes [27, 29, 381, 564, 565, 693, 694, 917, 1034, 1526, 1752, 1933, 1934]. For example, 1-hexene is converted to n-heptanal, styrene to o-phenylpropionaldehyde and 3-nitroprop-1-ene to 3-methyl-3-nitropropionaldehyde. The carbonyl–rhodium catalyst system has also been applied to reactions with ethyl acrylate and related vinyl esters [28, 30, 1751], cyclic alkenes such as cyclohexene [564, 823], acyclic dienes [63] and diketenes [1397]. The effects of temperature, pressure, CO/H_2 ratio, catalyst concentration and solvent on the behavior of this catalyst system have been determined [338, 565].

A comparative study of the use of metal carbonyls for the hydroformylation of octene, butadiene, α-methylstyrene and pinacol has established the following order of activity [820, 821]: $Ir < Co < Rh$. Another study indicates that $Rh_4(CO)_{12}$ is more effective than $Co_2(CO)_8$ by a factor of ca. 12 000 [693], whereas an independent investigation [1934] provides relative rates of 1:300–500:50–100 for Co/Rh/Ir in the metal carbonyl catalysed hydroformylation reactions. Similarities and differences in the behavior of the cobalt and rhodium catalysts are highlighted in a review by Pino [1394].

The structure of the alkene affects the rate of the hydroformylation reactions. With $Rh_4(CO)_{12}$ as the catalyst precursor, the order of relative rates is styrene $\gg$ linear terminal alkenes $>$ linear internal alkenes $>$ branched alkenes $>$ cyclic alkenes [694]. Some substituent effects have been observed with ring-substituted styrenes [1033].

In the hydroformylation of 1,1,1-d_3-butene with $Rh_4(CO)_{12}$ as the catalyst, an isotope effect on the product composition has been observed [346] (see Equation (4.5)).

$$CD_3CH{=}CHCH_3 \longrightarrow \begin{array}{l} CD_3CH_2CH_2CH_2CHO\ (75\%) \\ + CH_3CH_2CH_2CHDCHO\ (2\%) \\ + CH_3CH_2CH_2CD_2CHO\ (23\%) \quad (4.5) \end{array}$$

Investigation of the related hydroformylation of 5,5,5-d_3-1- and 5,5,5-d_3-2-pentene has established that intramolecular H-shifts decrease with an increase in CO pressure [1875].

Studies of the kinetics of the hydroformylation of cyclohexene with $Rh_6(CO)_{16}$ as the catalyst precursor have shown that the reaction is first order in alkene, but one-sixth order in [Rh] [380, 1786]. Similarly, with $Rh_4(CO)_{12}$, the rate equation is first order in cyclohexene and one-fourth order in [Rh] [380]. These results are consistent with the active catalyst being a compound of the type $HRh(CO)_a$ (see Scheme 4.7). $HRh(CO)_4$ has been positively identified by Fourier subtraction of the infrared spectra of $Rh_4(CO)_{12}$ and $Rh_2(CO)_8$ from that of a solution of $Rh_4(CO)_{12}$ in dodecane under 1542 atm pressure of CO/H_2 (4.5:1) [1861].

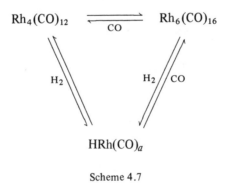

Scheme 4.7

All the kinetic results are consistent with the reactions depicted in Scheme 4.8. With terminal alkenes such as 1-heptene, most of the rhodium is transfomed to the relatively stable acyl complex $Rh(COR)(CO)_4$, and the rate-determining step in the hydroformylation reaction is the oxidative-addition of H_2 to $Rh(COR)(CO)_3$. With cyclohexene, however, the coordinative-addition of alkene to $HRh(CO)_3$ is the rate-determining step.

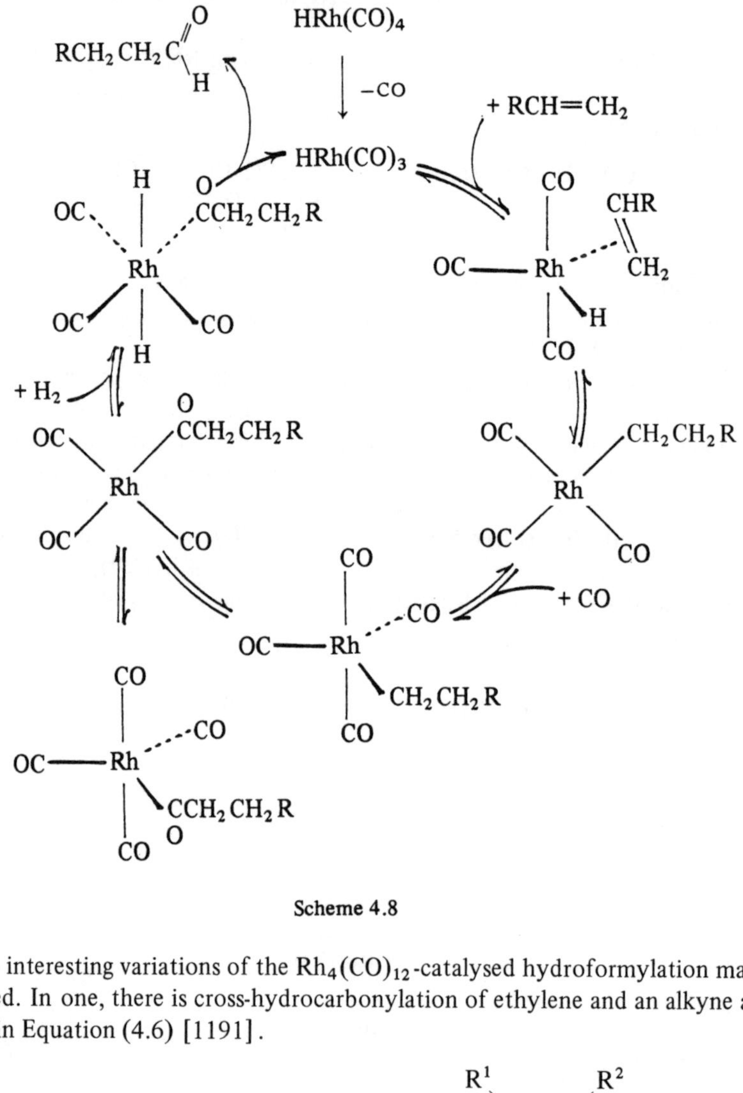

Scheme 4.8

Two interesting variations of the $Rh_4(CO)_{12}$-catalysed hydroformylation may be noted. In one, there is cross-hydrocarbonylation of ethylene and an alkyne as shown in Equation (4.6) [1191].

$$H_2C{=}CH_2 + R^1C{\equiv}CR^2 \xrightarrow{CO/H_2} \begin{array}{c} R^1 \qquad R^2 \\ \diagdown \diagup \\ C{=}C \\ \diagup \diagdown \\ H \qquad COEt \end{array} \qquad (4.6)$$

A variety of α,β-unsaturated ethylketones (e.g. $R^1 = R^2 = Ph$ or Me, $R^1 = H$ and $R^2 = Bu$) have been formed in this way. In the other, 2-propanol is the source of [H]. Thus, γ-ketopimelic acid esters and amides are obtained from acrylic acid esters and amides according to:

$$2 H_2C=CHCOY + CO + Me_2CHOH \longrightarrow OC(CH_2CH_2COY)_2 + Me_2CO$$

The reaction is achieved at 180 °C and 75 kg cm^{-2} pressure CO [754].

$HRh(CO)(PPh_3)_3$

With $Rh_4(CO)_{12}$ or $Rh_6(CO)_{16}$ as the catalyst source, a significant improvement in performance is obtained if an excess of triphenylphosphine is added to the system [321, 532, 1370], and a low partial pressure of CO is used. These conditions favor the formation of $HRh(CO)(PPh_3)_3$ in the catalyst equilibria:

$$HRh(CO)_3(PPh_3) \rightleftharpoons HRh(CO)_2(PPh_3)_2 \rightleftharpoons HRh(CO)(PPh_3)_3$$

Preformed $HRh(CO)(PPh_3)_3$ is, in its own right, a very effective hydroformylation catalyst for the conversion of alkenes such as 1-octene and styrene to aldehydes [150, 206, 223, 492, 613, 1155, 1331, 1405, 1408, 1410, 1435, 1917, 1919]. With this catalyst, steric factors strongly favor the formation of linear aldehydes. The selectivity is independent of catalyst concentration [1697].

Addition of extra PPh_3 to the $HRh(CO)(PPh_3)_3$ decreases the rate of the hydroformylation reaction [1520] but increases further the ratio of normal to branched aldehyde [205, 231, 232, 296, 403, 958, 1141, 1142, 1199, 1331, 1437, 1520, 1588, 1692, 1768, 1918]. Phosphorus/rhodium ratios as high as 50:1 or 100:1 provide the best conditions for obtaining high ratios of linear over branched aldehydes [1403, 1431, 1437]. In some instances, molten PPh_3 (m.p. 79 °C) has been used as the solvent for the reaction [536, 1696, 1918, 1922] – in this medium, the catalyst is very selective and is almost indefinitely stable. Only minor effects on selectivity are observed when the solvent is changed from benzene to THF, or when Et_3N is added to the system [1403].

There has been some examination of factors affecting the formation of by-products ('Oxo-bottoms') during the hydroformylation of propene with catalysts based on $\{HRh(CO)(PPh_3)_3 + PPh_3\}$. It has been established [618] that degradation of the catalyst produces mainly $[Rh(CO)_2(PPh_2)]_n$, Ph_3PO and $PrPh_2P$. Temperature, aldehyde concentration and residence time have the biggest effect on the amount of 'heavy ends', but under all conditions of practical interest the amount of bottoms is $< 1\%$ [1216]. The main constituents of these bottoms are C_8-aldols, 2-ethylhexenal, butyrates and isobutyrates of C_8-aldols.

Hydroformylation of cycloalkenes such as cyclooctene [205, 231, 1828, 1830] and of 1,5-hexadiene or related dienes [104, 231, 1159, 1188, 1200] has also been achieved with $HRh(CO)(PPh_3)_3$ or $\{HRh(CO)(PPh_3)_3 + PPh_3\}$ as the catalyst. With highly branched dienes such as 2,3-dimethylbutadiene, the formation of γ,δ-unsaturated aldehydes is more favored by steric factors than is the case with unsubstituted dienes [505]. The incorporation of secondary phosphines in the catalyst system results in an almost complete 1,4-hydroformyl addition to 1,3-butadiene [104]. Treatment of an unsaturated polymer with CO and H_2 in the presence of $HRh(CO)(PPh_3)_2$ has given a polymer with up to 20 mol % of CHO groups [1523]; this hydroformylation reaction is not accompanied by crosslinking or degradation.

Hydroformylation products have been obtained from a number of functionalized alkenes and dienes. As shown in Equation (4.7), treatment of the cyanoalkene 4-17 with CO/H_2 in toluene at 70 °C and 180 kg cm^{-2} pressure gives 2-formyl-2-methylglutaronitrile [1201].

$$
\begin{array}{ccc}
\underset{\text{(4-17)}}{\overset{\overset{\displaystyle CH_2}{\|}}{NC-C-CH_2CH_2CN}} & \longrightarrow & \underset{\overset{\displaystyle |}{CHO}}{\overset{\overset{\displaystyle CH_3}{|}}{NC-C-CH_2CH_2CN}}
\end{array}
\qquad (4.7)
$$

The catalyst for this reaction is obtained form $HRh(CO)(PPh_3)_3 + PPh_3$. Dienes of the type 4-18 yield heterocyclic ketones 4-19 (for example, with $R = CH_2Ph$ or CO_2Et) upon hydroformylation with $HRh(CO)(PPh_3)_3$ as the catalyst [570].

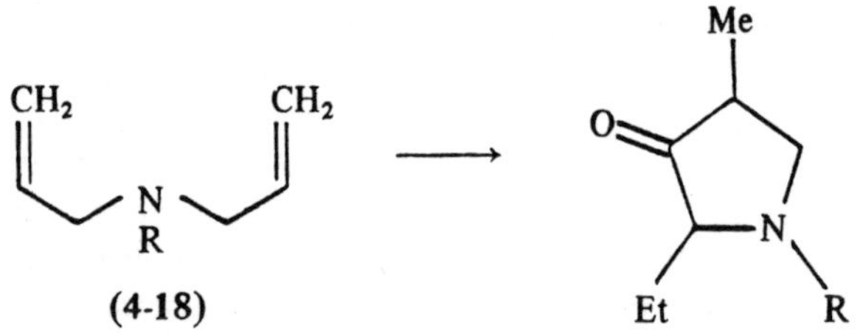

(4-18)   (4-19)

The hydroformylation of allylic alcohol to give 4-hydroxybutanal and 3-hydroxy-2-methylpropanal (Equation (4.8)) is complicated by isomerization and hydrogenation of the unsaturated alcohol [1402, 1508, 1591]:

$$
HOCH_2CH=CH_2 \longrightarrow HOCH_2CH_2CH_2CHO + \underset{\overset{\displaystyle |}{CH_3}}{HOCH_2CHCHO} \qquad (4.8)
$$

The addition of water and Ni to this reaction mixture followed by hydrogenation gives 1,4-butanediol and 2-methyl-1,3-propanediol as the major products [75, 1509]. The hydroformylation of allyl-t-butyl ether, $Me_3C-O-CH_2CH=CH_2$, with $HRh(CO)(PPh_3)_3 + PPh_3$ as the catalyst, proceeds in a similar manner [660, 661].

Several cyclic products useful in perfumes have been obtained *via* hydroformylation reactions [1190, 1748, 1749]. Typical examples are given in Equations (4.9) and (4.10).

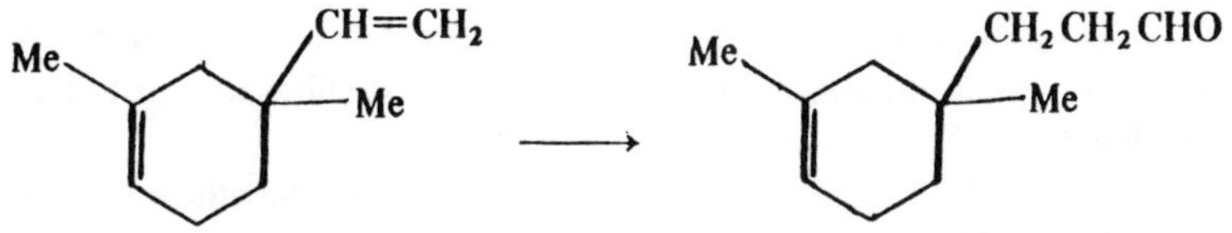

$$(4.9)$$

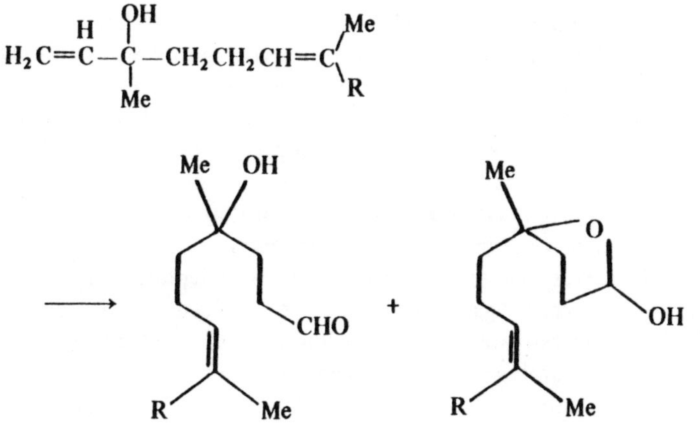

$$(4.10)$$

(minor)   (major)

Various factors affect the rates of hydroformylation reactions with $HRh(CO)$-$(PPh_3)_3$ as the catalyst. They include temperature, the partial pressures of CO and H_2 [232, 728, 1402, 1403, 1829], the CO/H_2 ratio [1402, 1692, 1917] and, as discussed previously, the addition of excess phosphine. With added PPh_3, there are conflicting claims about the effect of the partial pressures of CO and H_2. One suggests [1829] that the rate of hydroformylation depends on the partial pressure of H_2 but not that of CO; the other [296] indicates that the reaction is zero order in both CO and H_2 The hydroformylation rates are enhanced by γ-radiation and are about four times faster than with thermal conditions [887].

Hydroformylation catalysts have been obtained by adding phosphines other than PPh_3, or phosphites, to $HRh(CO)(PPh_3)_3$. The ligands added include $PPh_2(OR)$ (R = H, Et) [1143, 1144], PMe_2Ph, $PCyPh_2$ [780], PBu_3, $P(OPh)_3$

[1402] and bibenzophospholes [683]. Further understanding of the effects of added phosphine has come from the use of chelating phosphine ligands, $Ph_2P(CH_2)_aPPh_2$ ($a = 2-4$) [1400, 1405]. With added $Ph_2PCH_2CH_2PPh_2$, there is a large decrease in the ratio of normal to branched aldehyde formed from 1-pentene. Hydroformylation of 2-pentene with added $Ph_2PCH_2CH_2PPh_2$ also exhibits low normal/branched selectivities. Thus, the cis-chelated hydridorhodium complexes display a lower propensity for anti-Markovnikoff Rh–H addition to a terminal double bond than do trans-bis(phosphine)rhodium hydride complexes.

Some hydroformylation catalysts have been obtained by the addition of an equimolar amount of $Ph_2P(CH_2)_nPPh_2$ ($n = 2-4$) plus a large excess of PPh_3 to the rhodium compounds. In the absence of the bisphosphine, the hydroformylation of terminal alkenes such as vinyl acetate or ethyl acrylate does not proceed substantially at low pressures; the hydroformylation proceeds smoothly at 1 atm pressure when the bidentate ligand is present. It is suggested [1147] that the bisphosphine suppresses the formation of a metastable acyl–rhodium complex in which a six-membered chelate ring is formed through coordination of the acyl-O to the metal. The added $Ph_2P(CH_2)_nPPh_2$ also significantly reduces the extent of isomerization occuring in the hydroformylation of 1-octene [1146]. This is attributed to a decrease in the concentration of coordinatively unsaturated Rh complexes due to coordination of the bidentate ligand. 1,1'-Bis(diarylphosphino)ferrocene ligands have also been added to the catalyst system [1819]. The highest rates and higher linear/branched aldehyde ratios are observed when the aryl groups have electron-withdrawing substituents (e.g. p-Cl, p-CF$_3$). It is postulated that the active catalyst is a dirhodium complex with three phosphorus atoms attached to each rhodium atom.

Various other substances have been added to $HRh(CO)(PPh_3)_3$ in attempts to improve catalyst performance. The addition of weak acids such as phthalic acid or phosphoric acid is said to suppress hydrogenation of the alkenes during hydroformylation [1633]. The effects of adding metal complexes such as $Zn(acac)_2$ [1471], $Mo(CO)_6$, $Co_2(CO)_8$, $Ru_3(CO)_{12}$ and $Pt(PPh_3)_4$ [400, 716] have been explored. The activity of the catalyst is reduced in the presence of O_2 [1408].

Studies of the relative rates of hydroformylation of alkenes with $HRh(CO)(PPh_3)_3$ as the catalyst have established the following orders: terminal alkenes > corresponding internal alkenes [427]; styrene > 1,5-hexadiene > 1-pentene > cyclooctene > cis- and trans-2-pentenes > dl-limonene [231]. With {$HRh(CO)(PPh_3)_3$ + PPh_3} as the catalyst, the reactivity ratio of 1-, (E/Z)-2-, (E)-3-, (E)-4-, (E)-5- and (E)-6-dodecene towards hydroformylation at 120–140 °C is 40:1:1:1:1:1 when the alkenes are examined as mixtures [351].

There has been some interest in the stereochemistry of hydroformylation reactions with catalysts of the type $HRh(CO)(PPh_3)_3$. cis-Addition is observed with both the cis-and trans-isomers of 3-methyl-2-pentene [1655, 1656], but different regioselectivity is observed for the two enantiomers of $H_2C=CHCHPhMe$ [1657]. Another study has shown that the alkenes $MeCH=CHMe$, diethyl maleate, bicyclo[2.2.2]oct-2-ene and 2,3-dihydrofuran give fairly low optical yields of the hydroformylation products [212]. The optical yields are improved significantly when the hydroformylation catalyst is prepared from $HRh(CO)$-$(PPh_3)_3$ and an optically active phosphine ligand (see below).

The mechanism of the hydroformylation reaction with $HRh(CO)(PPh_3)_3$ as catalyst is presumably similar to that for $HRh(CO)_4$ (see Scheme 4.8). Indirect support for various steps in the reaction scheme has come from a detailed infrared study of the reactions between $HIr(CO)_3(PPr^i_3)$ and ethylene, $Ir(CH_2CH_3)$-$(CO)_3(PPr^i_3)$ and CO, and $HIr(CO)_3(PPr^i_3)$ and H_2 [1909], and from the isolation of trans-$Rh(CF_2CF_2H)(CO)(PPh_3)_2$ and related species from the reaction between $HRh(CO)(PPh_3)_3$ and tetrafluoroethylene [1928, 1930].

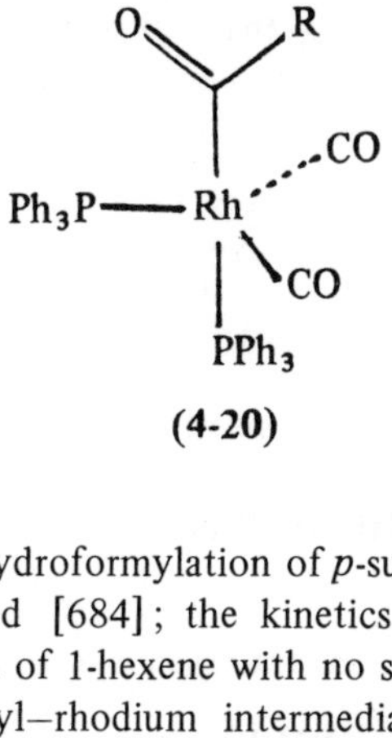

(4-20)

In the hydroformylation of p-substituted styrenes, pseudo-first-order kinetics are observed [684]; the kinetics are not diffusion-controlled in the hydroformylation of 1-hexene with no solvent present [1697]. The reaction pathway involves acyl–rhodium intermediates, and compounds of the type 4-20 in Equation (4.11) have been isolated and characterized [233, 241].

$$HRh(CO)(PPh_3)_3 \; \overset{CO}{\underset{}{\rightleftharpoons}} \; HRh(CO)_2(PPh_3)_2 \; \overset{\text{1-alkene}}{\underset{}{\rightleftharpoons}} \; 4\text{-}20 \qquad (4.11)$$

The acyl–rhodium complexes undergo inter- and intra-molecular rearrangements. With the species formed from styrene, for example, 4-21 and 4-22 co-exist in solution.

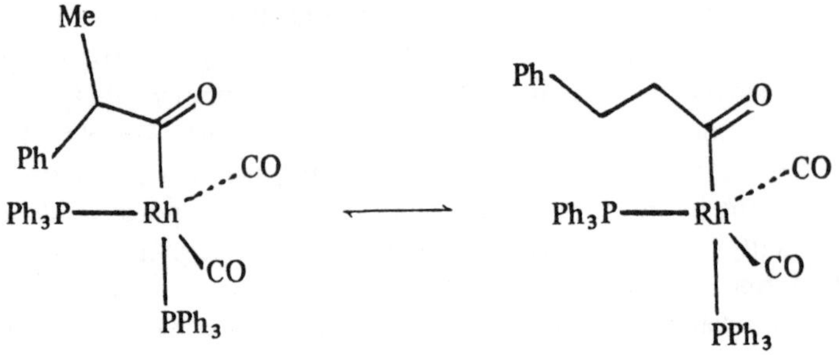

(4-21)   **(4-22)**

In benzene at 278 K, the ratio of **4-21** to **4-22** is 91:9, but the ratio changes to favor **4-22** over 2 h at 298 K. This is consistent with the observation that 2-phenylpropanal is the major hydroformylation product at ambient temperature, but 3-phenylpropanal predominates at higher temperatures. Thus the isomer ratio in hydroformylation reactions may be controlled by the kinetic lability of acyl–rhodium intermediates. It is interesting that the regioselectivity in the hydroformylation of trifluoropropene and pentafluorostyrene with $HRh(CO)(PPh_3)_3$ and other rhodium catalysts is opposite to that observed with hydrocarbon alkenes [552], for example

$$CF_3CH{=}CH_2 \xrightarrow[CO/H_2]{HRh(CO)(PPh_3)_3} \begin{array}{l} CF_3CH_2CH_2CHO\ (5\%) + \\ + CF_3(CH_3)CHCHO\ (95\%) \end{array}$$

This is attributed to stabilization by the electron-withdrawing substituents of intermediates containing $Rh{-}CH(Me)CF_3$.

Formation of $HRh(CO)_2(PPh_3)_2$ from $HRh(CO)(PPh_3)_3$ is less likely in systems with added PPh_3. The equilibrium

$$HRh(CO)(PPh_3)_3 + n\,PPh_3 \rightleftharpoons HRh(CO)(PPh_3)_2 + (n+1)\,PPh_3$$

has been investigated for $n = 0$–140 by ^{31}P NMR spectroscopy. In toluene at 5–105 °C, the equilibrium generally favors the tris- rather than the bis-phosphine complex [932].

Other catalyst sources

Many compounds closely related to $HRh(CO)(PPh_3)_3$ have been used as hydroformylation catalysts. These include $HRh(CO)(PPh_3)_2(AsPh_3)$ [1150]; $HRh(CO)(PPh_3)(PR_3)_2$ (e.g. R = 2-pyridyl [1024]) and $HRh(CO)(PPh_3)$-$(Ph_2PCH_2CH_2PPh_2)$ [1400]; $HRh(CO)L_3$ (L = $PPh_2(CH_2CH_2SiMe_3)$) [773],

$P(OR)_3$ [150, 939, 940, 1436], $As(OR)_3$, $Sb(OR)_3$ [939, 940], $N(CH_2Ph)_3$ [510], piperidino-$(PPh_2)_2$ or related ligands [1059], and $[Ph_2P(CH_2)_{14}P^+(Me) (CH_2CHMe_2)_2]$ $[4$-n-$C_{12}H_{25}C_6H_4SO_3]^-$ [1347]; $HRh(CO)L'(L-L)$ $(L' = R_3N,$ R_3P, R_3As and $L-L = $ **4-23** [779]); $[HRh(CO)(L-L)]_2(\mu$-$L-L)$ $(L-L = $ **4-23** [1822]) or $trans$-1,2-$(Ph_2PCH_2)_2C_4H_6$ [1145, 1821] and other 1,2-bisdiphenylphosphinocycloalkanes [781].

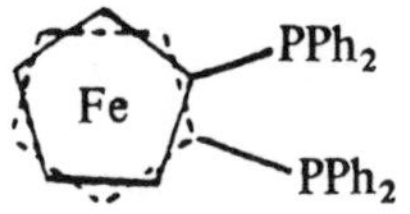

(4-23)

With the latter systems, small changes in the bisphosphine configuration and rigidity affect the selectivity of the hydroformylation reaction. Dicarbonyl complexes such as $HRh(CO)_2(PPh_3)(PR_3)$ [1521] and $HRh(CO)_2\{trans$-1,2-Ph_2P-cyclobutane$\}$ [1820] have also been tried as catalysts. Some of these systems may offer particular advantages in specialist applications over the $HRh(CO)(PPh_3)_3$ system.

Catalysts of similar type are presumably formed under hydroformylation conditions when appropriate ligands are added to $Rh_6(CO)_{16}$. For example, catalysts obtained from $Rh_6(CO)_{16}$ and bidentate ligands such as $Ph_2P(CH_2)_3PPh_2$ or $Ph_2PCH=CHPPh_2$ give high normal/branched chain isomer ratios when alkenes are hydroformylated [123]. The catalyst formed from $\{Rh_6(CO)_{16} + PBu_3\}$ achieves the conversion of allylic alcohols to butanediols [1629]; diols are also obtained from cyclic acetalaldehydes in the presence of $\{Rh_6(CO)_{16} + P(OMe)_3\}$ [155, 156]. The hydroformylation of cyclic acrolein acetals such as 2-vinyl-4-methyl-1,3-dioxan has been accomplished with $\{Rh_6(CO)_{16} + $ excess $P(OPh)_3\}$ as the catalyst source [389]. The addition of amines such as pyridine [822] or $PhCH_2NMe_2$ [1486] to $Rh_6(CO)_{16}$ provides related catalysts which hydroformylate 1,3-butadiene and 1-hexene respectively, but high pressures are required. A catalyst formed when small amounts of pyridine are added to $Rh_4(CO)_{12}$ has been used to convert $AcOCH_2CH=CH_2$ to $AcO(CH_2)_3CHO$ [1049]; again, harsh conditions (110–130 °C, 140–160 atm pressure) are required. Finally, α-alkenes are converted to alcohols under mild water–gas shift reaction conditions with catalysts formed from $\{Rh_6(CO)_{16} + (Me_2NCH_2)_2CH_2\}$ or 4-(dimethylamino)pyridine [916].

Other catalyst precursors of the types $Rh(CO)_aL_b$ (L = tertiary phosphine, amine or related ligand) [258, 507, 542, 597, 888, 1032, 1438] and [Rh-$(CO)_3L]^+$ (L = tertiary phosphine) [1212, 1226] are effective for the hydroformylation of alkenes.

Some catalysts have been formed from analogous iridium carbonyl systems.

With $\{Ir_4(CO)_{12} + PBu_3\}$ [1777], there is competition from alkene hydrogenation, but this is reduced if the CO/H_2 ratio is *ca.* 4 and high temperatures and pressures are maintained. The preformed hydrido complex $HIr(CO)(PPh_3)_3$ has also been used to catalyse the hydroformylation of alkenes [533].

Many complexes of the type $Rh(CO)_a XL_b$ are useful as catalyst sources for the hydroformylation of alkenes. These include:

$Rh(CO)_2(CO_2Et)L$ (L = pyridine, 2-picoline, or aniline) [226].

$Rh(CO)_2Clpy$ [956].

$Rh(CO)XL_2$ (X = halide, L = PR_3) [47, 150, 206, 269, 367, 492, 563, 646, 647, 714, 720, 1376,. 1411, 1487, 1760, 1768], AsR_3 [720], $P(OR)_3$ [150, 832], PPh_{3-n} (NRR')$_n$ (n = 1,2,3) [631, 632] or a phosphinoalkyl-organosilane ligand such as $Me_3SiCH_2PPh_2$ [502].

$Rh(CO)(CO_2R)L_2$ [226, 227, 228, 876, 1002, 1070, 1195, 1646].

$Rh(CO)(acac)(PPh_3)$ [447, 662, 1586], $\{Rh(CO)_2acac + PMePh_2$ [1586] or related ligand [515] $\}$.

$Rh(CO)X(PR_3)_2$ (X = SR) [47], $SnCl_3$ or BF_4 [1920].

$Rh(CO)ClL_2$ ($L_2 = Ph_2PCH_2CH_2PPh_2$ and related ligands) [1059, 1756].

$Ir(CO)Cl(PPh_3)_2$ [144, 687] and $[Rh(CO)\{P(OPh)_3\}Pz]_2$ or related pyrazolate complexes [1835].

With the $M(CO)XL_2$ catalyst sources, the hydroformylation reaction is subject to an inhibition period which is removed by the addition of hydrogen halide acceptors. This indicates that the principal catalytically active species may be a compound of the type $HRh(CO)_2L_2$ [492]. Hydroformylation reactions undertaken in the presence of cyclohexenyl hydroperoxide are greatly accelerated, and this is thought to be due to the formation of *cis*-$Rh(CO)_2Cl(PPh_3)$ [1782]. The addition of excess PPh_3 also activates $Rh(CO)Cl(PPh_3)_2$, and it has been suggested that this method of activation is most effective when the phosphine is added after the initial activity has dropped to about 80% [840]. Catalysts have also been formed from $\{Rh(CO)Cl(PPh_3)_2 + AsPh_3\}$ and $\{Rh(CO)Cl-(AsPh_3)_2 + SbPh_3\}$ [1514]. Strong acids [1099, 1100, 1596], O_2 and CS_2 strongly inhibit the reaction, and $HC\equiv CH$ [1001] has a lesser effect. The effect of added alkoxide [619] and of increased CO pressure [563] have also been investigated.

With catalysts formed from $Rh(CO)XL_2$, the nature of L has a significant influence on the specificity of the reaction. For example, when L = PAr_3 the linear/branched aldehyde ratio in the hydroformylation of 1-hexene is enhanced by the presence of *p*-alkyl groups. The optimum comprimise between the hydroformylation rate and the aldehyde ratio is observed when L = $P(p\text{-}BuC_6H_4)_3$. The highest total yield of aldehyde is found when L = PR_3(R = long-chain alkyl such as n-$C_{16}H_{33}$) [543].

Other rhodium and iridium compounds have been used as catalyst sources. The hydroformylation of alkenes is catalysed by $[Rh(CO)_2Cl]_2$ [563, 876, 956], but the reaction is accompanied by some isomerization of the alkene. Increasing the partial pressure of CO (up to a limit of 150 atm) favors the hydroformylation reaction relative to isomerization [563]. The formation of alcohols by the hydroformylation of alkenes is promoted when nitriles such as succinonitrile are added to the $[Rh(CO)_2Cl]_2$ [808, 809, 1136]. For example, the hydroformylation of undecene with a catalyst formed from $\{ [Rh(CO)_2Cl]_2 +$ succinonitrile$\}$ gives dodecanol (81%) + dodecanal (16%); in the absence of succinonitrile, the same products are obtained in 20% and 76% yields, respectively. Similar effects are observed [1265] in the hydroformylation of 1-hexene by the addition of N-methylpyrrolidine and polyalkylene glycol-600 to $[Rh(CO)_2Cl]_2$. There is some preference for α-formylation in reactions with α,β-unsaturated esters, and this preference is increased by the addition of $R_2P(CH_2)_aPR_2$ (R = Ph, Cy; a = 2—4) to the $[Rh(CO)_2Cl]_2$ [1755]. With acyclic dienes, hydroformylation in the presence of formaldehyde produces alkanediols [63].

When $RhCl_3$ is used as the catalyst precursor [63, 721, 722] it seems likely that $[Rh(CO)_2Cl]_2$ is formed *in situ* prior to generation of the active catalyst. A similar sequence is presumably followed in reactions which employ $\{RhCl_3 + SnCl_2\}$ [1915] and $IrBr_3$ [63].

Some related catalyst precursors include the carboxylato complex $[Rh(CO)_2-(O_2CR)]_2$ [695, 876], and the chelate complexes $Rh(CO)_2(L-L)$ (HL_2 = acetylacetone, hexafluoroacetylacetone [167, 563, 956, 1064] or salicyl-aldoximine [1069]) and $Rh(CO)(L-L)L'$ (HL_2 = acetylacetone, L' = PR_3 [881, 1066, 1068, 1071, 1435]; HL_2 = salicylaldoximine, L' = $P(OPh)_3$ [534]; HL_2 = dihydrobis(pyrazolyl)borate, L' = PBu_3 [1065]). It seems likely that the chelate ligand is cleaved from the rhodium atom in the hydroformylation reaction. Thus, optically inactive products are obtained when **4-24** is used as the catalyst source [1549]. The addition of PPh_3 or $PPrPh_2$ to $Rh(CO)(acac)(PPh_3)$ improves the stability of this catalyst [1223].

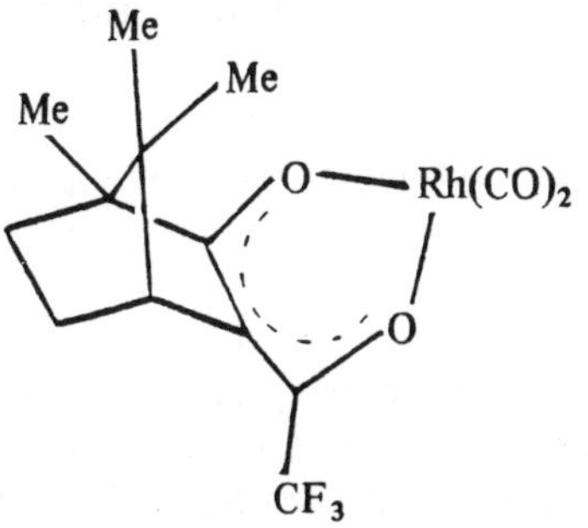

(4-24)

The versatile catalyst $RhCl(PPh_3)_3$ is also effective for the hydroformylation of alkenes [531, 1297, 1507, 1736, 1750, 1915] and cyclic dienes [1098]. Some related hydroformylation catalysts are $\{RhCl_3 + PR_3\}$ [203, 1188, 1620], $RhCl(Ph_2PCH_2CH_2PPh_2)$ [1159, 1283], $RhPh(PPh_3)_3$ [1159], $Rh_2(\mu\text{-}SR)_2L_4$ (e.g. $R = Bu^t$, $L = P(OMe)_3$) [900, 901] and iridium(III) compounds of the type H_3IrL_a ($L = PR_3$, AsR_3 or SbR_3 and $a = 2$ or 3) [334]. The rhodium(O) compounds $[RhL_3]_2$ ($L = PPh_3$, PBu_3 or $P(OPh)_3$), $[Rh(Ph_2PCH_2CH_2PPh_2)]_2$ [1158], $Rh_3(CO)_3(\mu\text{-}PPh_2)_3(PPh_3)_2$ and $Rh_4(CO)_5(\mu\text{-}PPh_2)_4(PPh_3)$ [168] are also active catalyst sources, and $[Rh(PPh_3)_3]_2$ in particular is useful in the formation of BuCHEtCHO. The addition of sodium acetate to catalysts which are obtained from iridium salts and tertiary phosphines enhances the formation of alcohols when alkenes undergo hydroformylation [1621].

The diene complex $[Rh(COD)Cl]_2$ catalyses the hydroformylation of alkenes, cyclic alkenes and dienes [72, 718, 719, 723, 724, 1232, 1540] but high pressures (600–700 atm) are often required. Phosphine or N-base ligands are included in some related systems such as $\{[Rh(COD)Cl]_2 + P(3\text{-}NaO_3SC_6H_4)_3\}$ [874], $\{Rh(COD)OAc + PPh_2(CH_2CH_2PPh_2)$ or related phosphines$\}$ [1846, 1914], $[Rh(COD)IL_2]^+$ ($L = NPh_3$, PPh_3, $AsPh_3$) [363, 1832], $\{[Rh(NBD)Cl]_2 + PPh_3\}$ [1339, 1499] and $\{[Rh(COD)$ (imidazole)$]_3 + PR_3$, NR_3 or piperidine$\}$ [1758]. The addition of chelating diphosphines to $[Rh(1,5\text{-hexadiene})Cl]_2$ or $[Rh(C_2H_4)_2Cl]_2$ reduces the ratio of β- to α-formylation product obtained from styrene [1283], but 2,2′-bipyridyl has no effect [1282]. Catalysts prepared from $[Rh(1,5\text{-hexadiene})Cl]_2$ or $[Rh(C_2H_4)_2Cl]_2$ and chiral ligands [1281, 1284, 1659] are more specific for the conversion of styrene to α-phenylpropionaldehyde. Some polyenyl complexes that have been used as hydroformylation catalysts are the rhodium(I) compound $(\eta\text{-}C_5H_5)Rh(CO)(PBu_3)$ [1067] and species formed by treating $Rh^{III}LCl_2$ (e.g. L = cyclododeca-1,5,9-trienyl) with $P(OPh)_3$ or related ligands and then hydrazine or a metal hydride such as NaH or $NaBH_4$ [1971].

A final catalyst system that has been used for the low-pressure hydroformylation of alkenes, $RR^1C{=}CHR^2$, to carboxylic esters, $RR^1CHCHR^2CO_2CHPhCOPh$, is Rh_2O_3 in benzil [211]. Some aldehyde, RR^1CHCHR^2CHO, is also formed in the reaction.

There is one instance of the use of a CO_2/H_2 mixture to convert ethylene to propionic acid. The catalyst for this reaction was $RhCl(PPh_3)_3$ [1043].

Supported catalysts

In some instances, the problem of catalyst recovery in hydroformylation reactions has been tackled by anchoring the catalyst to an appropriate support. The materials used include polymers with appended phosphine groups, porous alumina and functionalized silica.

Macroporous styrene–divinylbenzene resin is the most commonly used polymer support. Phosphine or related ligands are linked to the polymer, and a typical catalyst is formed by adding a solution of $HRh(CO)(PPh_3)_3$ [45, 591, 1398, 1401, 1405]. Sometimes PPh_3 [423] or a chelating phosphine [1405] is also added.

These polymer-supported catalysts have been used in the hydroformylation of alkenes such as propylene, 1-hexene and styrene. Generally, there is a high ratio of linear to branched aldehydes in the products. This selectivity is affected by temperature and pressure [1398, 1401, 1405], and is increased by the addition of small amounts of PPh_3 [1403].

The activity of the catalysts is affected somewhat by the method used to functionalize the support. Those prepared via chlorophosphonation are stable whereas those formed by chloromethylation deactivate slightly. The latter do, however, have a higher selectivity for normal aldehyde formation [424, 425]. The number of anchor sites involved in coordination to rhodium also affects the activity — the highest activity is achieved when only one phosphine ligand in $HRh(CO)(PPh_3)_3$ is substituted [1521]. The complexes formed can be square-planar or trigonal-bipyramidal, and the structure adopted depends on the separation between the anchor sites and the flexibility of the side-chain containing the anchor sites. There can be significant redistribution of the rhodium in the reactor beads during the course of the hydroformylation reaction. This may be due to the dissolution of some of the Rh:

$$P - Rh \rightleftharpoons P + [Rh \cdot soln]$$

Measurement of the concentration of rhodium in solution shows that it decreases with increasing temperature or H_2 pressure and with decreasing CO pressure [1041]; in general, the amount of metal leached during hydroformylation is extremely low [1763].

There has been a comparison of the performance of $HRh(CO)(PPh_3)_3$ as a homogeneous and a supported catalyst in the hydroformylation of 1-hexene [1684]. In the liquid phase (no solvent) at 120 °C, the activity of the homogeneous catalyst is 250 min^{-1} with a linear/branched product ratio of 5.4. In the gas phase with a supported liquid-phase catalyst, the activity is only 1.7 min^{-1}, but the linear/branched product ratio is 46.

A number of hydroformylation catalysts have been prepared by supporting rhodium compounds other than $HRh(CO)(PPh_3)_3$ on polymeric materials. Some examples are:

$(\eta\text{-}C_5H_5)Rh(CO)_2$ on polystyrene [225, 283, 1562] or poly(2,6-dimethyl-1,4-phenylene oxide) [1855] with attached cyclopentadienyl anions.
$Rh_2Co_2(CO)_{12}$ and related clusters on a crosslinked polymer matrix containing $-NR_2$ or $-NH(CH_2)_nCR_2$ groups [458, 569, 671, 672].

$Rh_4(CO)_{12}$ on poly(vinylpyridine) which is crosslinked or supported on glass [612].

$Rh_6(CO)_{16}$ on hydroxyl-terminated styrene–butadiene copolymer or Ir_4-$(CO)_{12}$ on hydroxyl-terminated polybutadiene – thermal decomposition of the bound cluster produces a dispersion of colloidal metal particles (size 10–200 Å) [1628].

$[Rh(CO)_2Cl]_2$ on polystyrene with attached phosphine groups – the polymer is coated over silica gel [74].

$[Rh(CO)_2Cl]_2$ attached to styrene–divinylbenzene copolymer which has been sulfonated and then stirred with a solution of $P(NMe_2)_3$ [952].

$Rh(CO)_2(acac)$ anchored to Ph_2P-functionalized polystyrene [46, 126] or polypropylene [668].

$RhCl(PPh_3)_3$ on a styrene–divinylbenzene copolymer functionalized with PBu_2Ph [643].

$[Rh(THF)Cl_2BH_4]$ on poly(vinylpyridine) cross linked with divinylbenzene [284, 287].

$RhCl_3$ or $Rh(CO)_2(acac)$ on chelate resins with iminodiacetic acid moieties [725]; the latter system is more active but gives lower normal/branched aldehyde ratios.

Porous Al_2O_3 or SiO_2 can also be used to support $HRh(CO)(PPh_3)_3$ [198, 422, 733]. Surface hydroxy groups on the support probably act as the anchoring sites. The stability of the $HRh(CO)(PPh_3)_3$ on SiO_2 catalyst is increased by incorporating excess PPh_3 into the catalyst – this also increases the selectivity for n-aldehyde formation and reduces the activity for alkene hydrogenation [732]. One way of achieving this is to prepare the catalyst from porous silica [421, 589] or Chromosorb [1695] and a solution of $HRh(CO)(PPh_3)_3$ in molten PPh_3. A catalyst formed from $HRh(CO)(PPh_3)_3$ and silica containing bound N, P, O or S ligands has also been used to hydroformylate 1-hexene [45] – it demonstrated high stability when tested under flow conditions.

Other hydroformylation catalysts have been prepared from $Rh_4(CO)_{12}$ or $Rh_6(CO)_{16}$ on oxide supports such as zeolites [1048, 1118, 1119], ZnO, ZrO_2 Cr_2O_3, MoO_2, CoO or SiO_2 [794, 795, 798, 1285]. In general, basic oxides (e.g. ZnO, MgO) produce active catalysts but the catalysts are ineffective when supported on acidic oxides such as Al_2O_3 [795]. Mixed Co–Rh clusters, including $Co_2Rh_2(CO)_{12}$, have also been supported on ZnO [794, 795] and MgO [795]. The hydroformylation rates and preference for linear aldehyde formation are sensitive to the structure of the catalyst source. With ZnO as the support, the rates are faster for $Rh_4(CO)_{12}$ than for $Rh_6(CO)_{16}$, but the proportion of linear isomer is lower [794]. With mixed Co–Rh tetranuclear clusters, the proportion of linear aldehyde increases with the Co content of the catalyst [794]. The acidic oxide Al_2O_3 has given an active catalyst with $Rh(CO)Cl(PPh_3)_2$ [1374, 1482].

An *in situ* preparation of $Rh(CO)_a(PPh_3)_b$ from Rh/Al_2O_3, a CO/H_2 mixture and PPh_3 has been described, and this catalyst dihydroformylates polycyclic dienes (such as norbornadiene) and nonconjugated dienes including 4-vinyl-cyclohexene [1433, 1434]. Catalysts obtained from Rh_2O_3 and PR_3 under hydroformylation conditions [503, 504, 508, 511, 512] may be of a similar type.

A few other rhodium compounds have been attached to oxide supports. Examples include $Rh(CO)Cl(PPh_3)_2$ on alumina or silica [1087], $RhCl(PPh_3)_3$ on diatomaceous earth [1210] and $[Rh(COD)Cl]_2$ on functionalized silica [216].

Activated carbon is another high surface area support for hydroformylation catalysts. It has been used to anchor $Rh_4(CO)_{12}$ or $Rh_6(CO)_{16}$ [799, 1285]. Decomposition of these carbonyl clusters gives metallic Rh on charcoal which has been used in commercial hydroformylation reactions. Hydroformylation of an alkene is achieved with synthesis gas at 10–20 atm pressure and *ca.* 100 °C in the presence of PPh_3 or $P(OPh)_3$. Under these conditions the selectivity for linear aldehyde is >90%, and there is little or no hydrogenation of the aldehyde to the alcohol.

An interesting recent development in the formation of supported hydroformylation catalysts is to apply the rhodium complex (e.g. $HRh(CO)(PPh_3)_3$) as a solution in molten PPh_3 [588, 592], $P(p\text{-}MeC_6H_4)_3$, $P(CH_2CH_2CN)_3$ or $PPh_2\{(S)\text{-}(+)\text{-}(2\text{-phenylbutyl})\}$ [590]. The catalyst is distributed uniformly by capillary condensation in the pores of the support material (e.g. macroreticular styrene–divinylbenzene copolymer or silica). These catalysts show high activity, selectivity and stability. Thus, the hydroformylation of propylene is achieved at 90 °C and 1.57 MPa total pressure, and there is no loss of activity over at least 800 h.

SOME SPECIAL APPLICATIONS OF THE HYDROFORMYLATION
REACTION

Asymmetric hydroformylation

Various chiral ligands have been incorporated into hydroformylation catalysts in attempts to achieve asymmetric hydroformylation. Catalysts can be formed, for example, by adding DIOP to $HRh(CO)(PPh_3)_3$ [136, 213, 345, 779, 1395, 1897] or $[Rh(CO)_2Cl]_2$ [1659, 1897]. In the hydroformylation of isoprene to 3-methylpentanal at 95 °C and 90 atm pressure with $\{HRh(CO)(PPh_3)_3 + (-)\text{-DIOP}\}$ as the catalyst, the optical yield is 32% [213]. The optical yields are typically 10–24% for the conversion of vinyl acetate to (S)-2-acetoxy-propanal with the { [Rh(CO)$_2$Cl]$_2$ + (−)-DIOP} catalyst [734, 1897]. Very high selectivity in the asymmetric hydroformylation of styrenes and butenes is obtained with the Rh(CO)Cl(DIOP) catalyst system.

Other asymmetric hydroformylation catalysts have been formed from:

HRh(CO)(PPh$_3$)$_3$ + 1,1′-bis(diphenylphosphino)ferrocene [780].

HRh(CO)(PPh$_3$)$_3$ + 1,1′-bis[bis(α,α,α-trifluoro-p-tolyl)phosphino] ferrocene [1818].

[Rh(CO)$_2$Cl]$_2$ + PPh$_2$(neomenthyl) [1897].

[Rh(CO)$_2$Cl]$_2$ + (R)-benzylmethylphenylphosphine [1897].

[Rh(CO)$_2$Cl]$_2$ + (1S, 2S)-1,2-bis(diphenylphosphino-oxy)cyclohexane, (1R, 3R)-1,2-bis(diphenylphosphinomethyl)cyclobutane and related ligands [681].

[Rh(CO)$_2$Cl]$_2$ or HRh(CO)(PPh$_3$)$_3$ + a copolymer of styrene, divinylbenzene and 4-25 [548].

Rh(CO)Cl(PPh$_3$)$_2$ + PPh$_2$(neomenthyl) [1760].

[Rh(COD)L$_2$] [BPh$_4$] (L = cyclohexylanisylmethylphosphine) [1783].

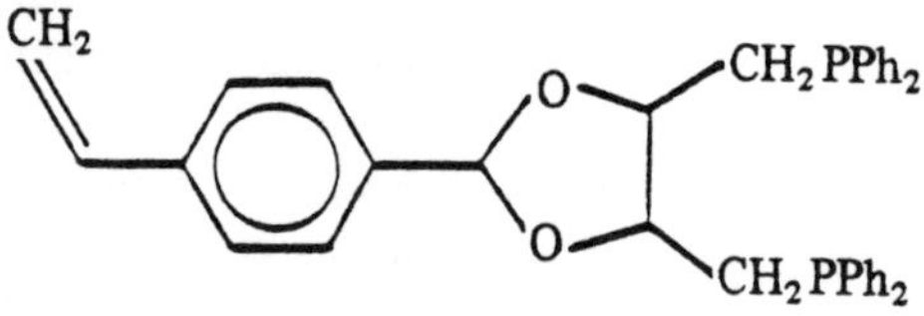

(4-25) 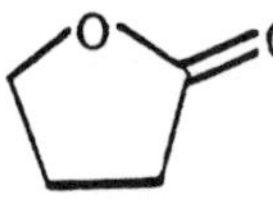   (4-26)

Use of silanes in 'hydroformylation' reactions

Alkenyloxysilanes are obtained by treatment of alkenes with CO and a silane such as HSiEt$_3$ or HSiMeEt$_2$. Appropriate catalysts include [Rh(CO)$_2$Cl]$_2$, RhCl(PPh$_3$)$_3$ [1239] and {RhCl(PPh$_3$)$_3$ + Et$_3$N} [1560]. The reaction of 1-hexene with CO and HSiMeEt$_2$ in the presence of {RhCl(PPh$_3$)$_3$ + Et$_3$N} is an example of this type of reaction, the observed distribution of products being (Z)-BuCMe=CHOSiMeEt$_2$ (8%), (E)-BuCMe=CHOSiMeEt$_2$ (13%), (Z)-BuCH$_2$CH=CHOSiMeEt$_2$ (56%) and (E)-BuCH$_2$CH=CHOSiMeEt$_2$ (23%).

Combined hydroformylation and oxidation, including hydroacylation

Some useful organic syntheses have been achieved by hydroformylation followed by oxidation. Conversion of the unsaturated ether Me$_3$COCH$_2$CH=CH$_2$ to butyrolactone (Equation (4.12)) is a good example [426].

$$H_2C=CHCH_2OH + Me_2CH=CH_2 \longrightarrow Me_3COCH_2CH=CH_2 \xrightarrow{H_2,CO}$$

$$Me_3CCOCH_2CH_2CH_2CHO \xrightarrow{O_2} Me_3CCOCH_2CH_2CH_2CO_2H \xrightarrow{H_2O}$$
4-26   (4.12)

A catalyst prepared from $\{HRh(CO)(PPh_3)_3 + PPh_3\}$ is used for the hydroformylation step. In another system, hydroformylation and oxidation are carried out simultaneously. Thus, treatment of $HC\equiv CH$ with CO, H_2 and O_2 over a mixture of $Rh(CO)_2(acac)$ and $\{Me(CH_2)_7\}_3PO$ gives $EtCO_2H$, $(EtCO)_2O$, EtCHO and Et_2CO [516].

The conversion of alkenes to carboxylic acids can also be achieved with $\{CO + H_2O\}$, thus

$$MeCH=CH_2 \longrightarrow Pr^iCO_2H + PrCO_2H$$

Appropriate catalysts for this type of reaction include $Ir(CO)Cl(PPh_3)_2$ or related complexes + HI promoter [368, 1224, 1225], $Ir(CO)_2ClL$ [1227], $HIr(CO)_2(PPh_3)_2$ [1224, 1225] and $[Rh\{P(OPh)_3\}_4]^+$ + EtI promoter [1926]. The rate of carboxylation with some of these catalysts is increased by adding minor amounts of compounds such as FeI_2 [1612]. With chelate complexes such as $Rh(acac)_3$, the major product obtained from ethylene and $\{CO + H_2O\}$ is Et_2CO; some $EtCO_2H$ and 3,6-octanedione are also obtained [38]. Dicarboxylic acids are obtained in similar reactions with conjugated dienes. For example, butadiene gives adipic acid in the presence of $[Rh(CO)_2Cl]_2$ + MeI [1876] or unsaturated monoesters [1955].

The water in these systems presumably serves two roles – to form an hydrido–metal species and to hydrolyse an acyl–metal intermediate (see Equation (4.13)).

$$M \xrightarrow{H_2O} H-M \xrightarrow[CO]{C_2H_4} C_2H_5\overset{\overset{\displaystyle O}{||}}{C}-M \xrightarrow{H_2O} C_2H_5CO_2H \quad (4.13)$$

These reactions generally require a solvent or a phase-transfer agent to bring the alkene and water into contact. Byproducts can be produced in these systems due to the formation of hydrogen via the water–gas shift reaction.

In some related reactions, secondary alcohols or other compounds with an active hydrogen are used in place of water. Thus, ethylene is converted to products such as diethylketone and ethylpropionate by treatment with CO and 2-propanol in the presence of rhodium complexes including $Rh(CO)Cl(PPh_3)_2$ and $Rh(CO)_a\{P(OR)_bR'_{3-b}\}_c$ [514, 782, 1583]. The treatment of nitrobenzene with CO and butanol in the presence of catalyst systems such as

$\{[Rh(CO)_2Cl]_2 + FeCl_2\}$ gives butyl-N-phenylurethane [813]. Similar reactions in methanol with catalysts such as $\{Ir(CO)Cl(PPh_3)_2 + FeCl_3\}$ give methylcarbanilate, and polyurethanes are obtained in a similar manner from polynitro compounds [815]. Vinyl-, allyl- and isopropenyl-o-carboranes are converted to alkyl-o-carboranes by treatment with CO and saturated alcohols in the presence of $Rh_4(CO)_{12}$ [1106]. Deuteration studies have established that there is H-transfer from the alcohol in these reactions. With carbonyl–rhodium compounds as catalysts, the carbonylation of alkyl halides of the type $C_nF_{2n+1}CH_2CH_2I$ in the presence of ROH gives $C_nF_{2n+1}CH_2CH_2CO_2R$ (n = 6 or 8, R = H or Et) [1483]. In the presence of $[Rh(CO)_2X]_2$ with promoters such as I_2, HI or MeI the compounds $RCO_2CH_2CO_2R'$ (e.g. R = Me, Et; R$'$ = H) are obtained [1022, 1050]. Treatment of 2-phenylaziridine with CO and $BuNH_2$ in the presence of carbonyl–rhodium compounds gives MePhC-$(NH_2)CH_2CONHBu$, and a similar carbonylation of 2-methylaziridine in MeSH with carbonyl–iridium compounds as catalysts gives $MeCH(NH_2)CH_2COSMe$ [827].

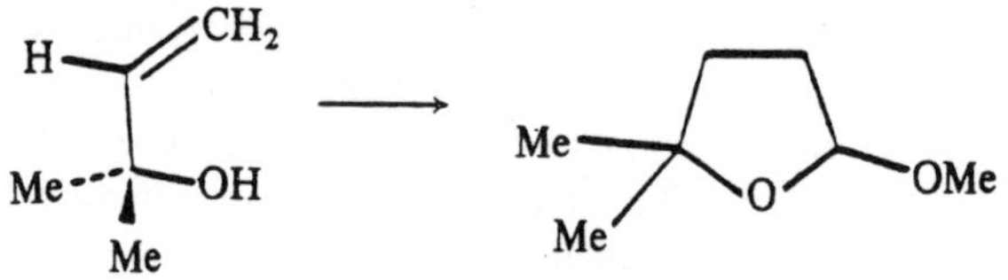

(4.14)

(4-27)

2-Alkoxytetrahydrofurans (4-27) can be obtained from allylic alcohols such as 2-methyl-3-buten-2-ol by treatment with CO, H_2 and MeOH (Equation (4.14)). A suitable catalyst for this conversion is obtained from $\{Rh_6(CO)_{16} + PPh_3 +$ phthalic acid$\}$ [1630, 1631, 1632]. 1,4-Diols are obtained by the hydrogenation of the products. A similar conversion of butenes such as $HOCH_2CH_2CMe{=}CH_2$ to dihydropyrans has also been achieved [1927].

The hydroacylation of C_4–C_5 unsaturated aldehydes provides a useful synthesis of cyclopentanones. The cyclization reaction is inhibited by alkyl substituents at the C–2 or C–5 positions, and disubstitution at the C–2 position results in the formation of ethylketones. The catalyst for the reaction is formed from $[Rh(C_8H_{14})_2Cl]_2$ by treatment with ethylene in CH_2Cl_2 and 2 equiv of PR_3 (R = p-MeC$_6$H$_4$, p-MeOC$_6$H$_4$ or p-Me$_2$NC$_6$H$_4$) [1056].

CHAPTER 5

OXIDATION REACTIONS

Catalytic oxidation is an important process in organic synthesis. Homogeneous catalysts are used in many large-scale processes for the oxidation of alkanes, alkenes and arenes. The industrial importance of these reactions, and a general overview of the chemistry, is presented in Parshall's book entitled *Homogeneous Catalysis* [1367].

One of the most extensively studied of these processes is the Wacker process [709, 710] for the oxidation of alkenes to aldehydes:

$$RCH{=}CH_2 + \tfrac{1}{2}O_2 \longrightarrow RCH_2CHO$$

Palladium(II) salts are used as the catalyst for this reaction. It is known that C$-$O bond formation occurs through intramolecular reaction between a coordinated alkene and a bound hydroxide, but there is some uncertainty about the detail of the final steps in the mechanism.

On an industrial scale, the Wacker process is now being phased out. Preference is being given to other processes, such as the *Halcon* glycol acetate synthesis and the *Oxirane* propylene oxide synthesis. The former involves the oxidative acetoxylation of ethylene with the metalloid tellurium being involved in one step of a complex series of organic reactions. In the latter process, interaction between an alkene and an alkyl hydroperoxide in the presence of a molybdenum catalyst gives an epoxide and an alcohol.

Rhodium and iridium compounds have been used to assist the oxidation of a number of organic compounds. These reactions are discussed below.

5.1. Direct Reaction with O_2

The oxygenation of terminal alkenes to give the corresponding methyl ketones (Equation (5.1)) is promoted by various complexes of rhodium and iridium.

$$RHC{=}CH_2 \xrightarrow{\ O_2\ } R\overset{\displaystyle O}{\overset{\|}{C}}CH_3 \tag{5.1}$$

Effective catalysts for the reactions with alkenes such as 1-hexene and 1-octene include $RhCl(PPh_3)_3$ [471, 1762], $Rh(CO)Cl(PPh_3)_2$ [1422], $HRh(CO)(PPh_3)_3$ [471], $[Rh(C_2H_4)_2Cl]_2$ [393] and $[RhL_4]^+A^-$ (L = $AsPh_3$, $AsMe_2Ph$: A = ClO_4, PF_6) [803]. The addition of hydroperoxides activates the former two catalysts [1812]. There is oxidation at the vinyl group when 3,7-dimethyl-1,6-octadiene is treated with O_2 in the presence of $RhCl_3/FeCl_3$ in DMF [1157]. The products of this reaction are $R(CH_2)_2CHMeCOMe$ plus cis- and trans-$R(CH_2)_2CMe{=}CHMe$ (R = $Me_2C{=}CH$). The corresponding reaction with 7-methoxy-3,7-dimethyl-oct-1-ene gives the same products (R = $Me_2C(OMe)CH_2$). An interesting catalyst system for the oxidation of 1-hexene to 2-hexanone is obtained by binding carbonyl—rhodium(I) and copper(II) species to organo-sulfide-functionalized silica gel (SG-SH) [1275]. The most active catalyst is obtained with site-separated organosulfide groups — in these systems, monomeric carbonyl—rhodium species are present.

The conversion of ethylene to acetaldehyde is achieved with $[RhCl_a$-$(H_2O)_{6-a}]$ (a = 4, 5 but not 6) as the catalyst, and the mechanism of this reaction is thought to be analogous to that of the Wacker process [853, 854]. The oxygenation of styrene can give a range of products. With $Ir(CO)Cl(PPh_3)_2$ as the catalyst, acetophenone is formed if toluene is the solvent; in acetic acid, benzaldehyde is the major product and smaller amounts of acetophenone are obtained [1746]. Both acetophenone and benzaldehyde are obtained when $[Rh(C_2H_4)_2Cl]_2$ is used as the catalyst in the presence of a radical inhibitor [501], and styrene oxide is an additional product when the catalyst is $Rh(CO)$-$Cl(PPh_3)_2$, $HIr(CO)(PPh_3)_3$ or $[Ir(Ph_2PCH_2CH_2PPh_2)_2]^+$ [1097, 1746].

Epoxides and alcohols are generally obtained from reactions of internal alkenes with oxygen. For example, 4-methyl-2-pentene gives 4-methyl-pentene-2,3-epoxide, 2-methyl-pent-3-en-2-ol and 4-methyl-pent-3-en-2-ol. These reactions have been catalysed by $RhCl(PPh_3)_3$ [41] and $M(CO)XL_2$ complexes (M = Rh, Ir: X = Cl, Br; L = PPh_3, $AsPh_3$) [41, 1096, 1810].

The oxygenation of various cycloalkenes has also been achieved. Cyclohexene, for example, yields 2-cyclohexen-1-one, 2-cyclohexen-1-ol, cyclohexene oxide and benzaldehyde. This reaction has been catalysed by $RhCl(PPh_3)_3$ [181, 342, 561, 914, 1023, 1469], $M(CO)X(PPh_3)_2$ (M = Rh, Ir) [342, 561] and $Ir(N_2)Cl(PPh_3)_2$ [342]. Again, the addition of hydroperoxides activates the catalysts [1675]. Greater specificity in the formation of α,β-unsaturated ketones is achieved with $[Rh_3O(OAc)_6(H_2O)_3]OAc$ as the catalyst in acetic acid in the presence of t-butyl hydroperoxide [1814]. The catalytic oxidation of (+)-carvomenthene (5-1) with $RhCl(PPh_3)_3$ yields two cyclic ketones, 5-2 and 5-3 [111], and this reaction is thought to proceed via an allyl radical.

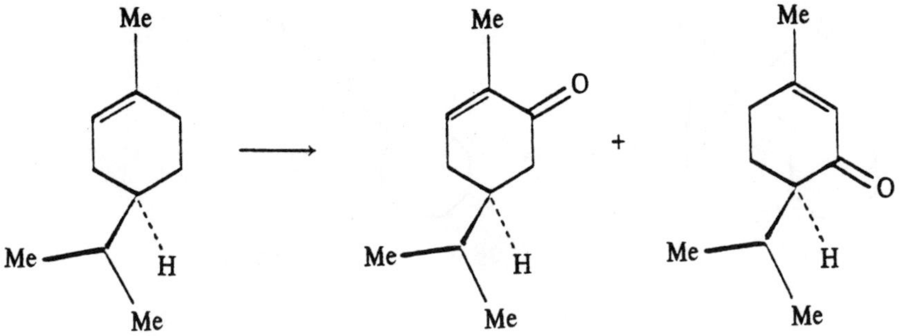

(5-1) (5-2) (5-3)

With catalysts formed from Ir(COD)Cl(DMA) or HIr(COD)Cl$_2$(DMA), an O$_2$/H$_2$ mixture is used to achieve the oxidation of cyclooctene to cyclooctanone [87]:

$$C_8H_{14} + O_2 + H_2 \longrightarrow C_8H_{14}O + H_2O$$

It seems likely that O-transfer occurs from an Ir hydroperoxide intermediate.

Mixed metal catalyst systems have been used in some oxidations of cycloalkenes. With catalysts formed from rhodium(II) carboxylates and molybdenum or vanadium compounds such as MoO$_2$(acac)$_2$ or VO(acac)$_2$, the oxidation of cyclohexene gives 1,2-epoxy-3-cyclohexanol [1269]. Various catalyst systems of the type { RhCl$_3$ · 3H$_2$O + Cu(ClO$_4$)$_2$ } have been used to form unsaturated ethers from cycloalkenes and oxygen in the presence of alcohols [1186].

There has been considerable interest in the mechanism of oxygenation of alkenes. With RhCl(PPh$_3$)$_3$ as the catalyst, initial oxygen uptake is known to give a dioxygen adduct of the type RhCl(O$_2$) (PPh$_3$)$_a$ [1187]; a related complex [RhL$_4$(O$_2$)]$^+$ is formed with the [RhL$_4$]$^+$ cations [803]. Oxygen-18 labelling experiments have established that there is direct transfer of this coordinated oxygen to the alkene. For the oxidation of coordinated cyclooctene to cyclooct-1-en-3-one [860], infrared evidence favors a process where there is formation and transformation of hydroperoxides. Scheme 5.1 shows the various steps proposed for this reaction. Although this particular reaction is stoichiometric, the oxidation of cyclooctene becomes catalytic with added styrene or in *N,N*-dimethylacetamide. Kinetic results indicate that the rate-limiting steps are the metal-centered insertion of O$_2$ into the allylic C—H bond, and subsequent reactions of the resulting cyclooct-1-en-3-ol [743].

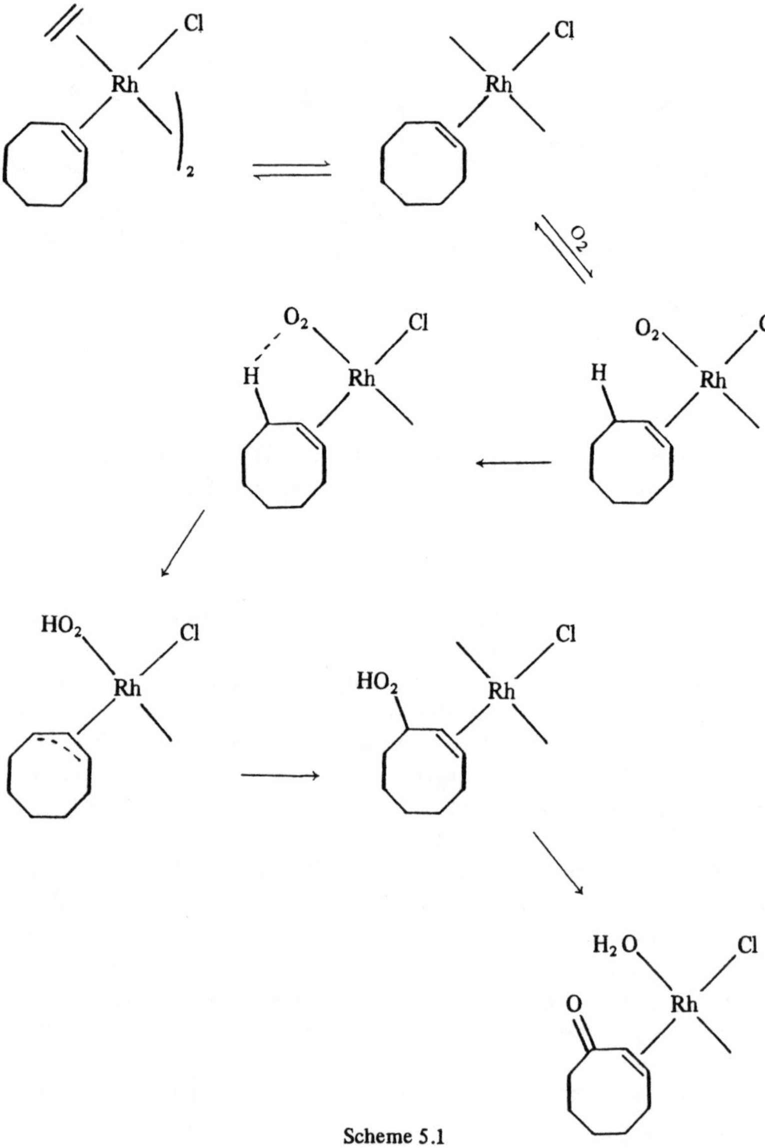

Scheme 5.1

Free-radical chain inhibitors have no significant influence on the oxygenation reactions just discussed. However, a free-radical pathway is indicated in some rhodium(I)-catalysed oxidations of alkenes [561, 1023, 1097] and of butadiene [1122]. For example, it is suggested that the reaction of 2,3-dimethyl-2-butene with O_2 in the presence of $M(CO)ClL_2$ compounds proceeds via a free-radical pathway to give an allylic hydroperoxide intermediate [1096]. A radical mechanism is supported also by EPR studies which indicate the formation of Rh^{II} intermediates of the types $Rh^{II}O_2^-$ and $Rh^{II}ROO\cdot$ during the oxidation of alkenes catalysed by $Rh(CO)Cl(PPh_3)_2$ or $RhCl(PPh_3)_3$ [1812]. It seems likely that the effectiveness of added hydroperoxides in activating some oxidation catalysts can be correlated with their ability to oxidize rhodium(I) to rhodium(II) (see Equations (5.2) and (5.3)) or possibly rhodium(III) [517, 1812].

$$Rh^I + ROOH \longrightarrow Rh^{II} + RO\cdot + OH^- \qquad (5.2)$$

$$Rh^{II} + ROOH \longrightarrow Rh^I + ROO\cdot + H^+ \qquad (5.3)$$

Presumably the hydroperoxide does not react with the organic substrate because the product distribution obtained from metal-catalysed decompositions of cyclohexene hydroperoxide differs from that from cyclohexene oxidation [914].

The co-oxygenation of terminal alkenes and PPh_3 has been achieved with several rhodium catalysts [86, 290, 1461]:

$$RHC{=}CH_2 + PPh_3 + O_2 \longrightarrow \overset{\displaystyle O}{\overset{\displaystyle \|}{R}}CCH_3 + OPPh_3$$

$RhCl(PPh_3)_3$ is the most effective catalyst, but $RhX(PPh_3)_{2 \text{ or } 3}$ (X = CN, OCN, SCN) can also be used. The concentration of added triphenylphosphine affects the rate of the reaction, and the ratio of the two products is sensitive to trace amounts of acid. The most probable paths for the formation of the products are shown in Scheme 5.2, and the suggested route to octan-2-one is detailed in Scheme 5.3.

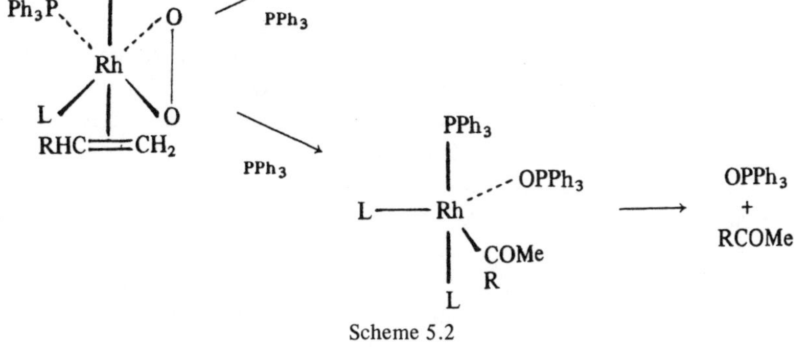

Scheme 5.2

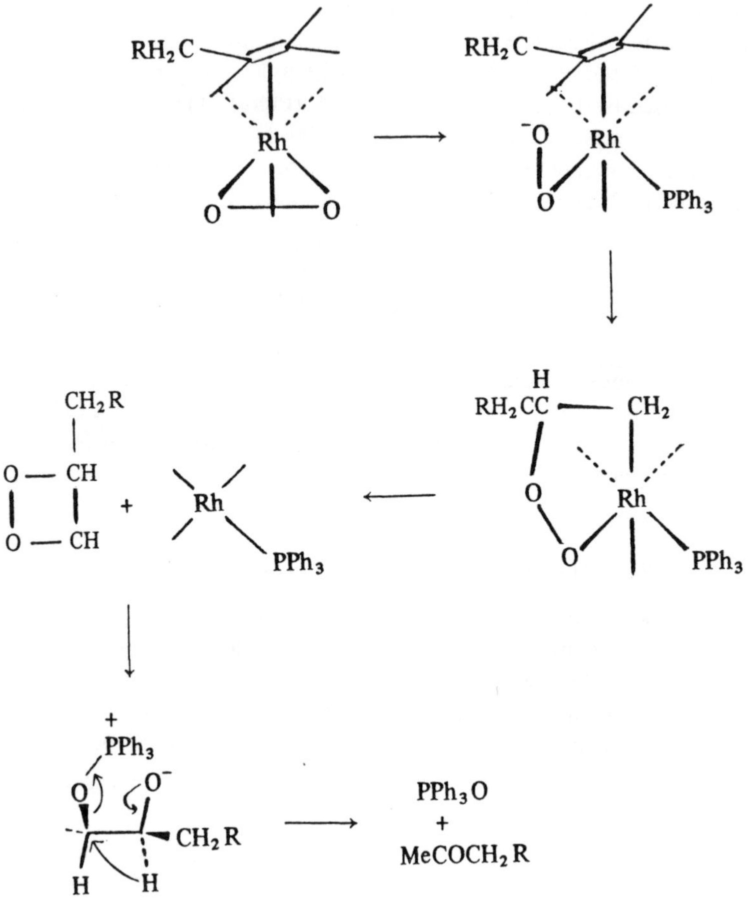

Scheme 5.3

Some homogeneous bimetallic catalysts have been used for the co-autoxidation of terminal alkenes and hydrocarbons with labile H atoms. A typical catalyst combination is $RhCl(PPh_3)_3$ and $MoO_5(HMPT)H_2O$ — the rhodium complex achieves oxidation of the hydrocarbon, and the molybdenum species assists the selective decomposition of the hydrocarbon hydroperoxides in the epoxidation step [81]. The hydrocarbons used include toluene and cumene, and typical alkenes are 1-hexene and 1-octene.

The reactions of vinyl esters with oxygen in the presence of $RhCl(PPh_3)_3$ sometimes involve $C=C$ bond cleavage. For example, styryl acetate in toluene yields benzaldehyde, acetic acid and methyl formate. The same reaction in ethanol produces benzyl acetate, methyl acetate and benzylmethylketone [1745]. Alkene substituents can affect the course of these reactions. The cleavage of C=C bonds also occurs in the oxygenation of cinnamaldehyde with RhCl(PPh$_3$)$_3$ as the catalyst. This reaction yields benzaldehyde, glyoxal, benzene and styrene [1747]. The decarbonylation reactions which occur in this system produce Rh(CO)Cl(PPh$_3$)$_2$ which is also an active oxygenation catalyst.

Some other substrates that have been oxidized in the presence of complexes of rhodium or iridium include:

alkylbenzenes, e.g. ethylbenzene (→ acetophenone, RhCl(PPh$_3$)$_3$ as catalyst) [184, 192];
diphenylmethane (→ benzophenone, Rh(CO)Cl(PPh$_3$)$_2$ as catalyst) [517];
arenes, e.g. 5-4; (with RhCl(PPh$_3$)$_3$ as the catalyst, 5-5 is formed in refluxing benzene and 5-6 with no solvent) [171];

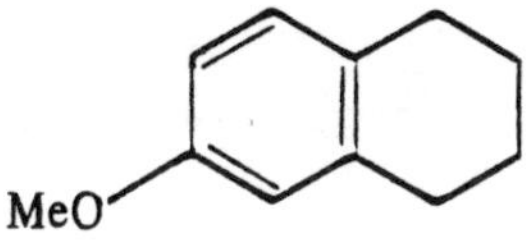

(5-4)

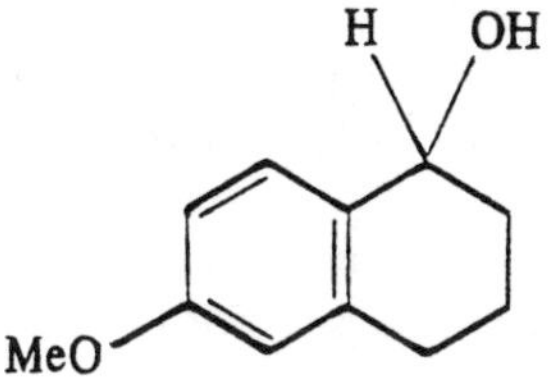

(5-5)

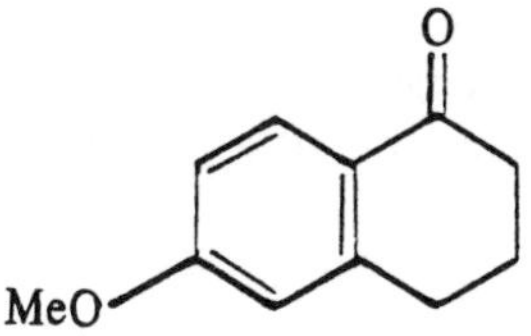

(5-6)

anthracene (→ anthraquinone, RhCl(PPh$_3$)$_3$ as catalyst; e.g. yield is 55% in dioxan at 70 °C for 4 d [1235] − reaction proceeds via solvent-derived hydroperoxides);
aldehydes (+ O$_2$ → nitriles such as acrylonitrile and crotonitrile, catalyst formed from [Rh(CO)$_2$Cl]$_2$ + NH$_3$) [1084];
benzaldehyde (→ benzoic acid and perbenzoic acid; order of catalytic activity is Rh(CO)ClL$_2$ > RhClL$_3$ > Ir(CO)ClL$_2$) [741];
ketones such as 2-butanone, cyclohexanone and 3-pentanone (→ corresponding carboxylic acids; acetophenone not oxidized; Rh$_6$(CO)$_{16}$ or Ir(CO)Cl(PPh$_3$)$_2$ as catalyst) [444, 1165, 1166];

tetrahydrofuran ($\rightarrow$ γ-butyrolactone, with $[\eta\text{-}C_5Me_5)_2 Rh_2(OH)_3]Cl \cdot 4H_2O$ in the presence of small amounts of water) [726];

cyclohexanol ($\rightarrow$ cyclohexanone $\rightarrow$ adipic acid, $Rh_6(CO)_{16}$; the effects of variation in the reaction conditions can be interpreted in terms of a lower nuclearity rhodium carbonyl as the reactive intermediate in the reaction cycle) [444, 1495];

ethanol ($\rightarrow$ acetaldehyde, slow; $[Rh(CO)_2Cl]_2$ as catalyst) [158];

sulfoxides ($\rightarrow$ sulfones; catalyst formed from $HIrCl_2(Me_2SO)_3$ in aqueous PrOH) [702];

dioxan ($\rightarrow$ di- and mono-formates of ethane-1,2-diol; catalyst source, $\{Rh\text{-}(acetate)_2 + LiCl\}$) [702];

CO ($\rightarrow$ CO_2; $Rh_6(CO)_{16}$ as catalyst $-$ solvent has a decisive effect) [320, 1165, 1166];

tertiary phosphines and arsines ($Rh_6(CO)_{16}$ as catalyst, e.g. $PPh_3 \rightarrow OPPh_3$, $Rh_6(CO)_{16}$ is re-formed when CO is added [1496]; the oxidation of phosphines is also accomplished with CO_2 in refluxing decalin with $RhCl(PPh_3)_3$ or $[Rh(C_8H_{14})_2Cl]_2$ as catalyst [1257] $-$ the rate of oxidation increases in the order $PPh_3 < PBuPh_2 < PEt_3$).

General discussions of these types of oxidation reactions are presented in recent reviews [1094, 1185, 1581].

5.2. Other Means of Forming C—O Bonds

Several reactions which involve the formation of new C—O bonds have H_2O or ROH as the source of [O]. A simple example is the hydration of acetylene (Equation (5.4)) which is catalysed by the complexes $[RhCl_a(H_2O)_{6-a}]^{(a-3)-}$ [861].

$$HC\equiv CH \xrightarrow{H_2O} \text{'}H_2C=CH(OH)\text{'} \longrightarrow CH_3CHO \qquad (5.4)$$

The corresponding reactions involving primary or secondary alcohols rather than water give acetals [812]. A similar reaction occurs between butadiene and alcohols in the presence of $RhCl_3 \cdot 3H_2O$ (see Equation (5.5)) [937].

$$H_2C=CHCH=CH_2 \xrightarrow{ROH} CH_3\overset{\overset{\displaystyle OR}{|}}{C}HCH=CH_2 \text{ (e.g. R = Me, Bu}^t\text{, allyl) (5.5)}$$

Binuclear allyl-rhodium intermediates are implicated in these reactions. Aryl-2,7-alkadienyl ethers are formed in the reaction of butadiene with phenol in the presence of the catalyst system $\{RhCl(PPh_3)_3 + MeMgI\}$ [437, 438].

There are some examples of the acetylization of ketones and aldehydes. Thus, the selective acetalization of steroid ketones is catalysed by $RhCl(PPh_3)_3$ (Equation (5.6)) [1867]. The conversion of crotonaldehyde to the dimethyl-acetal is achieved with $[Rh(CO)_2Cl]_2$ in methanol [735]; the rhodium catalysed reaction can be differentiated from related acid promoted addition reactions.

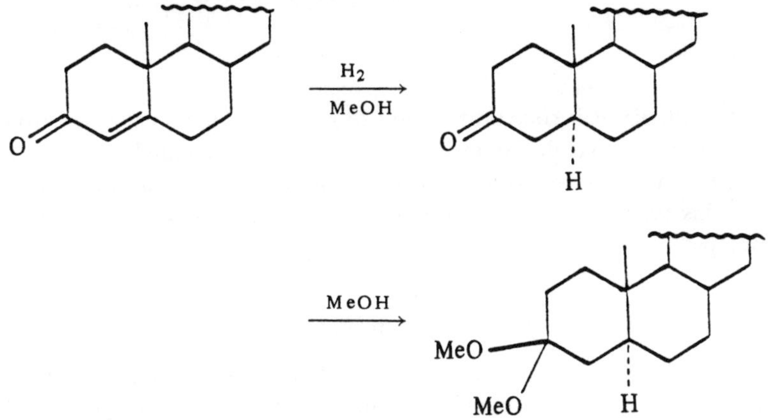

(5.6)

The hydration of nitriles to carboxamides has been achieved with $Rh(CO)$-$(OH)(PPh_3)_2$ as the catalyst [142]. This reaction is thought to proceed according to Equation (5.7):

$$RhOH \xrightarrow{RCN} RhN\!=\!\overset{\displaystyle OH}{\overset{|}{C}}R \longrightarrow RhNH\overset{\displaystyle O}{\overset{\|}{C}}R \xrightarrow{H_2O} RCONH_2 \quad (5.7)$$

Other reactions of a similar type are the conversion of acrolein to ethyl acrylate with $Rh(C_2H_4)_2(acac)$ in ethanol [834], the formation of $ROSiHMeR'$ from polysilanes and ROH in the presence of $Rh(CO)XL_2$ [1245], the production of vinyl acetate from ethylene with $\{RhCl_3(PPh_3)_3 + LiCl + Li(O_2CMe)\}$ as the catalyst source and acetic acid as the solvent [1652], and the formation of a mixture of $MeCOCH(CH_2CH\!=\!CMe_2)CO_2Et$ and $MeCOCH(CH_2CH_2CMe\!=\!CH_2)$ when isoprene and $MeCOCH_2CO_2Et$ are heated with $[Rh(1,5\text{-hexadiene})Cl]_2$, $P[3\text{-}(NaO_3S)C_6H_4]_3$ and Na_2CO_3 in water [1221].

CHAPTER 6

FUNCTIONAL GROUP REMOVAL

In organic synthesis, it is sometimes useful to remove a functional group from a molecule. For example, decarbonylation is a convenient method of removing formyl groups in steroid molecules. There are no large-scale commercial applications of this type of reaction, but numerous laboratory-scale syntheses have been developed.

6.1. Decarbonylation

The stoichiometric decarbonylation of acid halides is achieved by treatment with $RhCl(PPh_3)_3$ in a high boiling solvent such as benzonitrile [182, 187, 190, 193, 910, 1666, 1798, 1799, 1801, 1803]:

$$RCOX + RhCl(PPh_3)_3 \longrightarrow RX + Rh(CO)Cl(PPh_3)_2 + PPh_3$$

The $Rh(CO)Cl(PPh_3)_2$ which is formed in this reaction also decarbonylates acid halides [1293, 1705, 1795], but it is less effective than $RhCl(PPh_3)_3$. With PhCOBr and $PhCH_2COCl$ in the presence of $Rh(CO)X(PPh_3)_2$, decarbonylation starts at about 170 °C and becomes very rapid over the range 200–230 °C. Under these conditions, it is possible that the $Rh(CO)Cl(PPh_3)_2$ loses CO to form $RhClL_2$ and that this is the active decarbonylation catalyst. Although the product of the catalytic decarbonylation of an aliphatic acyl halide is generally an alkyl halide, an alkene can be obtained if the acid chloride contains β-hydrogens [187, 1798, 1799, 1803] (see, for example, Equation (6.1)).

$$PhCH_2CH_2COCl + RhCl(PPh_3)_3 \longrightarrow PhCH{=}CH_2 + HCl + \\ + Rh(CO)Cl(PPh_3)_2 \quad (6.1)$$

There is a report [1328] that aryl fluorides are obtained by the catalytic decarbonylation of aroyl fluorides. However, a detailed re-investigation [483] of this reaction using ^{18}F-labelled compounds has established that the product is ArH rather than ArF.

The mechanism of the decarbonylation reaction has been determined by Stille *et al.* [1060, 1662, 1664, 1665] and is shown in Scheme 6.1. When $Rh(CO)X(PPh_3)_2$ is the catalyst source, the rate of the decarbonylation reaction

168

depends on the nature of the X ligand in the rhodium complex; the order is Br > Cl > I [1705]. Considerable attention has been given to the detail of the reaction steps in this scheme.

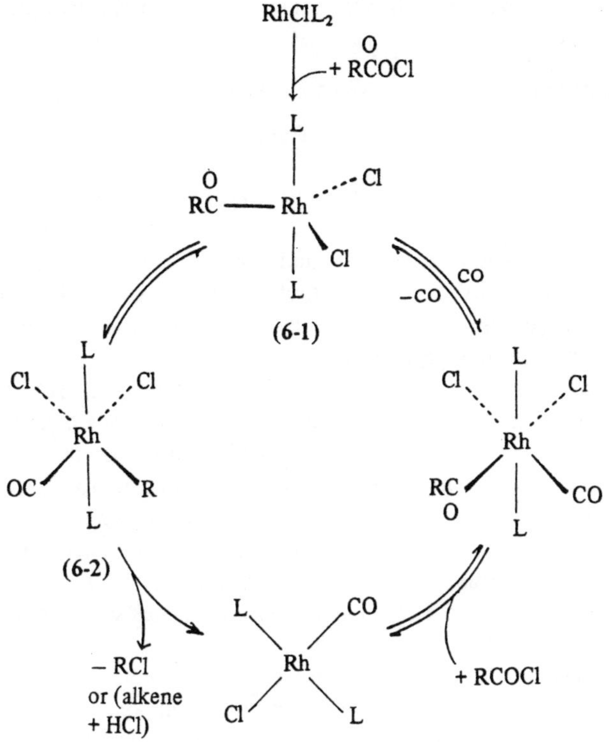
RhClL₂
O
+ RCOCl
L
O
RC
Rh
Cl
Cl
L
(6-1)
CO
–CO
L
Cl
Cl
Rh
RC
O
CO
L
L
Cl
Cl
Rh
OC
R
L
(6-2)
L
CO
Rh
Cl
L
– RCl
or (alkene
+ HCl)
+ RCOCl

Scheme 6.1

Both the acyl- and alkyl- (or aryl-) rhodium(III) complexes (**6-1** and **6-2**, respectively) have been isolated as intermediates in the decarbonylation of *p*-substituted benzoyl and phenylacetyl chlorides [1665]. The kinetics for the rearrangement of the acyl to the alkyl (or aryl) complex, and of the alkyl (or aryl) complex to RX and $Rh(CO)Cl(PPh_3)_2$, have been measured. Further, it has been established [452] that the equilibrium between complexes of the type **6-1** and **6-2** depends on the nature of the migrating group R. When R = Me, migration onto the carbonyl predominates to give the acyl complex. However, the reaction proceeds in the opposite direction to yield mainly the aryl complex when R = Ph.

Decarbonylation with $RhCl(PPh_3)_3$ of (*S*)-(–)$CF_3CHPhCOCl$ gives racemic α-$CF_3CHPhCl$ [1662]. The optically active form of the product is not racemized under the conditions of the decarbonylation reaction, and so racemization occurs either during the rearrangement of the acyl— to the benzyl—rhodium complex, or in the degradation of the benzyl complex to benzyl chloride. The former pathway is preferred since desulfonylation of a related (S)-α-trifluorobenzyl—rhodium complex occurs with 62% net conversion in the two steps shown in Equation (6.2).

$$Rh(CO)Cl(PEt_2Ph)_2 + (S)\text{-}\alpha\text{-}CF_3CHPhSO_2Cl$$

$$\longrightarrow \quad Rh(CO)(SO_2PhCHCF_3)Cl_2(PEt_2Ph)_2$$

$$\xrightarrow{-SO_2} \quad Rh(CO)(CHCF_3Ph)Cl_2(PEt_2Ph)_2 \quad (6.2)$$

The decarbonylation of *erythro-* and *threo*-2,3-diphenylbutanoyl chlorides gives exclusively *trans-* and *cis*-methylstilbenes, respectively [1664]. Two mechanistic sequences have been considered to account for the observed stereospecificity. In one, the acyl → alkyl rearrangement occurs with retention of configuration at carbon and there is then a *cis*-β-elimination. The other involves an acyl → alkyl rearrangement with inversion followed by a *trans*-β-elimination. The retention, *cis*-elimination pathway is favored in this system.

Styrene-d_0, -d_1 and -d_2 have been obtained from the decarbonylation with RhCl(PPh$_3$)$_3$ of *erythro-* and *threo*-2,3-dideuterio-3-phenylpropionyl chloride [475]. The nonstereospecific nature of this reaction is consistent with a reversible β-elimination mechanism involving the alkyl intermediate **6-1** (R = PhCHDCHD).

The decarbonylation of (S)-(+)-α-deuteriophenylacetyl chloride with RhCl-(PPh$_3$)$_3$ yields (S)-(+)-benzyl-α-deuterio chloride with an overall stereospecificity of 20—27% [1060]. With 3,3-dideuterio-3-perdeuteriophenylpropionyl chloride, stoichiometric decarbonylation gives a mixture of all possible d_0, d_1, d_2 and d_3 perdeuteriophenylethylenes, but catalytic decarbonylation gives exclusively 1-deuterio-1-perdeutereophenylethylene. Inter- and intra-molecular hydride-transfer mechanisms account for the hydrogen isotope exchange, and the overall results favor a mechansim in which a rapid equilibrium between the acyl— and alkyl—rhodium complexes is established initially, followed by a concerted *cis*-elimination of Rh—H. The latter is the rate-limiting step.

Both RhCl(PPh$_3$)$_3$ [105, 1421, 1797, 1800, 1888] and Rh(CO)Cl(PPh$_3$)$_2$ [1293, 1804, 1805, 1887] are also good catalysts for the decarbonylation of aldehydes. For example, the reaction with CH$_3$CH$_2$CDO in benzene at 75 °C gives CH$_3$CH$_2$D [1421]. With RhCl(PPh$_3$)$_3$ as the catalyst, it is possible to decarbonylate unsaturated aldehydes (in PhCN at 160 °C) without affecting the double bond (see Equation (6.3)) [1293].

$$\underset{Ph}{\overset{H}{\diagdown}}C=C\underset{Me}{\overset{CHO}{\diagup}} \quad \longrightarrow \quad \underset{Ph}{\overset{H}{\diagdown}}C=C\underset{Me}{\overset{H}{\diagup}} \quad (6.3)$$

A particularly interesting application of this type of reaction is the decarbonylation of an intermediate polycyclic aldehyde in the synthesis of α-linked disaccharides [805].

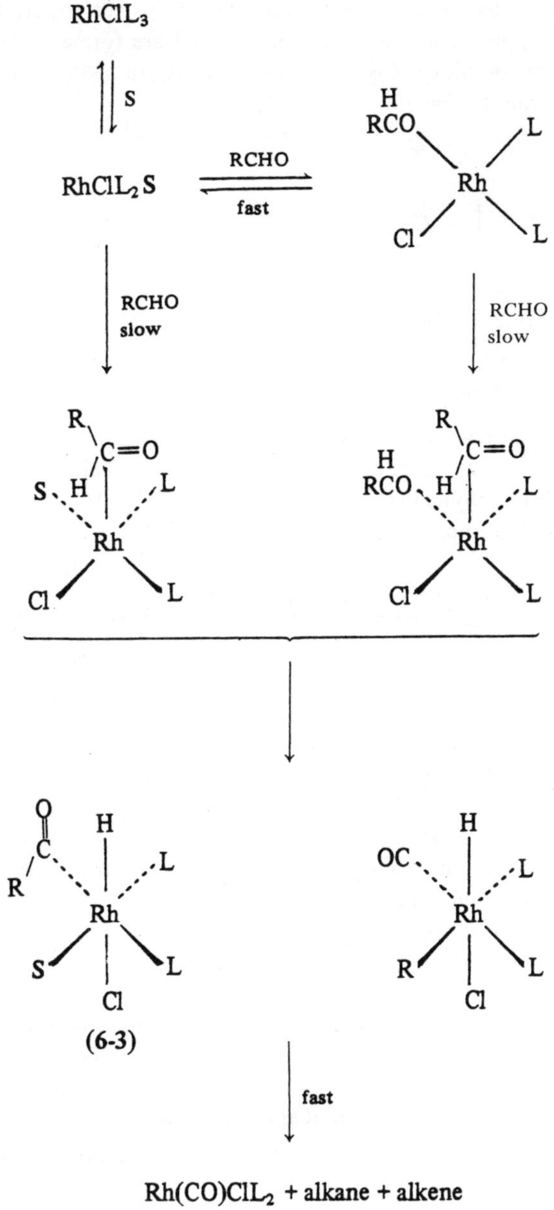

Scheme 6.2

The kinetics [105] for the decarbonylation of BuCHO with $RhCl(PPh_3)_3$ in CH_2Cl_2 at 20 °C are consistent with the mechanism outlined in Scheme 6.2. The possible involvement of free-radical intermediates in the pathway [1888] has been negated by deuterium-labelling and cross-over experiments [909]. There is some support that complexes of type **6-3** are formed from aldehydes. Thus, the iridium complex **6-4** is obtained when $Ir(CO)Cl(PPh_3)_2$ is treated with *o*-diphenylphosphinobenzaldehyde [1457].

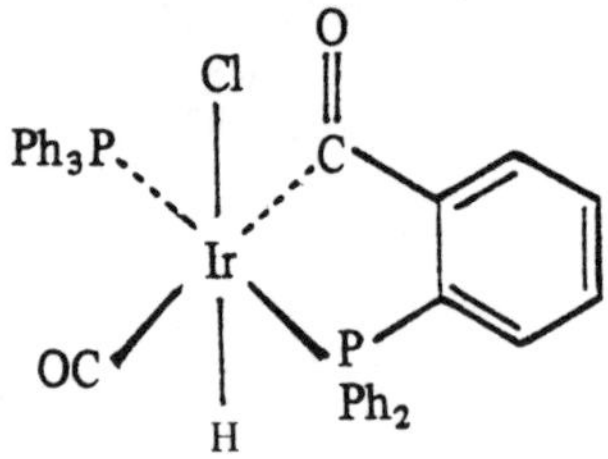

(6-4)

The stereochemistry of some of these reactions has been studied [1800, 1887, 1888] and there is predominant retention of optical activity. The decarbonylation of (*E*)-α-ethylcinnamaldehyde occurs with 100% retention of optical activity. A further example is shown in Equation (6.4).

$$\text{Et} \blacktriangleright \underset{\overset{|}{Ph}}{\overset{\overset{Me}{|}}{C^*}} \blacktriangleleft \text{CHO} \quad \xrightarrow[160\,°C]{RhCl(PPh_3)_3,\ PhCN,} \quad \text{Et} \blacktriangleright \underset{\overset{|}{Ph}}{\overset{\overset{Me}{|}}{C^*}} \blacktriangleleft \text{H} \qquad (6.4)$$

(81% optical yield)

Cationic complexes of the type $[Rh(L—L)_2]^+$ (e.g. $L—L = Ph_2P(CH_2)_nPPh_2$, $Ph_2PCH{=}CHPPh_2$) have been used to catalyse the decarbonylation of aldehydes such as PhCHO [455, 456, 457]. The complex $[Rh(Ph_2PCH_2CH_2CH_2PPh_2)_2]^+$ is a particularly effective catalyst with an activity exceeding that of $RhCl(PPh_3)_3$. With this catalyst, the decarbonylation of (−)-(*R*)-2-methyl-2-phenylbutanal is stereospecific under mild conditions.

Other organic substrates that have been decarbonylated by $RhCl(PPh_3)_3$ include:

ketones: with simple ketones, C—C bond cleavage sometimes occurs, e.g. acetone → $CH_4 + Rh(CO)Cl(PPh_3)_2$ [33]; alkynes are obtained from ketones of the type $RCOC{\equiv}CR'$ [1233, 1234]; some diketones give monoketones, e.g. acetylacetone → methylethylketone in refluxing toluene; the reactions with β-diketones can be complicated by the formation of stable chelate complexes [911].

acylphosphines: $RCOPPh_2 \longrightarrow RPPh_2 + PPh_3 + Rh(CO)Cl(PPh_3)_2$ (e.g. R = Me, CF_3) at 120 °C, xylene [1081].

ketenes: $Ph_2C\!=\!C\!=\!O \longrightarrow Ph_2C\!=\!C\!=\!CPh_2$ [755].

acid anhydrides: e.g. benzoic anhydride → fluorenone at 240 °C [188, 189].

secondary amides: $RCONHR' \longrightarrow RCN$ at 250–285 °C (e.g. *N*-benzyl-benzamide → benzonitrile), $ArCONH(CH_2R') \longrightarrow ArCN + R'CH_2OH$ [186].

allylic alcohols: $RCH\!=\!CHCH_2OH \longrightarrow RCH_2CH_3$ (major) + $RCH\!=\!CH_2$ (minor) at 110–115 °C in acetonitrile or benzonitrile (e.g. *cis*-but-2-en-1,4-diol → 1-propanol) [488].

With preformed $Rh(CO)Cl(PPh_3)_2$, the decarbonylation of fatty acids to alkenes (e.g. stearic acid → heptadecenes) [524], and of benzoic anhydride to fluorenones [189], has been achieved. The corresponding iridium compound, $Ir(CO)Cl(PPh_3)_2$, has also been used in some decarbonylation reactions. Aliphatic acid halides such as undecanoyl chloride give a mixture of alkene isomers (*cf.* $RhCl(PPh_3)_3$ and $Rh(CO)Cl(PPh_3)_2$ → terminal alkene selectively [187]), and 2,6-diphenyl-4-carboxypyrylum perchlorate (**6–5**) yields the stable cation radical **6-6** [151].

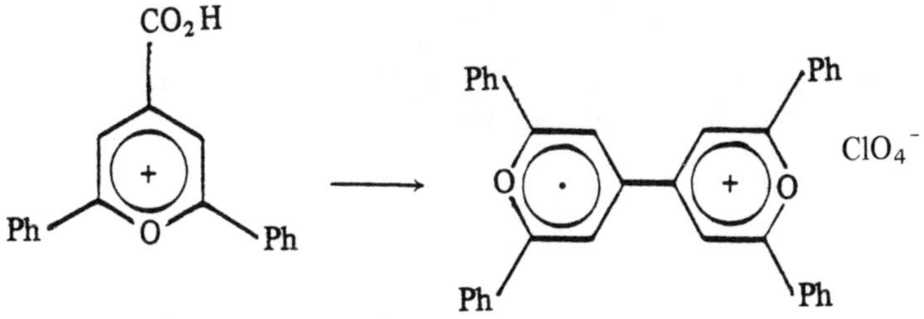

(6-5)   **(6-6)**

The decarbonylation of metal carbonyls can be achieved with $RhCl(PPh_3)_3$ and other rhodium(I) complexes. For example, treatment of $Mo(CO)_6$ in toluene at 80 °C with $RhCl(PPh_3)_3$ results in the rapid formation of phosphine-substituted complexes as shown in Equation (6.5) [1849].

$$Mo(CO)_6 \longrightarrow [Mo(CO)_5] \longrightarrow [Mo(CO)_4]$$

$$Mo(CO)_5(PPh_3) \qquad cis\text{-}Mo(CO)_4(PPh_3)_2 \qquad (65)$$

Similar reactions occur with $Fe(CO)_5$. In another recent example of this type of reaction, terminal CO and CS groups are both removed from the complex $[(\eta\text{-}C_5H_5)Fe(CO)_2(CS)][PF_6]$ [1015]. The decarbonylation of acyl–iron complexes of the type $(\eta\text{-}C_5H_5)Fe(CO)_2(COR)$ to give $(\eta\text{-}C_5H_5)Fe(CO)_2R$ is also effected by $MCl(PPh_3)_a$ complexes (M = Rh, Ir) [39, 40]; the use of $[RhCl(PPh_3)_2]_2$ in acetonitrile avoids the complication of PPh_3 transfer to the iron atom [1016]. In a related reaction with $(\eta\text{-}C_5H_5)Fe(CO)_2$ {trans-$C(O)CH{=}CHR$}, the vinyl complex $(\eta\text{-}C_5H_5)Fe(CO)_2(CH{=}CHR)$ is preferentially formed [1453].

Some rhodium(III) compounds have been used in decarbonylation reactions, and these include:

$RhCl_3(PEt_2Ph)_3$; saturated carboxylic acids $\longrightarrow$ alkenes (e.g. hexanoic acid $\longrightarrow$ 2-pentene) [1420].

$RhCl_3(AsPh_3)_3$ in the presence of PPh_3; aliphatic acid halides $\longrightarrow$ terminal alkenes selectively [187].

6.2. Desulfonylation

Arene sulfonyl halides are converted to aryl halides in the presence of $RhCl(PPh_3)_3$ or $Rh(CO)Cl(PPh_3)_2$ [183, 194]; thus

$$PhSO_2Cl \longrightarrow PhCl$$

In the corresponding reaction with $Ir(CO)Cl(PPh_3)_2$, the steps shown in Equation (6.6) have been identified [194].

$$Ir(CO)Cl(PPh_3)_2 + ArSO_2Cl$$
$$\longrightarrow \quad Ir(CO)(SO_2Ar)Cl_2(PPh_3)_2$$
$$\xrightarrow{-PPh_3} \quad Ir(CO)(SO_2Ar)Cl_2(PPh_3)$$
$$\longrightarrow \quad Ir(CO)(Ar)Cl_2(SO_2)(PPh_3) \qquad (6.6)$$

With the rhodium compounds, the final stage of the reaction sequence is presumably reductive-elimination of the aryl chloride.

6.3. Deoxygenation, including the Decomposition of Hydroperoxides

The catalytic decomposition of alkyl hydroperoxides to the appropriate alkanol is achieved with $RhCl(PPh_3)_3$ [164, 202] or $Ir(CO)Cl(PPh_3)_2$ [202]. The corresponding decomposition of cyclohexenyl hydroperoxide with $RhCl(PPh_3)_3$ in benzene yields cyclohexenol, cyclohex-1-en-3-one and polymers. A kinetic study of this reaction indicates a free-radical process [82]. The decomposition of di-t-butyl peroxide is induced by $RhX(PPh_3)_3$ (X = Cl, Br) at 120 °C under H_2 at 600 psi [950]. t-Butanol and acetone are formed in the reaction via a free-radical pathway.

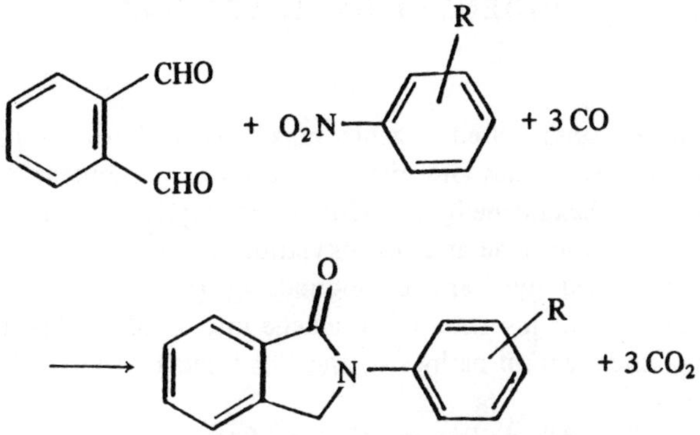

A complex deoxygenation reaction (Equation (6.7)) occurs when o-phthal-aldehyde is treated with aromatic nitro compounds under CO pressure. This reaction is catalysed by $Rh_6(CO)_{16}$ [830].

CHAPTER 7

OLIGOMERIZATION, POLYMERIZATION AND RELATED CONDENSATION REACTIONS

Some reactions of unsaturated organic substrates lead to the formation of new C—C bonds. Examples are the dimerization of ethylene to 1-butene, the formation of 1,4-hexadiene by the addition of ethylene to 1,4-butadiene, the polymerization of norbornene and the alkylation of ketones. These and related reactions are catalysed by many compounds of transition metals, including rhodium or iridium. The insertion of an alkene into an M—C bond plays a key role in many of the reaction pathways, and this topic has been reviewed [707].

7.1. Alkene Dimerization and Polymerization

The dimerization of ethylene to linear butenes is catalysed by $RhCl_3$ [36, 37, 369, 1325], $IrCl_3$ [36], $\{RhCl_3 + a$ reductant such as $LiAlH_4 + PR_3\}$ [819], $[Rh(C_2H_4)_2Cl]_2$ [1325] and $\{Rh_2Cl_2(SnCl_3)_4 + HCl + L\}$ [208]. A mechanism for the reaction with $RhCl_3$ as the catalyst source is shown in Scheme 7.1. Kinetic studies [369, 372, 1325] have established that formation of the butyl—rhodium(III) complex is the rate-determining step. Although 1-butene is formed initially, this can be isomerized to 2-butene; the isomerization reaction is retarded by high pressures of ethylene [1325]. A supported catalyst has also been used for the dimerization of ethylene to cis- and trans-2-butenes. The catalyst is obtained by the immobilization of $[RhCl_4(SnCl_3)_2]^{3-}$ on 'AV-17-8' anion-exchange resin [71]. The optimum yield (88%, 1:2.7 cis/trans ratio) is obtained after 6h at 75 °C and 42 atm pressure.

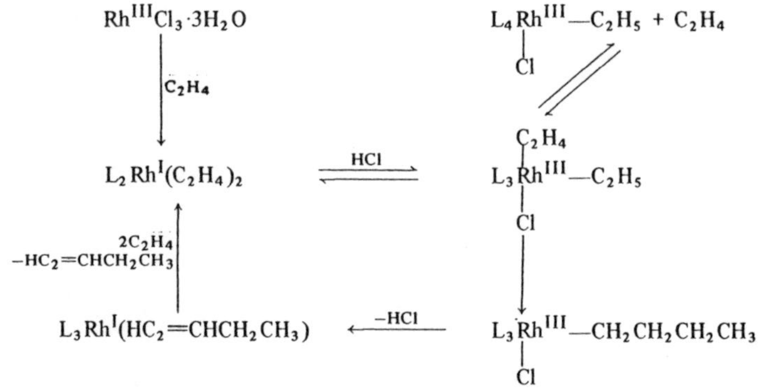

$$Rh^{III}Cl_3 \cdot 3H_2O \qquad\qquad L_4Rh^{III}\!-\!C_2H_5 + C_2H_4$$

$$\Big\downarrow C_2H_4 \qquad\qquad\qquad \overset{|}{Cl} \;/\!/$$

$$\qquad\qquad\qquad\qquad C_2H_4$$

$$L_2Rh^{I}(C_2H_4)_2 \xrightarrow{\;HCl\;} L_3Rh^{III}\!-\!C_2H_5$$

$$\qquad\qquad\qquad\qquad\qquad \overset{|}{Cl}$$

$$\Big\uparrow \overset{2C_2H_4}{\underset{-HC_2=CHCH_2CH_3}{}}$$

$$L_3Rh^{I}(HC_2=CHCH_2CH_3) \xleftarrow{\;-HCl\;} L_3Rh^{III}\!-\!CH_2CH_2CH_2CH_3$$

$$\qquad\qquad\qquad\qquad\qquad\qquad \overset{|}{Cl}$$

Scheme 7.1

Other alkenes undergo oligomerization reactions. A catalyst formed from $\{RhCl_3$ or $IrCl_3 + (EtAlCl_2)_2\}$ has been used for the dimerization of propylene [1387], and the rate of this reaction is increased by the addition of triphenyl-phosphine and related ligands. The oligomerization of branched alkenes such as 2-methylpropene is catalysed by the cationic nitrosyl complex $[Rh(NO)-(NCMe)_4]^+$ [344]. An interesting application of selective oligomerization involves the removal of methyl linoleate (which oligomerizes) from methyl oleate (which does not). The catalyst used for this separation is formed from $[Rh(C_8H_{14})_2Cl]_2$ and an organotin compound such as Bu_3SnCl [1605].

There are some examples of the codimerization of alkenes. The reaction of ethylene with propylene is catalysed by $RhCl_3$ in chlorinated solvents, with the rate in $CHCl_3$ being faster than that in CCl_4 [945]. Both $RhCl_3 \cdot 3H_2O$ and $[Rh(C_2H_4)_2Cl]_2$ are suitable catalysts for the codimerization of styrene with lower alkenes such as ethylene, propylene and 1-butene [1815]. The reaction between $H_2C=CHCH_2CO_2K$ and vinyl halides such as (E)-$PhCH=CHBr$ is catalysed by $RhCl(PPh_3)_3$ in EtOH at 85 °C [324]; conjugated dienoic acids (e.g. (E, E) − and (E, Z)-$Ph(CH=CH)_2CH_2CO_2H$) are obtained.

Some Ziegler-type catalysts for the low-pressure polymerization of alkenes have been obtained from compounds of rhodium and iridium. Appropriate catalyst combinations are $\{RhCl_3, Rh(acac)_3$ or $[Rh(CO)_2Cl]_2 + (EtAlCl_2)_2\}$ [1008] and $\{IrX_a + Et_3Al_2Cl_3$ or $Bu^i_3Al\}$ [470, 1272]. Another catalyst has been obtained by treatment of 7-1 (R, R^1, R^2 = H or a hydrocarbon group) with $(Et_3Al)_2$ [103].

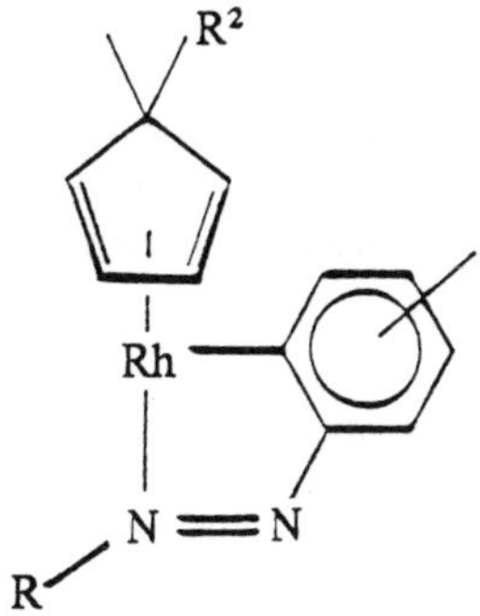

(7-1)

Diverse rhodium and iridium compounds have been used as catalysts for the polymerization of the following substrates:

styrene: radical polymerization initiated by $Ir(CO)Cl(PPh_3)_2$ and t-BuOOH [903].

$RCH=CH_2$: catalyst formed from $\{RhCl_3 \cdot 3H_2O) +$ sodium alkylarene sulfonate$\}$ [1582].

$RCH=CH_2$: polymerization or copolymerization, $Rh(CO)_2(acac)$ or Rh-$(acac)_3$ in polar solvents [896] or $\{Rh(COD)ClL$ (L = 4-picoline, p-toluidine or PPh_3) + a peroxide$\}$ [772] as catalyst.

acrylonitrile: anionic mechanism, catalyst obtained from $Ir(CO)Cl(PPh_3)_2$ + $AgClO_4$ [1364].

methylmethacrylate: catalyst obtained from $\{Rh(CO)Cl(PPh_3)_2 + CCl_4,$ C_6H_5Cl or other RX$\}$ [906]; or from $\{HM(CO)(PPh_3)_3$ (M = Rh [904] or Ir [905]) + CCl_4 or related RX$\}$.

7.2. Diene Oligomerization and Polymerization

The oligomerization of butadiene has given several products depending upon the catalyst used. With a catalyst formed from $RhCl_3$ and potassium acetate in ethanol, butadiene is transformed to 2,4,6-octatriene at 100 °C [36, 37]. A mixture of 3-methyl-1,4,6-heptatriene, 4-vinylcyclohexane, 1,5-cyclooctadiene, and various trimers and higher polymers is obtained when the catalyst source is $\{RhCl_3 + H_2C{=}CHCN\}$ in ethanol [790]. Similarly, with $RhCl_3 \cdot 3H_2O$ in Me_3COH as the catalyst source, butadiene is converted to 4-vinylcyclohexene, 1,5-cyclooctadiene, cyclododecatriene and polymers [1427]. With catalysts of the type $Rh(CO)_2ClL$ (e.g. L = acridine or benzotriazole), butadiene forms 4-vinylcyclohexane, 1,5-cyclooctadiene and 1,4-*trans*-polybutadiene; the actual products obtained depend on the catalyst and the solvent [313]. 4-Vinyl-cyclohexene and 1,5-cyclooctadiene are the major products obtained from butadiene in DMF or pyridine with catalysts formed from $Rh(CO)Cl(PPh_3)_2$ and Et_2AlCN [997]. The oligomers obtained from butadiene in allylic alcohol with $Rh(NO_3)_3$ and related compounds as the catalyst source have carbonyl and hydroxyl functionalities [79].

On supported catalysts of type **7-2**, the sequential cyclooligomerization and hydrogenation of butadiene gives products such as ethylcyclohexane, cyclooctane and cyclododecane [1406]. Similar reactions with catalysts such as **7-3** result in sequential cyclooligomerization and hydroformylation.

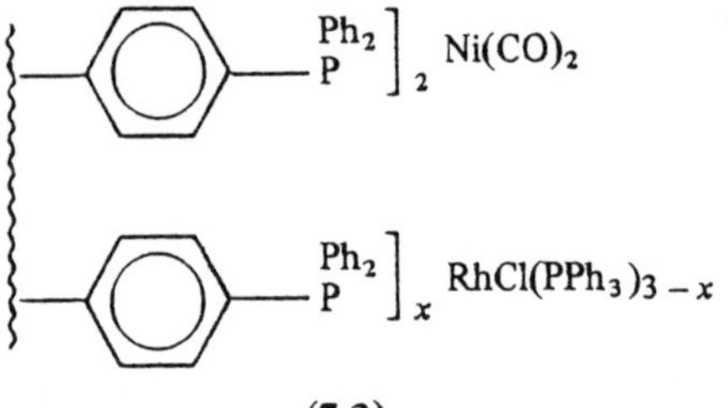

(7-2)

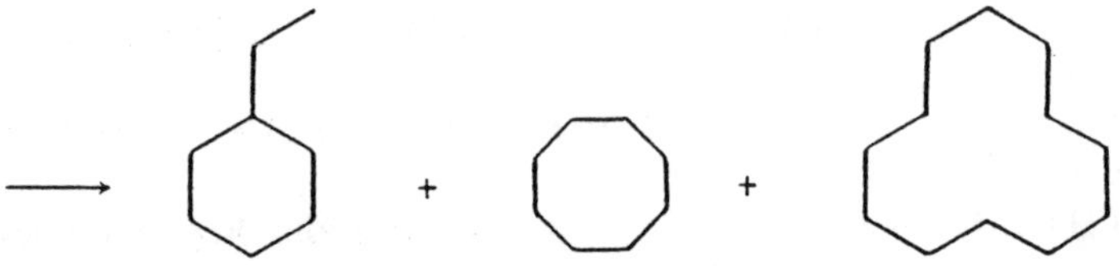

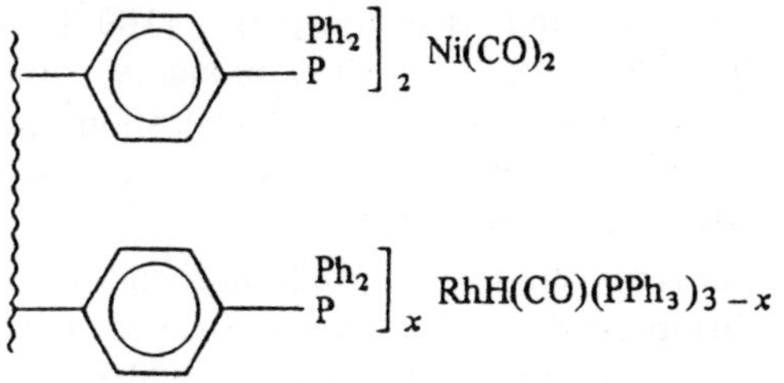

(7-3)

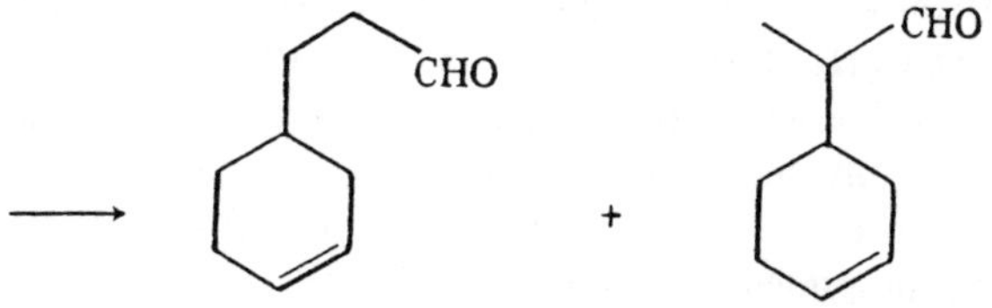

The stereospecific polymerization of budadiene to *trans*-1,4-polybutadiene is achieved in aqueous media with catalyst sources such as $RhCl_3 \cdot 3H_2O$, $Rh(NO_3)_3$, $[Rh(CO)_2Cl]_2$ and $IrCl_3 \cdot xH_2O$ [276, 1181, 1182, 1479, 1582, 1627]. The influence of factors such as catalyst concentration and emulsifier structure have been determined [399]. Tracer studies with T_2O have established that tritium is incorporated into the polymer [99], and that water is involved in both the initiation and termination steps [1637]. It seems likely that the active catalyst in these reactions is a metal(I) complex with a coordinated diene ligand [1771]. Certainly, the polymerization reaction is strongly accelerated by the addition of 1,3-dienes such as cyclohexadiene [1772, 1773], and preformed diene—rhodium(I) complexes including $[Rh(COD)Cl]_2$ and $[Rh(NBD)Cl]_2$ [100. 276, 1475, 1946] are active catalysts. Allyl—rhodium species may be involved in the reaction pathway [998, 1952]. The addition of pyridine, PPh_3, or other Lewis bases to these systems reduces the yield of polymer, presumably by competing for a coordination site in the active catalyst [1771, 1773]. The catalysts obtained from $RhCl_3$ or $[Rh(1,4\text{-hexadiene})Cl]_2$ also assist the polymerization of conjugated, fluorinated dienes such as 1,1,2-trifluoro-3-chlorobutadiene [1626]. However, the polymerization and cyclic dimerization of 2-chloro-1,3-butadiene is not readily catalysed by rhodium compounds [215]. The polymerization of isoprene is achieved with $[Rh(COD)Cl]_2$ as the catalyst source, and the active catalyst is thought to be a π-allyl complex [1139].

Some butadiene polymerization catalysts have been formed by the treatment of rhodium compounds with aluminum alkyls. For example, *trans*-1,4-polybutadiene is obtained in the presence of $\{RhCl_3, Rh(acac)_3, [Rh(COD)Cl]_2$ or $Rh_a(CO)_b + (Et_2AlCl)_2\}$ [863, 1856, 1951, 1953, 1954]. Other catalysts have been obtained by adding $(Et_2AlCl)_2$ to allyl—rhodium complexes including

Rh(allyl)$_3$, Rh(allyl)(2-butenyl) and [Rh(allyl)$_2$Cl]$_2$ [1007]. The catalyst obtained from Rh(CO)Cl(PPh$_3$)$_3$ and Et$_2$AlCN produces polybutadiene of *cis*-1,4- and *trans*-1,4-configuration when the polymerization reaction is performed in ethanol [997]. Active catalysts are also obtained by the addition of AlCl$_3$, SnCl$_4$, or better, HCl to Rh(1-methylallyl)$_3$ [1338].

The relative activities of some catalysts formed from complexes of rhodium(I) have been investigated. It appears that the catalyst is effective only if it has three 'active' ligands (e.g. Cl, CO, alkene), or possibily two if they are mutually *cis* [633]. Thus catalysts derived from [Rh(CO)$_2$X]$_2$ (X = Cl, Br), [Rh(CO)$_2$X$_2$]$^-$ and Rh(CO)$_2$(acac) are active, but [Rh(CO)$_2$(SCN)]$_2$, RhCl(PPh$_3$)$_3$, Rh(CO)-Cl(PPh$_3$)$_2$, (η-C$_5$H$_5$)Rh(COD) and (η-C$_5$H$_5$)Rh(C$_2$H$_4$)$_2$ are not effective catalyst sources. RhCl(PPh$_3$)$_3$ can be used as a hydrogenation catalyst to modify the unsaturated polymers obtained from 1,3-butadiene [1781].

Dienes of the type R$_2$C(CH$_2$CH=CH$_2$)$_2$ (e.g. R = COMe, CO$_2$Et) undergo cyclization in the presence of RhCl(PPh$_3$)$_3$ in boiling chloroform. The methylenecyclopentanes 7-4 are obtained in 60−90 % yields [624]. The co-oligomerization of dienes has been observed with catalysts formed from RhCl$_3$ and PPh$_3$ in ethanol [1219, 1220]. For example, AcOCH$_2$CH=CH(CH$_2$)$_3$CH=CH$_2$ and H$_2$C=CHCH=CH$_2$ give a mixture of AcOCH$_2$CH=CH(CH$_2$)$_3$C(=CH$_2$)-CH$_2$CH=CHMe, AcOCH$_2$C(CH$_2$CH=CHMe)=CHCH$_2$CH=CH$_2$ and AcOCH$_2$-CH=CHCH$_2$CH$_2$CH=CHCH$_2$CH$_2$CH=CHMe.

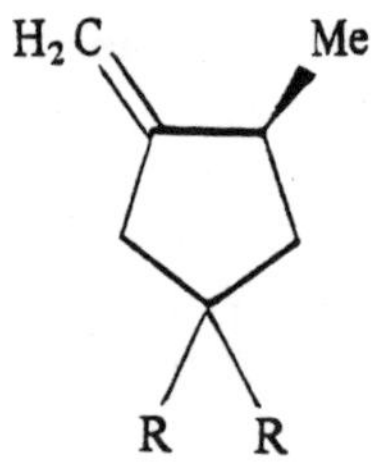

(7-4) 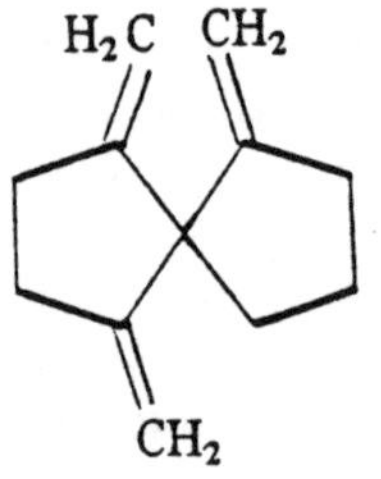   (7-5)

The tetramerization of allene to 1,4,7-trimethylenespiro[4.4]nonane (7-5) is catalysed by a complex formed from [Rh(C$_2$H$_4$)$_2$Cl]$_2$ and 2−4 equiv of PPh$_3$ [1351, 1353]. Crystalline polymers have also been obtained with { [Rh(C$_2$H$_4$)$_2$-Cl]$_2$ or [Rh(CO)$_2$Cl]$_2$ + PPh$_3$} as the catalyst [864], and soluble ladder spiro-polyallenes are obtained when the catalyst is [Rh(COD)Cl]$_2$ in benzene containing some PrCHO [1477]. Various other complexes of rhodium(I) (e.g. [Rh(CO)$_2$Cl]$_2$ or Rh(CO)$_2$ClL with L = PPh$_3$ or *m*-(H$_2$N)$_2$C$_6$H$_4$) [897, 1352, 1537, 1538] and Rh(CO)$_2$acac [1740]) or rhodium(III) (e.g. RhCl$_3$ · 3H$_2$O) [1214, 1352] have been used as allene polymerization catalysts. However, Rh(CO)Cl(PPh$_3$)$_2$ and RhCl(PPh$_3$)$_3$ are inactive. It has been shown that the polymerization of allene by complexes of rhodium(I) involves the formation of allylic–rhodium(III) complexes [593].

The cyclic diene norbornadiene is oligomerized with $RhCl(PPh_3)_3$ as the catalyst [19]. Four dimers (64% yield) and four trimers (11%) are obtained, and all are structurally complex.

7.3. Alkene–Diene Co-oligomerization

The formation of 1,4-hexadiene from ethylene and butadiene is catalysed by $RhCl_3$ [37, 371]. There is an induction period in this reaction which is probably associated with the reduction of the rhodium(III) chloride to a rhodium(I) complex. Certainly, the induction period is eliminated when preformed rhodium(I) compounds such as $[Rh(C_2H_4)_2Cl]_2$ [373, 1728, 1729, 1765], $(\eta\text{-}C_5H_5)\text{-}Rh(C_2H_4)_2$ [376, 378], $[Rh(C_4H_6)Cl]_2$ [1765] and $[Rh(\pi\text{-crotyl})Cl_2]_2 \cdot C_4H_6$ [1730] are used as the catalyst. *trans*-1,4-Hexadiene is the most desirable product of these reactions [1727] – it is used together with ethylene and propylene to form elastomers. There is a tendency for the *trans*-1,4-hexadiene to isomerize when it is produced from a 1:1 mole ratio of ethylene and butadiene, but high concentrations of butadiene suppress the isomerization reaction [1727]. Similar additions of other alkenes such as propene to 1,3-dienes are observed with some of the catalysts. When $[Rh(CO)_2Cl]_2$ is the catalyst, the reaction between ethylene and butadiene produces 1-*trans*,4-*trans*-8-decatriene [1766].

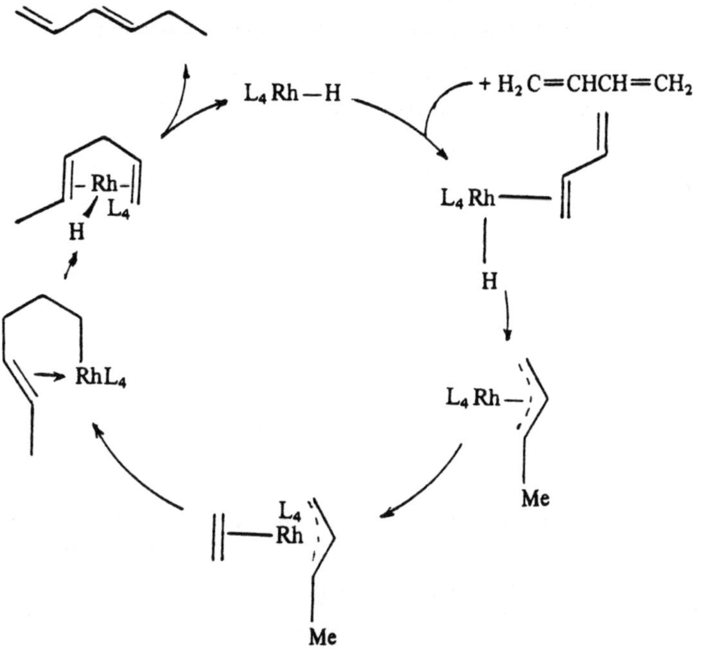

Scheme 7.2

The reaction involving ethylene and butadiene does not give dimers of the individual alkenes, and this is attributed to thermodynamic control [371]. Scheme 7.2 shows a pathway for this addition reaction, and it is proposed that the relatively high stability of the η^3-crotyl–rhodium(III) complex leads to the observed specificity of reaction. Comparison of Schemes 7.1 and 7.2 reveals an interesting difference. Thus, the oxidation number of rhodium is invariant during hexadiene synthesis, but a cyclic valency change is invoked for butene synthesis. The suggestion that rhodium(III) compounds are catalytically active in 1,4-hexadiene synthesis is supported by the observation that catalytic activity is maintained in the presence of excess crotyl chloride or $MeOCH_2Cl$ [1729]. With $[Rh(\pi\text{-crotyl})Cl_2]_2 \cdot C_4H_6$ as the catalyst, the addition of weak O-donors such as ethers, ketones or esters increases the rate of reaction but has no effect on the ratio of *trans*- to *cis*-1,4-hexadiene. The reaction rate is also increased by stronger donors such as ROH, H_2O, DMF and OPR_3, and these additives give increased selectivity in the formation of the *trans* product [1730].

The oligomerization of butadiene with allyl alcohol yields an oligomer with a predominantly *trans*-1,4-configuration terminated by —OH and —CO groups. Suitable catalysts for this reaction are formed by adding alkali metal salts of allyl alcohol to $RhCl_3$ or $Rh(NO_3)_3$ [698, 1163]. The addition of butadiene to acetoxyoctadiene is catalysed by $[Rh(C_2H_4)_2Cl]_2$ [675].

7.4. Oligomerization and Polymerization of Cyclic Alkenes
and Polyenes, including Ring-opening Reactions

A crystalline polymer has been obtained from cyclobutene in the presence of $RhCl_3 \cdot 3H_2O$, and this polymerization proceeds by *opening* the double bond rather than ring-cleavage [1256]. Several studies of ring-opening polymerization reactions have involved bicycloalkenes such as norbornene and norbornadiene. Catalysts for the polymerization of norbornene have been obtained from $\{IrX_3 + COD$ or cyclooctene in ethanol$\}$ [1816], cyclooctene complexes including $[Ir(C_8H_{14})_2Cl]_2$, $Ir(CO)(C_8H_{14})_2Cl$, $Ir(C_8H_{14})(F_3\text{-acac})$ [1413, 1414] and $\{IrCl_3 \cdot xH_2O$ or $(NH_4)_2IrCl_6$ + a reductant such as butyraldehyde or $SnCl_2\}$ [1478, 1817]. The ring-opening copolymerization of pairs of cyclic alkenes such as cyclopentene and norbornene, or cyclooctene and 1,5-cyclooctadiene, is promoted by $IrCl_3(COD)$ [839]. The products of these reactions can be characterized by ^{13}C NMR spectroscopy. Low yields of the *cis*-1,3-dialkenyl-cyclopentanes (7-6) have been obtained from the metathetical codimerization of norbornene with vinylidene or internally disubstituted alkenes; these reactions are assisted by $[Ir(C_8H_{14})_2Cl]_2$ [1489]. In another coupling reaction, (E)-2-styrylnorbornane is formed from norbornene and (Z)-PhCH=CHBr in $PhCH_2OH$ in the presence of $RhCl(PPh_3)_3$ and HOAc [294].

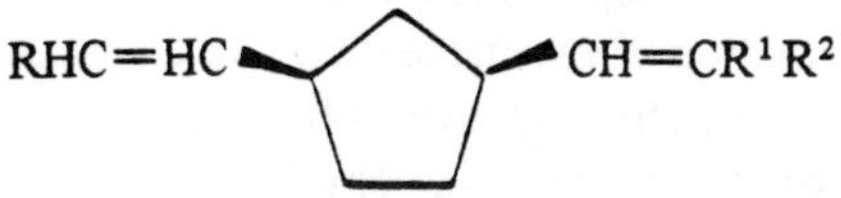

$RHC=HC$ ⟋⟍ $CH=CR^1R^2$

(7-6)

$(R = R^1 = Me, Et; R^2 = H, R = Me; R^1 = H, R^2 = Et)$

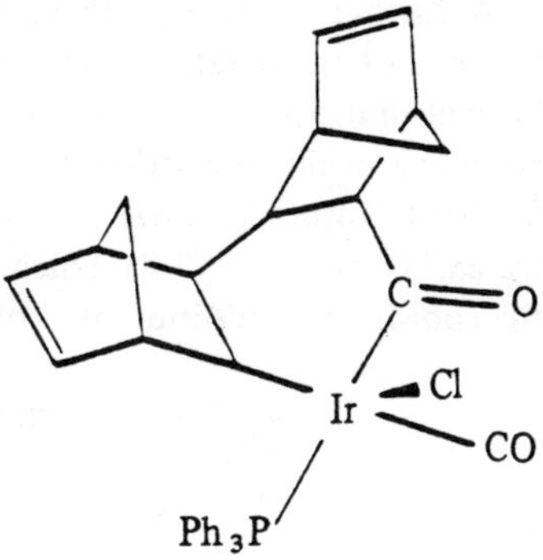

$C=O$

Ir — Cl / CO

Ph_3P

(7-7)

The dimerization and trimerization of norbornadiene gives structurally complex products. Suitable catalysts for these reactions include $RhCl(PPh_3)_3$ [19], $[Rh(NBD)_2]^+$ [615] and $\{Rh(acac)_3 + R_2AlCl\}$ [1535]. The head-to-tail dimerization product '*Binor-S*' is obtained with a catalyst prepared from $RhCl$-$(PPh_3)_3$ and $BF_3 \cdot Et_2O$ [1539]. The spectroscopic characterization of the iridium complex 7-7 has established that coupling of the NBD in these reactions may proceed via metallocyclic intermediates [544].

At 250 °C in the presence of $[Rh(NBD)Cl]_2$, biphenylene undergoes a ring-opening dimerization reaction to produce tetraphenylene [1447]. Some polyquinolines have been formed with NBD–Rh complexes as catalysts [1462, 1660]. Biphenylene is the crosslinking site in these polymers.

7.5. The Oligomerization, Polymerization
and Addition Reactions of Alkynes

The dimerization of some 1-alkynes has been achieved with $RhCl(PPh_3)_3$ as the catalyst. The reactivity of the alkyne $RC\equiv CH$ increases in the order $R=H < Me < Pr < Bu < MeOCH_2 < Bu^t < Ph$, and the ratio of branched (i.e. $RC\equiv CCR=CH_2$) to linear (i.e. $RC\equiv CCH=CHR$) product increases in the order $R = Ph \simeq Bu^t < Bu < Pr < Me < MeOCH_2$ [1944]. The dimerization of 1-octyne gives the branched-chain isomer, 7-methylenepentadec-8-yne, as the main product [289]. The α-hydroxyalkynes, 3-methyl-but-1-yne-2-ol and propargyl alcohol, have produced, 2,7-dimethyl-oct-3-en-5-yne-2,7-diol [1609] and 2-penten-4-yne-1,5-diol [1533], respectively. In some cases, the reactions with phenylacetylene have yielded polyphenylacetylene [944, 1609], and the polymer is also obtained when the catalyst source is $RhCl_3 \cdot 3H_2O$ or $Rh(NO_3)_3$ [825, 1774, 1794]. Linear polymers of $HC\equiv CR$ (R = H, Me or Ph) are obtained when $Rh(COD)(Ph)(PPh_3)$ is used as the catalyst [1753].

A further study of the dimerization of phenylacetylene by $RhCl(PPh_3)_3$ at ambient temperature and 1 atm pressure has shown that two products are formed initially in about equal amounts [289]. One is *trans*-1,4-diphenyl-butenyne while the other is the unstable 1,3-diphenylbutenyne which reacts further to form oligomeric products. Formation of the stable dimer may proceed via an intermediate **(7-8)** which is obtained by the oxidative-addition of one and the coordinative-addition of another alkyne to the rhodium complex [943].

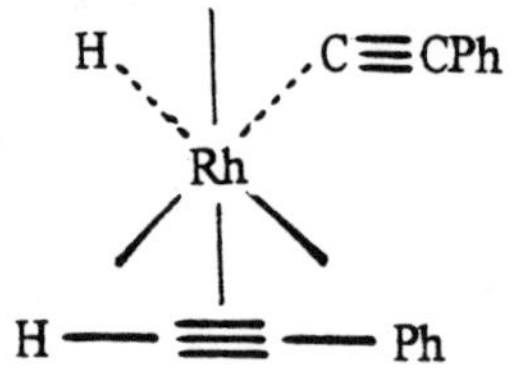

(7-8)

The cyclotrimerization of both phenylacetylene and diphenylacetylene is catalysed by $Rh_4(CO)_{12}$ [843], and the cyclotrimerization of hexafluoro-but-2-yne is assisted by $Rh(CO)Cl(AsCy_3)_2$ [1109]. The cationic complex $[Rh(CO)_2(Ph_2PCH_2CH_2PPh_2)]^+$ is an effective catalyst for the cyclic oligo-merization of propyne [34]. The formation of trimethylbenzene is enhanced if CO_2 is present in the reaction mixture, but 4,6-dimethyl-2-pyrone is also formed. With a mixture of $HC{\equiv}CCO_2Me$ and $MeO_2CC{\equiv}CCO_2Me$ in the presence of $RhCl(PPh_3)_3$, the major product is $1,2,4,5\text{-}(CO_2Me)_4C_6H_2$ [1202].

Addition reactions between aryl- or heteroaryl-acetylenes and $H_2C{=}CHCH_2\text{-}CO_2H$ are catalysed by $RhX(PR_3)_nL_m$ and $[Rh(PR_3)_nL_p]^+$ complexes in the presence of potassium acetate or related salts. For example, $PhC{\equiv}CH$ and $H_2C{=}CHCH_2CO_2H$ give *trans,trans*-$PhCH{=}CHCH{=}CHCH_2CO_2H$ when heated in EtOH in the presence of $RhCl(PPh_3)_3$ and KOAc [323].

There are some examples of the cycloaddition of alkynes to cyclohexene. The reactions with 2-butyne, 4-octyne and 2,8-decadiyne in the presence of $[Rh(C_2H_4)_2Cl]_2$ have given the bicyclic products **7-9** [1611].

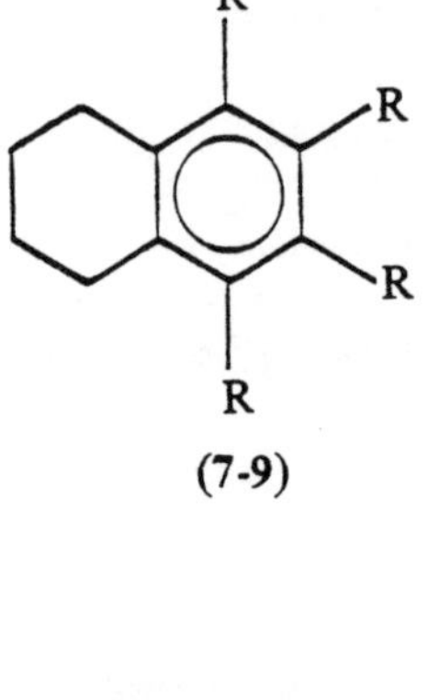

(7-9)

Various catalysts, including $RhCl(PPh_3)_2$, have been used in the formation of bis(silyl)ethylenes from acetylene and disilanes [1790]. For example, Cl_2-$MeSiCH=CHSiMeCl_2$ and $ClMe_2SiCH=CHSiMe_3$ are obtained by the addition of $Cl_2MeSi-SiMeCl_2$ and $ClMe_2Si-SiMe_3$, respectively.

7.6. The Formation of Functionalized Oligomers and Polymers

Some functionalized oligomers and polymers have been obtained by the co-condensation of alkenes, dienes or alkynes with other substrates. The formation of polyketones, $H(CH_2CH_2CO)_nR$ (R = Et, OMe, OEt, depending on the solvent used), from ethylene and CO in the presence of carbonyl—rhodium catalysts derived from Rh_2O_3 is an example [842]. Unsaturated polysulfones have been obtained in a similar manner from the reactions of conjugated dienes with alkenes and SO_2 in the presence of $RhX(PPh_3)_3$ compounds [1584].

In several patents [844, 845, 1456], the formation of polymers with amine groups attached to the chain is described. As an example, such a polymer is obtained when a mixture of piperazine, water, 4-methylpiperidine and 1,5-hexadiene is heated under CO pressure with $[Rh(NBD)(PMe_2Ph)_3][PF_6]$ as the catalyst. The polymers are soluble in dilute aqueous HCl due to the presence of the functional groups.

Reactions between butadienes and secondary amines or primary aromatic amines are catalysed by rhodium compounds such as $RhCl_3 \cdot 3H_2O$, $RhCl(PPh_3)_3$ and $Rh(CO)Cl(PPh_3)_2$, and give the products **7-10** to **7-12** [109].

(7-10)

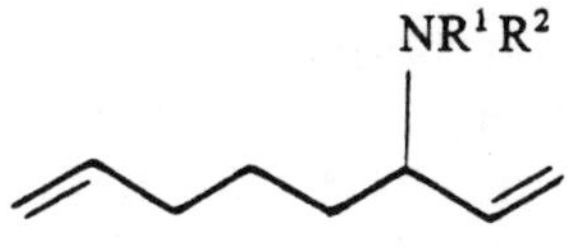

(7-11)

(7-12)

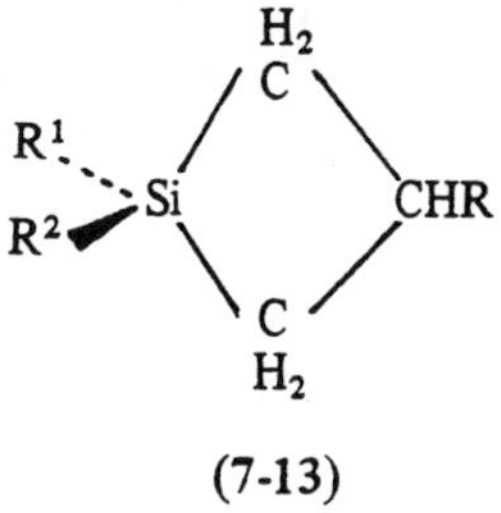

(7-13)

Polymers of the type $[-R^1R^2Si-CH^2-CHR-CH_2-]_n$ are obtained by ring-opening of the silacyclobutanes (**7-13**); suitable catalysts for the reaction are $[Rh(CO)_2Cl]_2$, $RhCl(PPh_3)_3$ and $RhMe(PPh_3)_3$ [390]. It is thought that this reaction proceeds by insertion of Rh into an Si—C ring bond followed by ring-opening.

7.7. Other Reactions Leading to the Formation of New C—X bonds

ALKYLATION (AND ARYLATION) REACTIONS

There is a recent review of transition metal catalysed additions of organic halides to alkenes, alkynes and other organic substrates [689]. In some of these reactions, rhodium compounds act as the catalyst.

The methylation of alkenes is achieved by treatment with iodomethane in the presence of $\{RhCl_3 + SnCl_2\}$ as the catalyst source [1076]. A good yield of $MeCH{=}CHPh$ is obtained from $PhCH{=}CH_2$, but the reactions with ethylene and 1-octene give low yields of the methylation products. The same catalyst system assists the methylation of cyclohexene by dimethyl sulfate. Similar reactions occur when alkenylmercurials, and the related alkynyl- and aryl-mercury compounds, are treated with CH_3I in the presence of $MeRhI_2$-$(PPh_3)_2$ [1055]:

$$RHgX + MeI \quad \longrightarrow \quad R{-}Me$$

With $Rh\{(-)\text{-}DIOP\}Cl$ as the catalyst, the asymmetric addition of CCl_3Br to styrene gives $(S)\text{-}(-)\text{-}PhCHBrCH_2CCl_3$ in greater than 32% ee [1240]. Treatment of this product with NaN_3 followed by reduction with $LiAlH_4$ yields $(R)\text{-}(+)\text{-}PhCHEtNH_2$.

In the reactions between $H_2C{=}CHCN$ and carboxylic acids such as $MeCO_2H$ and $HO_2C(CH_2)_4CO_2H$, formation of the cyanoethylation products $MeCO_2$-CH_2CH_2CN and $NCCH_2CH_2O_2C(CH_2)_4CO_2CH_2CH_2CN$ occurs with high selectivity (ca. 98 %) when the catalyst is $RhCl(PPh_3)_3$ [1409]. When the ester $MeCOCH_2CO_2Et$ is added to isoprene in the presence of a catalyst formed from $[Rh(COD)Cl]_2$, $P\{C_6H_4(SO_3Na)\}_3$ and Na_2CO_3 in water, a mixture of $MeCOCH(CH_2CH{=}CMe_2)CO_2Et$ and $MeCOCH(CH_2CH_2CMe{=}CH_2)CO_2Et$ is obtained [1221]. The reaction between ethylene and aniline to form $PhNHEt$ and 2-methylquinoline is catalysed by $RhCl_3 \cdot 3H_2O$ or $\{RhCl_3 \cdot 3H_2O + PPh_3\}$ [439, 440, 441]. Typical reaction conditions are 100 °C and 100 atm pressure. Low ethylene pressure or the addition of PPh_3 increases the proportion of the N-alkylation product. The amount of heterocyclic product is increased when $IrCl_3$ is used in place of $RhCl_3$. The addition of arenes HC_6H_4R (R = H, Me, OMe, F) to $Ph_2C{=}CO$ is catalysed by the carbonyl compounds $Rh_4(CO)_{12}$ and $Rh_6(CO)_{16}$ [757, 979]. These reactions yield $Ph_2CHCOC_6H_4R$, and the reaction is enhanced when R is an electron-withdrawing substituent.

There is transfer of the R group to ethylene and styrene in the reactions with $Rh(R)(DMG)_2py$ (R = Me, CH_2Ph, Ph) [1874]. Addition of diphenylacetylene to a reaction mixture obtained from $\{RhCl_3 + MeMgCl$ in THF at $-70°C\}$ results in the formation of methylation, hydrogenation and oligomerization products [1080]. Two products obtained from this reaction are *cis*-dimethylstilbene and *trans*-stilbene. The corresponding reaction between diphenylacetylene and $\{RhBr_3 + MeMgBr\}$ gives mainly *cis*-α-methylstilbene [1180]. *trans*-α-Methylstilbene is the main product from the reaction between diphenylacetylene and preformed $Rh(Me)(PPh_3)_3$ [1180].

Aryl halides can be added to $Rh(Me)(PPh_3)_3$, and a subsequent elimination reaction yields the appropriate methylarene [1567]:

$$Rh(Me)L_3 + ArX \longrightarrow Rh(Me)(Ar)XL_3$$
$$\longrightarrow RhXI_3 + MeAr$$

In this way, toluene can be obtained from iodobenzene and *p*-methylbiphenyl from *p*-bromobiphenyl.

Ketones are obtained from the alkylation or arylation of acid chlorides. For example, $MeCOCl$ and *trans*-$PhCH=CHCOCl$ give $MeCOPh$ and *trans*-$PhCH=CHCOPh$, respectively, when treated with $Rh(CO)(Ph)L_2$ compounds [690]. Polymer-bound catalysts can be used for this reaction, and these are formed by attachment of $Rh(CO)Cl(PPh_3)_2$ to a Ph_2P—styrene/divinylbenzene resin; subsequent treatment with $R'Li$ and $RCOCl$ gives the appropriate $RCOR'$ compound [1399].

The alkylation of hydrocarbons has been achieved in other ways. One approach uses $CO + H_2$ or $CO + H_2O$ as the source of alkyl groups. For example, a mixture of alkylbenzenes, $Ph(CH_2)_nH$ (n = 1–5, but n = 2 predominant) is formed by treating benzene with CO and H_2 at 200 °C. Catalysts for this reaction were formed by treating metal carbonyls, including $Rh_6(CO)_{16}$, with $AlCl_3$ and a chelating phosphine such as $Ph_2P(CH_2)_3PPh_2$ [708]. Similarly, methyl iodide is converted to ethyl iodide by treatment with CO and H_2 at 200 °C under pressure in the presence of $RhCl_3 \cdot 3H_2O + PPh_3$ [1197]. In another example, the reaction between pyridine, CO and H_2O at 150 °C and 800 psi with $Rh_6(CO)_{16}$ as the catalyst source gave **7-14** as the major product [1036]. The *N*-alkylation of amines is achieved in similar manner [1901].

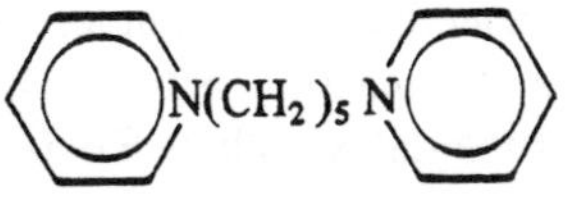

(7-14)

A more complicated alkylation system was used to convert $R'CH_2CO_2R$ to $R'CHMeCO_2R$. This involved treatment with HCHO, CO, H_2O and a tertiary amine such as N-methylmorpholine in the presence of $RhCl_3$ or some other rhodium compound [23]. Alkylpyridines containing mostly 2-ethyl-3,4-dimethylpyridine are obtained when a mixture of ethylene, CO, H_2 and NH_3 is heated at 145 °C and 226 atm pressure with a phthalocyanine-tetrasulfonate-rhodium complex as catalyst [751]. In a related reaction, the aminomethylation of alkenes is achieved under hydroformylation condition [806, 810] or via the water–gas shift reaction [846, 1035]:

$$RCH{=}CH_2 + 3\ CO + H_2O + HNR'_2 \longrightarrow R(CH_2)_3NR'_2 + 2\ CO_2$$

Rhodium carbonyl clusters such as $Rh_4(CO)_{12}$ and $Rh_6(CO)_{16}$ are effective catalysts. The occurrence of various side-reactions, such as the formation of 1-piperidinecarboxaldehyde, is minimized when mixed metal catalysts (e.g. $Rh_4(CO)_{12}/Fe_3(CO)_{12}$) are used.

In another approach, propylbenzenes are obtained from the interaction of cyclopropane with platinum metal chlorides under CO pressure in benzene [877]. With $IrCl_3$ as the catalyst pure $PhCHMe_2$ is obtained, whereas mixtures of the 2-propylbenzenes are formed when $RhCl_3$ is the catalyst; $PdCl_2$ gives pure $PhCH_2CH_2CH_3$.

REACTIONS INVOLVING CARBENE PRECURSORS

There are many accounts of the use of carbene precursors in the formation of new C—C bonds. Rhodium compounds of the type $[Rh^{II}(OAc)_2]_2$ and $Rh^{III}(TPP)I$ catalyse the decomposition of aryldiazoalkanes in inert solvents; *cis*-rather than *trans*-1,2-diarylethylenes are obtained [1459]. Azines are obtained from secondary aryldiazoalkanes in the presence of $[Rh(OAc)_2]_2$.

Carbenes generated from diazo esters can be inserted into the C—H bonds of aliphatic and cyclic alkanes. Rhodium carboxylates such as $Rh_2(O_2CCF_3)_4$ or $Rh_2\{O_2C(CF_2)_6CF_3\}_4$ are efficient catalysts for these reactions [418]. As an example, ethylcyclopentyl acetate is formed regioselectively when cyclopentane is treated with N_2CHCO_2Et at 22 °C in the presence of $Rh_2(O_2CCF_3)_4$.

Reports of the addition of carbenes to alkenes and dienes are fairly common. Reactions between N_2CHCO_2Et and aliphatic, alicyclic, heterocyclic and aromatic alkenes are catalyzed by $Rh_6(CO)_{16}$ [460, 462]; the formation of the cyclopropanation products (7-15) is highly selective. For example, 7-15 (R = Ph, R′ = OMe) is obtained in 87% yield from the reaction between $PhCH{=}CH_2$ and N_2CHCO_2Et in toluene. Rhodium carboxylates such as $Rh_2(OAc)_4$ [462, 1383] or $Rh^{II}(pivalate)$ [744] are alternative catalysts for these reactions, and they are effective under mild conditions. For example, the addition of

N_2CHCO_2Et to *trans,trans*-Cl_2C=CHCH=CMe_2 in the presence of rhodium pivalate occurs at 20 °C and gives a 66% yield of a 60:40 mixture of *cis-* and *trans*-(**7-15**) (R = R' = Me, R" = CH=CCl_2). With Rh^{III}(TPP)I as the catalyst, steric control of the reactions with alkenes gives predominantly the *cis*-isomer of the ethoxycarbonylcyclopropanes [270].

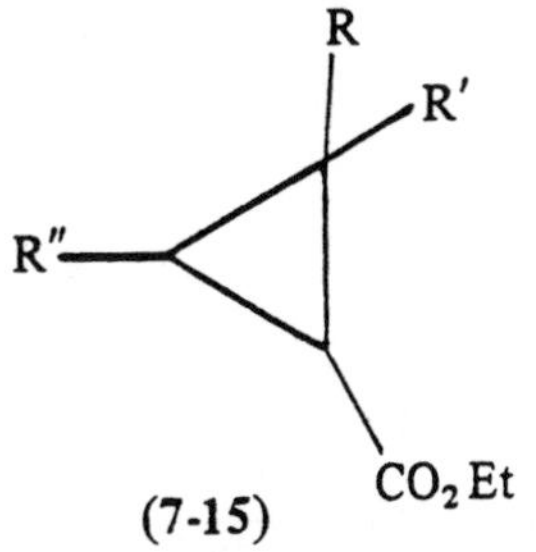

(7-15)

The cyclopropenation of alkynes is achieved by treatment with N_2CHCO_2R in the presence of rhodium acetate [451, 833, 1382]. The interaction between $AcOCH_2C$≡CCH_2OAc and $N_2CHCO_2CMe_3$, for example, gives a 60% yield of the cyclopropene **7-16** (R = $OCMe_3$). The sequential treatment of this compound with CF_3CO_2H, $(COBr)_2$ and CH_2N_2 produces **7-16** (R = CHN_2). The diazoketone function in this product undergoes an intramolecular addition to the cyclopropene C=C bond to form **7-17**; this cyclization is catalysed by $Rh_2(OAc)_4$ [833].

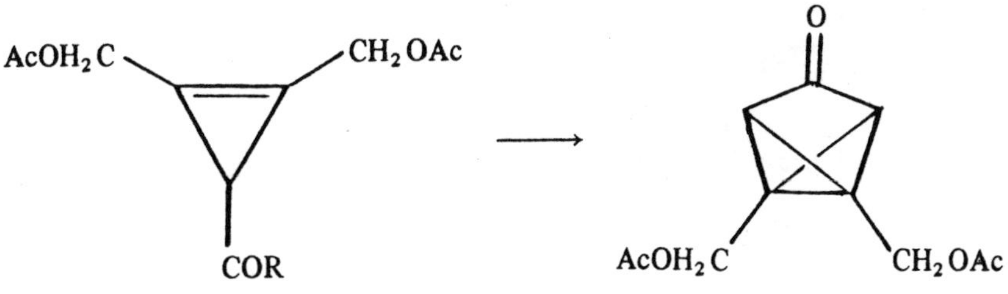

(7-16)   (7-17)

In the transition metal catalysed addition reactions of diazo esters with allylic compounds such as allylmethyl sulfide and allyl halides [459, 1754], two product types are detected. One is the cyclopropanation product **7-18**; the other, **7-19**, is derived from the [2,3]-sigmatropic rearrangement of ylide intermediates. Formation of the two products is competitive and is influenced by the nature of X and by the catalyst. For example, with $Rh_2(OAc)_4$ as the catalyst, exclusive formation of the ylide-derived product is observed when X = I, cyclopropanation is dominant when X = Cl and significant amounts of both products are obtained when X = Br. When the catalyst is a copper complex rather than rhodium acetate, there is an increased tendency to form the ylide-derived products.

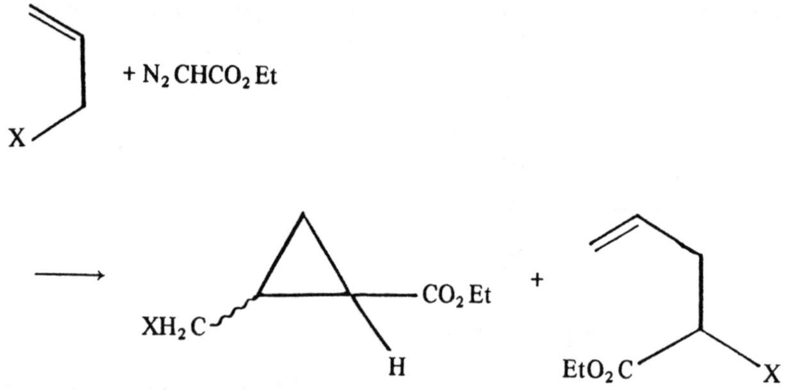

$$+ N_2 CHCO_2 Et$$

(7-18)   (7-19)

The formation of cycloheptatrienes can be achieved by the addition of diazo esters to aromatic molecules. For example, 1-carbalkoxycyclohepta-2,4,6-trienes (7-20) are formed specifically from toluene and $N_2 CHCO_2 Me$. Rhodium(II) salts formed from strong carboxylic acids (e.g. $Rh_2(O_2CCF_3)_4$) are good catalysts for these reactions [57].

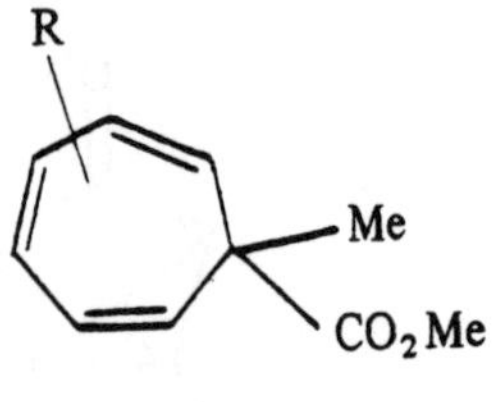

(7-20)

ADDITION, CYCLIZATION AND RELATED REACTIONS INVOLVING ALDEHYDES

The addition of aldehydes to ethylene results in hydroacylation (see Equation (7.1)):

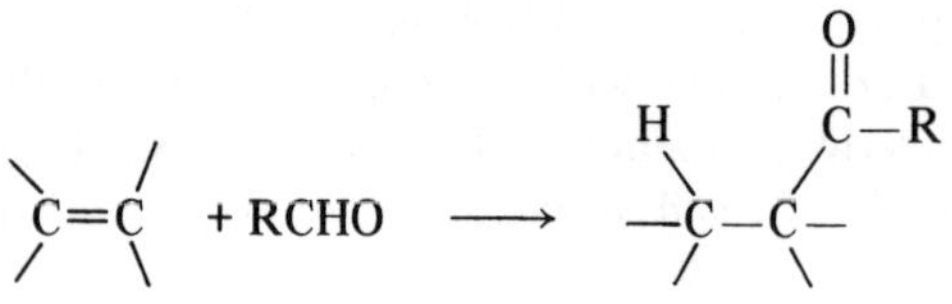

$$\diagdown C=C \diagup + RCHO \longrightarrow -\underset{|}{C}-\underset{|}{C}- \qquad (7.1)$$

As an example, 6-octen-3-one is obtained from ethylene and 4-hexenal in the presence of $Rh(C_2H_4)_2(acac)$ [1878]. Deuterium labelling has been used to follow the course of this reaction [1731]. Intramolecular alkene—aldehyde additions, which convert 4-pentenal to cyclopentanone or 4-hexenal to 2-methylcyclopentanone, are catalysed by $RhCl(PPh_3)_3$ [271, 272, 1085] or $\{ [Rh(C_8H_{14})_2Cl]_2 + P(p\text{-}MeC_6H_4)_3 \}$ [1052]. In these reactions, there is some decarbonylation of the aldehyde with the formation of $Rh(CO)Cl(PPh_3)_2$. Extensive deuterium-labelling experiments indicate a hydroacylation mechanism involving an acyl—rhodium(III) hydride intermediate. The related cyclization of (+)-citronellal produces a 3:1 mixture of (+)-neoisopulegol (7-21) and (−)-isopulegol (7-22) [1510]. With $Rh(C_2H_4)_2(acac)$ as the catalyst, 4-pentenal yields isomerization products which are formed by double-bond and ketone migrations [1085].

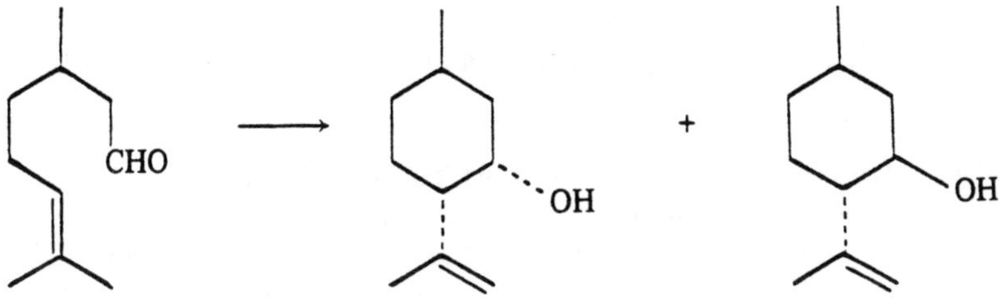

(7-21)   (7-22)

Aldimines are formed in the reactions of aldehydes with aminopyridines, and this provides an indirect way of activating the $C-H$ bond of the aldehyde towards oxidative-addition. Thus, treatment of an aromatic aldehyde with 3-methyl-2-aminopyridine in the presence of $RhCl(PPh_3)_3$ produces an aldimine which reacts with alkenes in the presence of $RhCl(PPh_3)_3$; ketones can then be obtained by hydrolysis [1731]. When aldehydes are treated with secondary amines and H_2 with a catalyst formed from $RhCl_3$ and CO/H_2, tertiary amines are obtained [807].

Esters have been obtained from reactions between aldehydes and alcohols with $HRh(CO)(PPh_3)_3$ as the catalyst [626]. For example, $PhCH_2OEt$ (42%) and $PhCH_2OH$ (42%) are obtained when PhCHO and EtOH are refluxed together for 3 d. The addition of an efficient proton-acceptor increases the proportion of ester in the products. Interactions between EtCHO and EtOH occur in the presence of $RhCl_3 \cdot 3H_2O$ as the catalyst. After 5 min at 25 °C, the major product is $EtCH(OEt)_2$; the products are $EtCH(OEt)_2$ plus $EtCH=CMeCHO$ after 16 h at 100 °C [745].

The α-methylation of cyclohexanone and acetophenone to give mono- and di-methylated ketones has been achieved with aqueous formaldehyde and CO under pressure in the presence of RhCl$_3$ as the catalyst source [1899]; an example is shown in Equation (7.2).

$$
\begin{array}{ccc}
\overset{\textstyle O}{\underset{\textstyle \|}{}} & & \overset{\textstyle O}{\underset{\textstyle \|}{}} \\
R^1CCH_2R^2 & \longrightarrow & R^1CC(Me)HR^2
\end{array}
\qquad (7.2)
$$

The reductive transformation of nitroarenes into (dialkylamino) arenes and 2,3-dialkyl-substituted quinolines is achieved using aliphatic aldehydes under CO. The N-alkylation reaction is catalysed by [Rh(COD)Cl]$_2$, and the N-heterocyclization by RhCl(PPh$_3$)$_3$ [1900]. The alkyl-substituted quinolines can also be obtained from aminoarenes and aliphatic aldehydes with catalysts such as [Rh(NBD)Cl]$_2$ [1898].

ALKYLATION, CYCLIZATION AND RELATED REACTIONS INVOLVING ALCOHOLS

The alkylation of ketones such as acetone, ethylmethylketone and cyclohexanone occurs when alcohols are added in the presence of NaOH or KOH and RhCl-(PPh$_3$)$_3$ or other catalysts [300]. Alkyl-substituted quinolines can be obtained from nitrobenzene and aliphatic alcohols with { [Rh(CO)$_2$Cl]$_2$ + MoCl$_5$} as the catalyst [218].

The oxidation of alcohols provides a route to the selective monoalkylation of arylacetonitriles. One catalyst for these reactions is HRh(PPh$_3$)$_4$. It has been used to form BuNHMe in 98% yield from BuNH$_2$ in refluxing methanol, and N-butyltetrahydropyrrole in 56% yield from BuN(CH$_2$)$_4$OH in refluxing dioxan [622, 627]. Another catalyst prepared *in situ* from RhCl$_3$, PPh$_3$ and Na$_2$CO$_3$ has been used for the selective monoalkylation of arylacetonitriles by alcohols [628]. For example, treatment of o-(NCCH$_2$)$_2$C$_6$H$_4$ with ROH (R = Me, Et) gives the indenes 7-23, whereas 2-NCC$_6$H$_4$CH$_2$CN with MeOH gives the isoquinolines 7-24 (R = OMe, R' = NH$_2$; R = NH$_2$, R' = OMe).

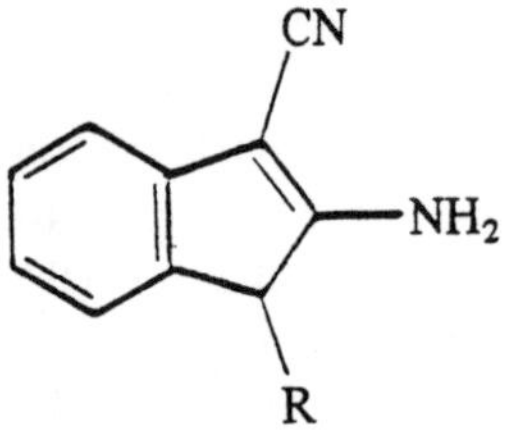

(7-23) 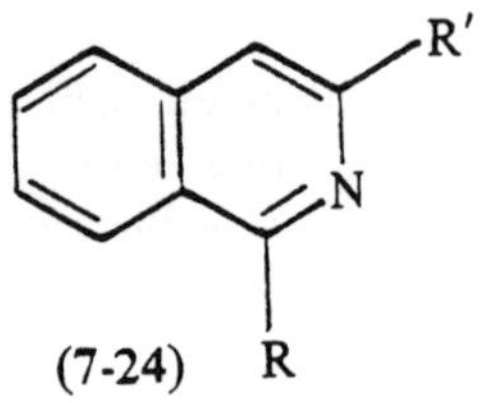   (7-24)   R

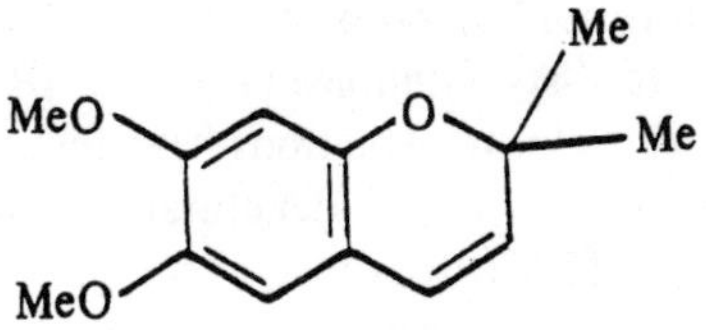

(7-25)

Some reactions of alcohols have led to the formation of O-heterocyclics. In the presence of a catalyst formed from $[Rh(COD)L]_3$ (L = imidazolato), 3,4-dimethoxyphenol condenses with isoprene in benzene at 140 °C to give 2-(3-methyl-2(and 3)-buten-1-yl)-4,5-dimethoxyphenol [947]. This product is converted to 7-25 by heating with chloranil in benzene. Cyclization of the fluorophenylpropanols 7-26 (e.g. R = H or CH_2OH, R' = H) gives some related chromenes (7-27); the cations $[(\eta^5-C_5Me_4Et)(\eta^6-C_6H_6)Rh]^{3+}$ are efficient catalysts for these conversions [767].

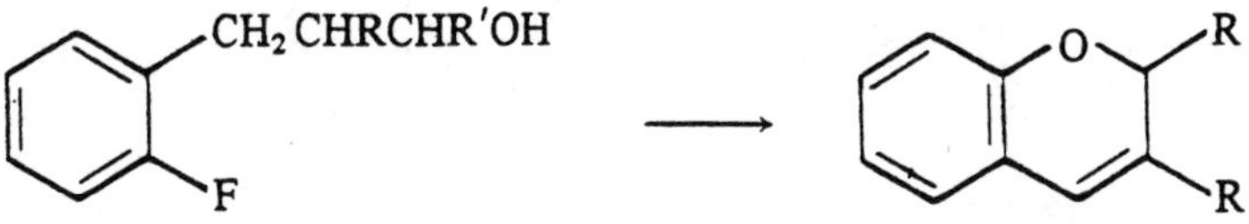

(7-26)   (7-27)

OTHER CYCLIZATION REACTIONS

A mixture of ethylene and benzene reacts with CO under pressure to give good yields of 3-pentanone and styrene. A suitable catalyst for this reaction is Rh_4-$(CO)_{12}$ [756]. In a related reaction between diphenylacetylene, benzene and CO in the presence of $Rh_4(CO)_{12}$, indenone (7-28) and $Ph_2C{=}CHPh$ are formed [752]. Similar reactions are observed with monosubstituted and disubstituted benzenes.

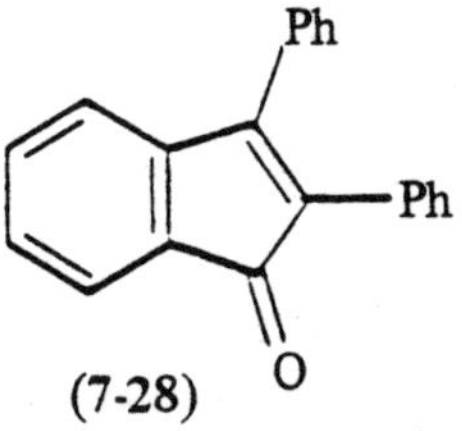

(7-28)

The formation of some heterocyclic compounds has been accomplished by ring-closure reactions in the presence of rhodium compounds. The cyclization of

7-29 to **7-30** is achieved with RhCl$_3$ under reflux conditions [1534] and 2-arylazirines (**7-31**) are transformed to 2-styrylindoles (**7-32**) by [Rh(CO)$_2$Cl]$_2$ or Rh(CO)Cl(PPh$_3$)$_2$ [54]. Other ring-closure reactions have involved oxygen-containing substrates. Tetrahydro-3-methyl-4-methylenefuran is obtained from diallyl ether with RhCl$_3$ as the catalyst [224].

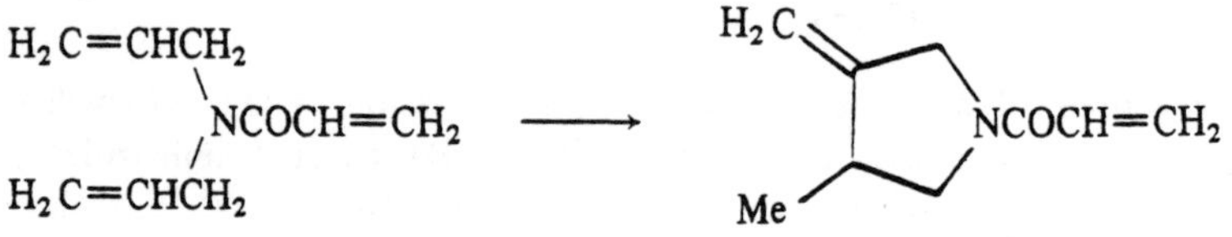

(7-29)   (7-30)

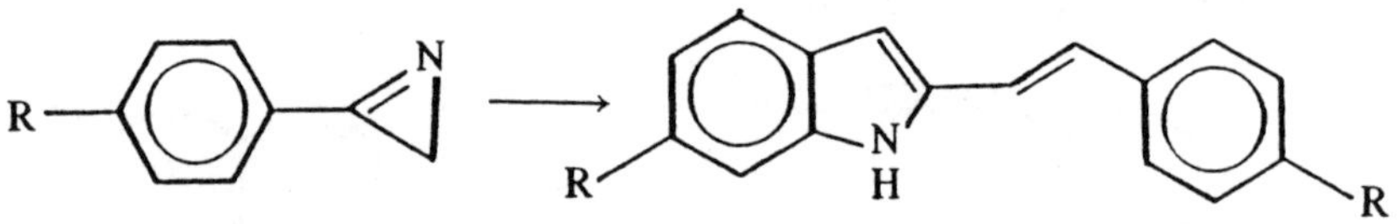

(7-31)   (7-32)

The thermal transformation of **7-33** to the β-lactam **7-34** is achieved in benzene at 80 °C with Rh$_2$(OAc)$_4$ as the catalyst [907]; the yield is 93% after 0.5 h.

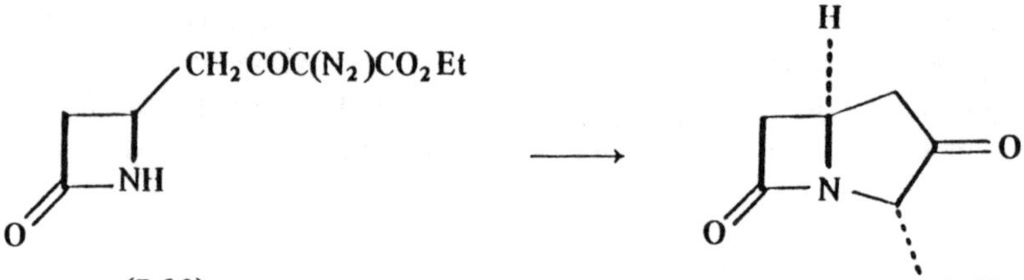

(7-33)   (7-34)   CO$_2$Et

COUPLING REACTIONS

At 25 °C in (Me$_3$N)$_3$PO, the system { [Rh(CO)$_2$Cl]$_2$ + LiCl} catalyses the conversion of *trans*-BuCH=CHHgCl to *trans,trans*-BuCH=CHCH=CHBu with greater than 98% stereospecificity [1051]. Similar reactions occur with *trans*-ClCH=CHHgCl [1587] and other vinylmercurials [1053]. Related reactions with arylmercurials yield biaryls [1053].

APPENDIX

TABULATION OF CATALYSTS AND THEIR APPLICATIONS

Catalyst or catalyst source	Type of reaction	Page*
(a) *Neutral metal carbonyls and related compounds*		
(i) $Rh_4(CO)_{12}$	aminomethylation	188
	arylation	186
	carbonylation	132^S, 135, 138
	CO → alcohols	123^A, 124^S
	CO → hydrocarbons	122^A, 122^S
	cyclization	184, 193
	hydroformylation	140^A, 140–143, 143–148^A, 149^A, 154^S, 155^S
	hydrogenation	79^S, 85^S
	hydrosilation	112
	polymerization	179^A
	reduction	158
	ring opening	33
	water-gas shift	120–121^A
(ii) $Ir_4(CO)_{12}$	carbonylation	131, 132^S
	CO → hydrocarbons	122, 122^A, 122^S
	H/D exchange	12
	hydroformylation	150^A, 154^S
	hydrogenation	79^S, 85^S
	water-gas shift	121
(iii) $Rh_2Co_2(CO)_{12}$	hydroformylation	153^S, 154^S
(iv) $Rh_6(CO)_{16}$	alkylation, arylation	186, 187^A, 188
	carbene addition	188
	carbonylation	135, 136, 138
	CO → alcohols	123^A, 124^S
	CO → hydrocarbons	122^S
	deoxygenation	175

* Superscript A signifies co-catalyst, promoter, or other additive; superscript S signifies catalyst attached to solid support material.

195

Catalyst or catalyst source	Type of reaction	Page
	hydrogenation	69
	hydrosilation	114
	polymerization	181, 182[A]
(ii) $M_a(\text{alkene})_b X_c$	aldehyde addition	191, 191[A]
	aldehyde cyclization	191
	carbonylation	137
	hydroformylation	152[A]
	hydrogenation	78[S], 82[S], 84[S]
	hydrosilation	114, 114[S]
	isomerization	16[A], 16[S]
	oxidation	160, 166
	polymerization	176, 177, 180[A], 181, 182, 184
	ring opening	30
(iii) $\{M_a(\text{alkene})_b X_c + L\} \to MXL_2$		
	alcohol dehydrogenation	11
	aldehyde addition	191
	asymmetric hydrogenation	90, 91
	H-transfer	75[A]
	hydrogenation	61, 78[S], 82[S], 84[S]
	hydrosilation	114[S], 115[S]
(iv) $[M(\text{diene})X]_2$	alcohol → aldehyde	12
	carbonylation	130, 130[A]
	hydroformylation	152, 155[S]
	hydrogenation	83[S], 84[S]
	isomerization	16, 16[AS], 17
	polymerization	179, 179[A], 180[A], 181
	ring opening	29, 30, 33, 34, 36
(v) $\{[M(\text{diene})X]_2 + L\}$ (→ "MXL_2" or "$[H_2ML_2S_2]$"[+])		
	asymmetric H-transfer	108
	asymmetric hydrogenation	90, 92
	H-transfer	74
	hydroformylation	152
	hydrogenation	56[A], 62–67, 80[S], 83[S]
	hydrosilation	114
	oxidation	167[A]
(vi) $M(\text{alkene})_a XL_b$ and $M(\text{diene})_a XL_b$		
	carbonylation	130
	H-transfer	73, 75
	hydroformylation	152

Catalyst or catalyst source	Type of reaction	Page
(f) *Metal(III) halides, and their derivatives*		
(i) RhX_3	aldehyde + alcohol	191
	alkylation	186[A], 187[A], 188, 191
	alkylation/hydrogenation/ polymerization	187[A]
	alkylation by alcohols	192[A]
	asymmetric hydrogenation	91[A]
	carbonylation	130[A], 132[A], 132[S], 133[A], 134[A], 138
	$CO \rightarrow$ alcohols	126
	dehydrogenation	10
	dimerization	176, 176[A], 177, 177[A], 179[A]
	disproportionation	22, 23[A]
	H/D exchange	12
	H-transfer	74, 74[A]
	hydroformylation	151[A], 154[S]
	hydrogenation	53[A], 71[A], 80[AS], 81[S]
	hydrosilation	114
	isomerization	17[A], 18, 18[A]
	oxidation	161[A]
	ring closure	194
	ring opening	33[A], 182
	polymerization	177[A], 178[A], 179, 179[A], 181, 182
	water-gas shift	121[A]
(ii) IrX_3	alkylation	186, 188
	carbonylation	130[A], 132[A], 133[A]
	dimerization	176, 177[A]
	hydrosilation	114
	isomerization	18[A]
	polymerization	177[A], 179, 182[A]
(iii) $[MX_6]^{3-}$	$CO \rightarrow$ alcohols	124[S]
	H/D exchange	16
	isomerization	18, 18[A]
	polymerization	176[S], 182[A]
(iv) MX_3L_3 and $[M^{III}X_a L_b]X_c'$		
	amines + carbonyl compounds	117[A]
	asymmetric hydrogenation	91
	carbonylation	132[S], 133[S]
	decarbonylation	174
	dehydrogenation	10
	H-transfer	73

Catalyst or catalyst source	Type of reaction	Page
(ii) $[MX_6]^{2-}$	H-transfer	72A
	hydrogenation	40A
(iii) $H_5IrL_{2\ or\ 3}$	disproportionation	22
	H/D exchange	11
	H-transfer	74
	hydrogenation	71, 71A
(iv) Carborane complexes		
	alcohol → aldehyde	12
	hydrogenation	53, 60
	hydrosilation	114, 115
	isomerization	17
(v) Nitrosyl complexes		
	H-transfer	74
	hydrogenation	60A
	isomerization	17
	polymerization	177
	reduction of NO by CO	128
(vi) Oxygen compounds, e.g. $[RhCl(O_2)L_2]_2$		
	hydrogenation	56
	oxidation	161
(vii) Miscellaneous		
Alkyl-Rh complexes, e.g. $RhMeL_3$, $RhMeX_2L_2$		
	alkylation	186, 187
	polymerization	185
Borohydride complexes		
	asymmetric hydrogenation	90
	hydroformylation	154S
	hydrogenation	53
Porphyrin and related complexes		
	carbene addition	189
	decomposition, diazo compounds	188
	H-transfer	73
COD $M^I(\mu\text{-}Cl)_2 M^{III}H_2L_2$		
	hydrogenation	67

REFERENCES

1. P. Abley, I. Jardine and F. J. McQuillin, *J. Chem. Soc. C*, 840 (1971).
2. P. Abley and F. J. McQuillin, *Chem. Commun.*, 1503 (1969).
3. P. Abley and F. J. McQuillin, *J. Catal.*, **24**, 536 (1972).
4. K. Achinami, *Jpn. Pat.* 119 414 (1979); *C. A.* **92**, 76 282 (1980).
5. K. Achiwa, *J. Am. Chem. Soc.*, **98**, 8265 (1976).
6. K. Achiwa, *Tetrahedron Lett.*, 3735 (1977).
7. K. Achiwa, *Chem. Lett.*, 561 (1978).
8. K. Achiwa, *Chem. Lett.*, 905 (1978).
9. K. Achiwa, *Tetrahedron Lett.*, 1475 (1978).
10. K. Achiwa, *Tetrahedron Lett.*, 2583 (1978).
11. K. Achiwa, *Jpn. Pat.* 65 872 (1978); *C. A.* **89**, 197 962 (1978).
12. K. Achiwa, *Jpn. Pat.* 122 219 (1979); *C. A.* **92**, 110 545 (1980).
13. K. Achiwa, *Fundam. Res. Homogeneous Catal.*, **3**, 549 (1979).
14. K. Achiwa, Y. Nakamoto and N. Ishizuka, *Jpn. Pat.* 158 494 (1979); *C. A.* **93**, 46 192 (1980).
15. K. Achiwa, Y. Ohga, Y. Iitaka and H. Saito, *Tetrahedron Lett.*, 4683 (1978).
16. K. Achiwa, I. Ojima and T. Kogure, *Jpn. Pat.* 70 257 (1979); *C. A.* **91**, 211 248 (1979).
17. K. Achiwa and T. Soga, *Tetrahedron Lett.*, 1119 (1978).
18. G. J. K. Acres, G. C. Bond, B. J. Cooper and J. A. Dawson, *J. Catal.*, **6**, 139 (1966).
19. N. Acton, R. J. Roth, T. J. Katz, J. K. Frank, C. A. Maier and I. C. Paul, *J. Am. Chem. Soc.*, **94**, 5446 (1972).
20. G. Adames, C. Bibby and R. Crigg, *J. Chem. Soc., Chem. Commun.*, 491 (1972).
21. R. O. Adlof and E. A. Emken, *J. Labelled Comp. Radiopharm.*, **18**, 419 (1981).
22. R. O. Adlof, W. R. Miller and E. A. Emken, *J. Labelled Comp. Radiopharm.*, **15**, 625 (1978).
23. Agency of Industrial Sciences and Technology, *Jpn. Pat.* 71 041 (1981); *C. A.* **95**, 168 802 (1981).
24. A. M. Aguiar, C. J. Morrow, J. D. Morrison, R. E. Burnett, W. F. Masler and N. S. Bhacca, *J. Org. Chem.*, **41**, 1545 (1976).
25. N. Ahmad, J. J. Levison, S. D. Robinson and M. F. Uttley, *Inorg. Synth.*, **15**, 59 (1974).
26. Z. Aizenshtat, M. Hausmann, Y. Pickholtz, D. Tal and J. Blum, *J. Org. Chem.*, **42**, 2386 (1977).
27. Ajinomoto Co., Inc., *Fr. Pat.* 1 341 874 (1963); *C. A.* **60**, 4061 (1960).
28. Ajinomoto Co., Inc., *Fr. Pat.* 1 361 797 (1964); *C. A.* **61**, 11 894 (1961).
29. Ajinomoto Co., Inc., *Jpn. Pat.* 1419 (1964); *C. A.* **60**, 11 950 (1960).
30. Ajinomoto Co., Inc., *Jpn. Pat.* 3020 (1964); *C. A.* **60**, 15 741 (1960).

31. Ajinomoto Co., Inc., *Jpn. Pat.* 61 937 (1980); *C. A.* **94**, 65 844 (1981).
32. Ajinomoto Co., Inc., *Jpn. Pat.* 121 837 (1980); *C. A.* **94**, 72 288 (1981).
33. I. S. Akhrem, R. S. Vartanyan, L. E. Kothyar and M. E. Volpin, *Isv. Akad. Nauk SSSR, Ser. Khim.*, 253 (1977); *Bull. Acad. Sci. USSR (Engl. Transl.)*, **26**, 227 (1977).
34. P. Albano and M. Aresta, *J. Organomet. Chem.*, **190**, 243 (1980).
35. N. W. Alcock, J. M. Brown, J. A. Conneely and D. H. Williamson, *J. Chem. Soc., Perkin Trans. 2*, 962 (1979).
36. T. Alderson, *U.S. Pat.* 3 013 066 (1961); *C. A.* **57**, 11 016 (1957).
37. T. Alderson, E. L. Jenner and R. V. Lindsey, *J. Am. Chem. Soc.*, **87**, 5638 (1965).
38. T. Alderson and J. C. Thomas, *U.S. Pat.* 3 040 090 (1962); *C. A.* **57**, P16 407 (1957).
39. J. J. Alexander and A. Wojcicki, *J. Organomet. Chem.*, **15**, P23 (1968).
40. J. J. Alexander and A. Wojcicki, *Inorg. Chem.*, **12**, 74 (1973).
41. K. Allison, M. R. Chambers and G. Foster, *Br. Pat.* 1 206 166 (1970); *C. A.* **74**, P3307 (1971).
42. K. G. Allum and R. D. Hancock, *Ger. Pat.* 2 003 294 (1970); *C. A.* **74**, 54 804 (1971).
43. K. G. Allum, R. D. Hancock, I. V. Howell, T. E. Lester, S. McKenzie, R. C. Pitkethly and P. J. Robinson, *J. Organomet. Chem.*, **107**, 393 (1976).
44. K. G. Allum, R. D. Hancock, I. V. Howell, S. McKenzie, R. C. Pitkethly and P. J. Robinson, *J. Organomet. Chem.*, **87**, 203 (1975).
45. K. G. Allum, R. D. Hancock, I. V. Howell, R. C. Pitkethly and P. J. Robinson, *J. Catal.*, **43**, 322 (1976).
46. K. G. Allum, R. D. Hancock and M. J. Lawrenson, *Ger. Pat.* 2 022 710 (1971); *C. A.* **75**, 5199 (1971).
47. K. G. Allum, R. D. Hancock, S. McKenzie and R. C. Pitkethly, *Ger. Pat.* 2 062 352 (1971); *C. A.* **75**, 129 339 (1971).
48. K. G. Allum, R. D. Hancock and R. C. Pitkethly, *Br. Pat.* 1 295 675 (1972); *C. A.* **78**, 71 379 (1973).
49. K. G. Allum and I. V. Howell, *Br. Pat.* 1 378 747 (1974); *C. A.* **82**, 155 268 (1975).
50. H. Alper (Ed.), *Transition Metal Organometallics in Organic Synthesis*, Academic Press, New York, Vol. 1 (1976) and Vol. 2 (1978).
51. H. Alper and K. Hachem, *J. Org. Chem.*, **45**, 2269 (1980).
52. H. Alper and K. Hachem, *Transition Met. Chem.*, **6**, 219 (1981).
53. H. Alper, K. Hachem and S. Gambarotta, *Can. J. Chem.*, **58**, 1599 (1980).
54. H. Alper and J. E. Prickett, *Chem. Commun.*, 483 (1976).
55. L. Alus, A. Sen and J. Halpern, *J. Am. Chem. Soc.*, **100**, 2915 (1978).
56. J. P. Amma and J. K. Stille, *J. Org. Chem.*, **47**, 468 (1982).
57. A. J. Anciaux, A. Demonceau, A. J. Hubert, A. F. Noels, N. Petiniot, R. Warin and Ph. Teyssie, *Chem. Commun.*, 765 (1980); *J. Org. Chem.*, **46**, 873 (1981).
58. J. R. Anderson, P. S. Elmes, R. F. Howe and D. E. Mainwaring, *J. Catal.*, **50**, 508 (1977).
59. J. R. Anderson and R. F. Howe, *Nature (London)*, **268**, 129 (1977).
60. J. R. Anderson and D. E. Mainwaring, *J. Catal.*, **35**, 162 (1974).
61. S. L. T. Anderson and M. S. Scurrell, *J. Catal.*, **71**, 233 (1981).
62. S. L. T. Anderson, K. L. Watters and R. F. Howe, *J. Catal.*, **69**, 212 (1981).
63. T. Anderson and R. V. Lindsey, *U.S. Pat.* 3 081 357 (1963); *C. A.* **59**, 8594 (1959).
64. A. Andreetta, G. Barberis and G. Gregorio, *Chim. Ind. (Milan)*, **60**, 887 (1978).
65. K. A. Andrianov, G. K. I. Magomedov, O. V. Shkol'nik, B. A. Izmailov, L. V. Morozova and V. N. Kalinin, *Dokl. Chem. (Engl. Transl.)*, **228**, 402 (1976).

66. K. A. Andrianov, G. K. I. Magomedov, O. V. Shkol'nik, V. G. Syrkin, B. A. Kamaritskii and L. V. Morozova, *Zh. Obshch. Khim.*, **46**, 2048 (1976); *C. A.* **86**, 106 705 (1977).

67. K. A. Andrianov, J. Soucek, J. Hetflejs and L. M. Khananashvili, *Zh. Obshch. Khim.*, **45**, 2215 (1975); *C. A.* **84**, 59 629 (1976).

68 J. Andrieux, D. H. R. Barton and H. Patin, *J. Chem. Soc., Perkin Trans. 1*, 359 (1977).

69. A. H. Andrist and J. E. Graas, *Prostoglandins*, **18**, 631 (1979); *C. A.* **93**, 25 941 (1980).

70. E. P. Antoniades, *U.S. Pat.* 4 194 056 (1980); *C. A.* **93**, 7665 (1980).

71. P. G. Antonov, N. V. Borunova, V. I. Anufriev and V. M. Ignatov, *Izv. Vyssh. Uchebn. Zaved., Khim. Khim. Tekhnol.*, **22**, 952 (1979); *C. A.* **92**, 41 220 (1980).

72. W. Aquila, W. Himmele and W. Hoffmann, *Ger. Pat.* 2 050 677 (1972); *C. A.* **77**, 19 184 (1972).

73. H. Arai, *J. Catal.*, **51**, 135 (1978).

74. H. Arai, T. Kaneko and T. Kunugi, *Chem. Lett.*, 265 (1975).

75. S. Arai, T. Saito, H. Matsunaga, M. Sumida and Y. Tsutsumi, *Toyo Soda Kenkyu Hokoku*, **25**, 71 (1981); *C. A.* **96**, 51 741 (1982).

76. H. Arakawa and Y. Sugi, *Chem. Lett.*, 1323 (1981).

77. M. Aresta, *Inorg. Chim. Acta*, **44**, L3 (1980).

78. V. Aris, J. M. Brown and B. T. Golding, *Chem. Commun.*, 1206 (1972).

79. K. Arlt and W. Heitz, *Makromol. Chem.*, **180**, 41 (1979).

80. M. Arthurs, M. Sloan, M. G. B. Drew and S. M. Nelson, *J. Chem. Soc., Dalton Trans.*, 1794 (1975).

81. H. Arzoumanian, H. Bitar and J. Metzger, *J. Mol. Catal.*, **7**, 373 (1980).

82. H. Arzoumanian, A. A. Blanc, J. Metzger and J. E. Vincent, *J. Organomet. Chem.*, **82**, 261 (1974).

83. R. W. Ashworth and G. A. Berchtold, *Tetrahedron Lett.*, 339 (1977).

84. R. W. Ashworth and G. A. Berchtold, *Tetrahedron Lett.*, 343 (1977).

85. J. G. Atkinson and M. O. Luke, *Can. J. Chem.*, **48**, 3580 (1970).

86. M. T. Atlay, L. R. Gahan, K. Kite, K. Moss and G. Read, *J. Mol. Catal.*, **7**, 34 (1980).

87. M. T. Atlay, M. Preece, G. Strukul and B. R. James, *Chem. Commun.*, 406 (1982).

88. C. J. Attridge and S. J. Maddock, *J. Organomet. Chem.*, **26**, C65 (1971).

89. C. J. Attridge and P. J. Wilkinson, *Chem. Commun.*, 620 (1971).

90. R. L. Augustine and R. J. Pellet, *J. Chem. Soc., Dalton Trans.*, 832 (1979).

91. R. L. Augustine and J. F. Van Peppen, *Ann. N.Y. Acad. Sci.*, **158**, 482 (1969).

92. R. L. Augustine and J. F. Van Peppen, *Chem. Commun.*, 571 (1970).

93. R. L. Augustine and J. F. Van Peppen, *Chem. Commun.*, 495 (1970).

94. R. Aumann and J. Knecht, *Chem. Ber.*, **111**, 3927 (1978).

95. V. A. Avilov, O. N. Eremenko and M. L. Khidekel, *Izv. Akad. Nauk SSSR, Ser. Khim.*, 2781 (1967); *Bull. Acad. Sci. USSR*, (*Engl. Transl.*), 2655 (1967).

96. V. A. Avilov, M. L. Khidekel, O. N. Eremenko, O. N. Efimov, A. G. Ovcharenko and P. S. Chekrii, *Br. Pat.* 1 262 885 (1972); *C. A.* **76**, 11 220 (1972).

97. V. A. Avilov, M. L. Khidekel, O. N. Eremenko, O. N. Efimov, A. G. Ovcharenko and P. S. Chekrii, *U.S. Pat.* 3 755 194 (1973); *C. A.* **79**, 129 711 (1973).

98. J. Azran, O. Buchman and J. Blum, *Tetrahedron Lett.*, 1925 (1981).

99. B. D. Babitskii, V. A. Kormer, I. Ya. Poddubnyi, V. N. Sokolov and N. N.

Chesnokova, *Dokl. Akad. Nauk SSSR,* **162,** 1060 (1965); *Dokl. Chem. (Engl. Transl.),* **162,** 561 (1965).

100. B. D. Babitskii, V. A. Kormer, I. Ya. Poddubnyi, V. N. Sokolov and N. N. Chesnokova, *Dokl. Adad. Nauk SSSR,* **167,** 1295 (1966); *Dokl. Chem. (Engl. Transl.),* **167,** 432 (1966).
101. G. L. Bachman and B. D. Vineyard, *Ger. Pat.* 2638070 (1977); *C. A.* **88,** 6562 (1978).
102. G. L. Bachman and B. D. Vineyard, *U.S. Pat.* 4194051 (1980); *C. A.* **93,** 94824 (1980).
103. Badische Anilin and Soda-Fabrik A-G., *Belg. Pat.* 670029 (1966); *C. A.* **66,** 66007 (1966).
104. H. Bahrmann and B. Fell, *J. Mol. Catal.,* **8,** 329 (1980).
105. M. C. Baird, C. J. Nyman and G. Wilkinson, *J. Chem. Soc.,* 348 (1968).
106. E. C. Baker, D. E. Hendriksen and R. Eisenberg, *J. Am. Chem. Soc.,* **102,** 1020 (1980).
107. G. L. Baker, S. J. Fritschel and J. K. Stille, *J. Org. Chem.,* **46,** 2960 (1981).
108. G. L. Baker, S. J. Fritschel, J. R. Stille and J. K. Stille, *J. Org. Chem.,* **46,** 2954 (1981).
109. R. Baker and D. E. Halliday, *Tetrahedron Lett.,* 2773 (1972).
110. J. Bakos, I. Toth and L. Marko, *J. Org. Chem.,* **46,** 5427 (1981).
111. J. E. Baldwin and J. C. Swallow, *Angew. Chem., Int. Ed. Engl.,* **8,** 601 (1969).
112. R. G. Ball, B. R. James, D. Mahajan and J. Trotter, *Inorg. Chem.,* **20,** 254 (1981).
113. R. Bardet, J. Thivolle-Cazat and Y. Trambouze, *C.R. Acad. Sci., Ser. B,* **292,** 883 (1981).
114. E. K. Barefield, G. W. Parshall and F. N. Tebbe, *J. Am. Chem. Soc.,* **92,** 5234 (1970).
115. K. W. Barnett, D. L. Beach, D. L. Garin and L. A. Kaempfe, *J. Am. Chem. Soc.,* **96,** 7127 (1974).
116. M. Bartholin, C. Graillat and A. Guyot, *J. Mol. Catal.,* **10,** 99 (1980).
117. M. Bartholin, C. Graillat and A. Guyot, *J. Mol. Catal.,* **10,** 361 (1981).
118. M. Bartholin, C. Graillat, A. Guyot, G. Coudurier, J. Bandiera and C. Naccache, *J. Mol. Catal.,* **3,** 17 (1977).
119. C. M. Bartish, *Belg. Pat.* 862828 (1978); *C. A.* **89,** 214906 (1978).
120. C. M. Bartish, *Ger. Pat.* 2800986 (1978); *C. A.* **89,** 163081 (1978).
121. C. M. Bartish, *U.S. Pat.* 4102920 (1978); *C. A.* **90,** 54480 (1979).
122. C. M. Bartish, *U.S. Pat.* 4102921 (1978); *C. A.* **90,** 5926 (1979).
123. C. M. Bartish, *U.S. Pat.* 4230641 (1980); *C. A.* **94,** 120864 (1981).
124. D. Baudry, M. Ephritikhine and H. Felkin, *Chem. Commun.,* 694 (1978).
125. D. Baudry, M. Ephritikhine and H. Felkin, *Nouv. J. Chim.,* **2,** 355 (1978).
126. E. Bayer and V. Schurig, *Angew. Chem., Int. Ed. Engl.,* **14,** 493 (1975).
127. B. Bayerl, M. Wahren and J. Graefe, *Tetrahedron Lett.,* 1837 (1973).
128. V. Bazant, M. Capka, I. Dietzmann, H. Fuhrmann, J. Hetflejs and H. Pracejus, *East Ger. Pat.* 103902 (1972); *C. A.* **81,** 49816 (1974).
129. V. Bazant, M. Capka, I. Dietzmann, H. Fuhrmann, H. Hetflejs and H. Pracejus, *East Ger. Pat.* 103903 (1972); *C. A.* **81,** 49815 (1974).
130. D. L. Beach and K. W. Barnett, *J. Organomet. Chem.,* **142,** 225 (1977).
131. D. L. Beach, D. L. Garin, L. A. Kaempfe and K. W. Barnett, *J. Organomet. Chem.,* **142,** 211 (1977).
132. D. Beaupère, P. Bauer, L. Nadjo and R. Uzan, *J. Organomet. Chem.,* **238,** C12 (1982).

133. D. Beaupère, P. Bauer and R. Uzan, *Can. J. Chem.*, **57**, 218 (1979).
134. D. Beaupère, L. Nadjo, R. Uzan and P. Bauer, *J. Mol. Catal.*, **14**, 129 (1982).
135. W. Beck and H. Menzel, *J. Organomet. Chem.*, **133**, 307 (1977).
136. Y. Becker, A. Eisenstadt and J. K. Stille, *J. Org. Chem.*, **45**, 2145 (1980).
137. P. D. Beirne, *U.S. Pat.* 3 524 898 (1970); *C. A.* **73**, 87 617 (1970).
138. R. W. Beisner and S. C. Winans, *U.S. Pat.* 4 225 530 (1980); *C. A.* **94**, 3751 (1981).
139. J. Benes and J. Hetflejs, *Collect. Czech. Chem. Commun.*, **41**, 2264 (1976).
140. M. A. Bennett and D. L. Milner, *J. Am. Chem. Soc.*, **91**, 6983 (1969).
141. M. A. Bennett and T. R. B. Mitchell, *J. Organomet. Chem.*, **70**, C30 (1974).
142. M. A. Bennett and T. Yoshida, *J. Am. Chem. Soc.*, **95**, 3030 (1973).
143. M. J. Bennett and P. B. Donaldson, *J. Am. Chem. Soc.*, **93**, 3307 (1971).
144. L. Benzoni, A. Andreetta, C. Zanzottera and M. Camia, *Chim. Ind.* (*Milan*), **48**, 1076 (1966).
145. A. S. Berenblyum, M. V. Ermolaev, I. V. Kalechits, L. I. Lakhman and M. L. Khidekel, *U.S.S.R. Pat.* 413 978 (1974); *C. A.* **81**, 127 334 (1974).
146. D. E. Bergbreiter and M. S. Bursten, *J. Macromol. Sci., Chem.*, **A16**, 369 (1981).
147. W. Bergstein, A. Kleemann and J. Martens, *Synthesis*, 76 (1981).
148. W. Bergstein, A. Kleemann and J. Martens, *Ger. Pat.* 3 000 445 (1981); *C. A.* **95**, 187 420 (1981).
149. G. Bernard, Y. Chauvin and D. Commereuc, *Bull. Soc. Chim. Fr.*, 1163 (1976).
150. J. Berthoux, J. P. Martinaud and R. Poilblanc, *Ger. Pat.* 2 039 938 (1971); *C. A.* **74**, 99 461 (1971).
151. V. V. Bessenov, O. Yu. Oklobystin, T. P. Panova and L. Yu. Ukhin, *Teor. Eksp. Khim.*, **12**, 829 (1976); *C. A.* **86**, 154 984 (1977).
152. S. Bhaduri, B. F. G. Johnson, C. J. Savory, J. A. Segal and R. H. Walter, *Chem. Commun.*, 809 (1974); S. Bhaduri and B. F. G. Johnson, *Transition Met. Chem.*, **3**, 156 (1978).
153. S. Bhaduri, H. Khwaja and V. Khanwalkar, *J. Chem. Soc., Dalton Trans.*, 445 (1982).
154. M. M. Bhagwat and D. Devaprabhakara, *Tetrahedron Lett.*, 1391 (1972).
155. K. K. Bhatia and C. C. Cumbo, *Ger. Pat.* 2 523 839 (1976); *C. A.* **84**, 180 235 (1976).
156. K. K. Bhatia and C. C. Cumbo, *Ger. Pat.* 2 523 889 (1976); *C. A.* **84**, 180 234 (1976).
157. C. E. Bibby, R. Grigg and J. N. Grover, *Tetrahedron Lett.*, 1783 (1975).
158. C. E. Bibby, R. Grigg and R. Price, *J. Chem. Soc., Dalton Trans.*, 872 (1977).
159. J. F. Biellmann, *Bull. Soc. Chim. Fr.*, 3055 (1968).
160. J. F. Biellmann and M. J. Jung, *J. Am. Chem. Soc.*, **90**, 1673 (1968).
161. J. F. Biellmann, M. J. Jung and W. R. Pilgrim, *Bull. Soc. Chim. Fr.*, 2720 (1971).
162. J. F. Biellmann and H. Liesenfelt, *C.R. Acad. Sci., Ser. C*, **263**, 251 (1966).
163. J. F. Biellmann and H. Liesenfelt, *Bull. Soc. Chim. Fr.*, 4029 (1966).
164. B. Bierling, K. Kirschke, H. Oberender and M. Schultz, *Z. Chem.*, **9**, 105 (1969).
165. J. L. Bilhou, V. Bilhou-Bougnol, W. F. Graydon, J. M. Basset and A. K. Smith, *J. Mol. Catal.*, **8**, 411 (1980).
166. J. L. Bilhou, V. Bilhou-Bougnol, W. F. Graydon, J. M. Basset, A. K. Smith, G. M. Zanderighi and R. Ugo, *J. Organomet. Chem.*, **153**, 73 (1978).
167. E. Billig and D. L. Bunning, *Eur. Pat. Appl.* 28 378 (1981); *C. A.* **95**, 132 293 (1981).
168. E. Billig, J. D. Jamerson and R. Lavelle, *Eur. Pat. Appl.* 28 892 (1981); *C. A.* **95**, 186 643 (1981).
169. P. Biloen and W. M. H. Sachtler, *Adv. Catal.*, **30**, 165 (1981).
170. D. Bingham, D. E. Webster and P. Wells, *J. Chem. Soc., Dalton Trans.*, 1514 (1974).
171. A. J. Birch and G. S. R. Subba-Rao, *Tetrahedron Lett.*, 2917 (1968).

172. A. J. Birch and G. S. R. Subba-Rao, *Tetrahedron Lett.*, 3797 (1968).
173. A. J. Birch and K. A. M. Walker, *J. Chem. Soc. C*, 1894 (1966).
174. A. J. Birch and K. A. M. Walker, *Tetrahedron Lett.*, 1935 (1967).
175. A. J. Birch and K. A. M. Walker, *Tetrahedron Lett.*, 3457 (1967).
176. A. J. Birch and K. A. M. Walker, *Aust. J. Chem.*, **24**, 513 (1971).
177. A. J. Birch and D. H. Williamson, *Org. React.*, **24**, 1 (1978).
178. K. C. Bishop, *Chem. Rev.*, **76**, 461 (1976).
179. D. St. C. Black, W. R. Jackson and J. M. Swan, *Comprehensive Organic Chemistry*, Vol. 3 (D. N. Jones, Ed.), Parts 15 and 16, Pergamon, Oxford, 1979.
180. S. N. Blackburn, R. N. Haszeldine, R. V. Parish and J. H. Setchfield, *J. Organomet. Chem.*, **192**, 329 (1980).
181. A. A. Blanc, H. Arzoumanian, E. J. Vincent and J. Metzger, *Bull. Soc. Chim. Fr.*, 2175 (1974).
182. J. Blum, *Tetrahedron Lett.*, 1605 (1966).
183. J. Blum, *Tetrahedron Lett.*, 3041 (1966).
184. J. Blum, J. Y. Becker, H. Rosenman and E. D. Bergmann, *J. Chem. Soc. B*, 1000 (1969).
185. J. Blum and S. Biger, *Tetrahedron Lett.*, 1825 (1970).
186. J. Blum and A. Fisher, *Tetrahedron Lett.*, 1963 (1970).
187. J. Blum, S. Kraus and Y. Pickholtz, *J. Organomet. Chem.*, **33**, 227 (1971).
188. J. Blum and Z. Lipskes, *J. Org. Chem.*, **34**, 3076 (1969).
189. J. Blum, D. S. Milstein and Y. Sasson, *J. Org. Chem.*, **35**, 3233 (1970).
190. J. Blum, E. Oppenheimer and E. D. Bergmann, *J. Am. Chem. Soc.*, **89**, 2338 (1967).
191. J. Blum and Y. Pickholtz, *Isr. J. Chem.*, **7**, 723 (1969).
192. J. Blum, H. Rosenman and E. D. Bergmann, *Tetrahedron Lett.*, 3665 (1967).
193. J. Blum, H. Rosenman and E. D. Bergmann, *J. Org. Chem.*, **33**, 1928 (1968).
194. J. Blum and G. Scharf, *J. Org. Chem.*, **35**, 1895 (1970).
195. J. Blum, C. Zlotogorski, H. Schwartz and G. Hoehne, *Tetrahedron Lett.*, 3501 (1978).
196. J. Blum, C. Zlotogorski and A. Zoran, *Tetrahedron Lett.*, 1117 (1975).
197. M. Boldt, G. Gubitosa, H. H. Brintzinger and F. Wild, *Ger. Pat.* 2 727 245 (1978); *C. A.* **90**, 168 738 (1979).
198. G. C. Bond, *Ger. Pat.* 2 055 539 (1971); *C. A.* **75**, 48 429 (1971).
199. G. C. Bond and R. A. Hillyard, *Faraday Discuss. Chem. Soc.*, **46**, 20 (1968).
200. R. Bonnaire, L. Horner and F. Schumacher, *J. Organomet. Chem.*, **161**, C41 (1978).
201. P. Bonvicini, A. Levi, G. Modena and G. Scorrano, *Chem. Commun.*, 1188 (1972).
202. B. L. Booth, R. N. Haszeldine and G. R. H. Neuss, *J. Chem. Soc., Perkin Trans. 1*, 209 (1975).
203. F. B. Booth; *U.S. Pat.* 3 511 880 (1970); *C. A.* **73**, 14 191 (1970).
204. F. B. Booth, *U.S. Pat.* 3 641 076 (1972); *C. A.* **76**, 101 934 (1972).
205. F. B. Booth, *U.S. Pat.* 3 965 192 (1976); *C. A.* **85**, 123 590 (1976).
206. F. B. Booth, *U.S. Pat.* 4 267 383 (1981); *C. A.* **95**, 114 803 (1981).
207. N. V. Borunova, P. G. Antonov, Y. N. Kukushkin, L. K. Freidlin, N D. Trink, V. N. Ignatov and I. K. Shirshikova, *Zh. Obshch. Khim.*, **50**, 1862 (1980); *C. A.* **94**, 30 265 (1981).
208. N. V. Borunova, V. M. Ignatov, A. F. Lunin, L. K. Freidlin and J.-C. Ir, *U.S.S.R. Pat.* 650 983 (1979); *C. A.* **90**, 186 337 (1979).
209. B. Bosnich and M. D. Fryzuk, *J. Am. Chem. Soc.*, **100**, 5491 (1978).

210. B. Bosnich and M. D. Fryzuk, *Top. Stereochem.,* **12**, 119 (1981).
211. K. Bott, *Angew. Chem., Int. Ed. Engl.,* **12**, 851 (1973).
212. C. Botteghi, M. Branca, G. Micera, F. Piacenti and G. Menchi, *Chim. Ind. (Milan)*, **60**, 16 (1978).
213. C. Botteghi, M. Branca and A. Saba, *J. Organomet. Chem.,* **184**, C17 (1980).
214. M. Bottrill and M. Green, *J. Organomet. Chem.,* **111**, C6 (1976).
215. K. Bouchal, J. Skramouska, J. Coupek, S. Pokorny and F. Hrabak, *Makromol. Chem.,* **156**, 225 (1972).
216. L. J. Boucher, A. A. Oswald and L. L. Murrell, *Prepr. Pap. Nat. Meet., Div. Pet. Chem., Am. Chem. Soc.,* **19**, 162 (1974); *C. A.* **83**, 198 255 (1975).
217. J. Bourson and L. Oliveros, *J. Organomet. Chem.,* **229**, 77 (1982).
218. W. J. Boyle and F. Mares, *Organometallics,* **1**, 1003 (1982).
219. K. A. Brady and T. A. Nile, *J. Organomet. Chem.,* **206**, 299 (1981).
220. P. S. Braterman, *Top. Curr. Chem.,* **92**, 149 (1980).
221. A. Brenner and D. A. Hucul, *J. Am. Chem. Soc.,* **102**, 2484 (1980).
222. J. L. Brewbaker, *U.S. Pat.* 3 367 961 (1968); *C. A.* **69**, 2522 (1968).
223. E. A. V. Brewester and R. L. Pruett, *U.S. Pat.* 4 247 486 (1981); *C. A.* **95**, 24 271 (1981).
224. A. Bright, J. F. Malone, J. K. Nicholson, J. Powell and B. L. Shaw, *Chem. Commun.,* 712 (1971).
225. H. H. Brintzinger, *Brit. Appl.* 2 000 153 (1979); *C. A.* **91**, 213 674 (1979).
226. British Petroleum Co. Ltd., *Fr. Pat.* 1 549 414 (1968); *C. A.* **72**, 2995 (1970).
227. British Petroleum Co. Ltd., *Fr. Pat.* 1 558 222 (1969); *C. A.* **72**, 31 226 (1970).
228. British Petroleum Co. Ltd., *Fr. Pat.* 1 573 158 (1969); *C. A.* **72**, 100 035 (1970).
229. D. Brodzki, B. Denise and G. Pannetier, *J. Mol. Catal.,* **2**, 149 (1977).
230. D. Brodzki, C. Leclere, B. Denise and G. Pannetier, *Bull. Soc. Chim. Fr.,* 61 (1976).
231. C. K. Brown and G. Wilkinson, *Tetrahedron Lett.,* 1725 (1969).
232. C. K. Brown and G. Wilkinson, *J. Chem. Soc. A.,* 2753 (1970).
233. J. M. Brown, L. R. Canning, A. G. Kent and P. J. Sidebottom, *Chem. Commun.,* 721 (1982).
234. J. M. Brown and P. A. Chaloner, *Chem. Commun.,* 321 (1978).
235. J. M. Brown and P. A. Chaloner, *J. Am. Chem. Soc.,* **102**, 3040 (1980).
236. J. M. Brown, P. A. Chaloner, G. Descotes, R. Glaser, D. Lafont and D. Sinou, *Chem. Commun.,* 611 (1979).
237. J. M. Brown, P. A. Chaloner, R. Glaser and S. Geresh, *Tetrahedron,* **36**, 815 (1980).
238. J. M. Brown, P. A. Chaloner, A. G. Kent, B. A. Murrer, P. N. Nicholson, D. Parker and P. J. Sidebottom, *J. Organomet. Chem.,* **216**, 263 (1981).
239. J. M. Brown, P. A. Chaloner and P. N. Nicholson, *Chem. Commun.,* 646 (1978).
240. J. M. Brown, P. A. Chaloner and D. Parker, *Adv. Chem. Ser.,* **196**, 355 (1982).
241. J. M. Brown, and A. G. Kent, *Chem. Commun.,* 723 (1982).
242. J. M. Brown and H. Molinari, *Tetrahedron Lett.,* 2933 (1979).
243. J. M. Brown and B. A. Murrer, *Tetrahedron Lett.,* 581 (1980).
244. J. M. Brown and R. G. Naik, *Chem. Commun.,* 348 (1982).
245. J. M. Brown and D. Parker, *Chem. Commun.,* 342 (1980).
246. J. M. Brown and D. Parker, *Organometallics,* **1**, 950 (1982).
247. J. M. Brown and D. Parker, *J. Org. Chem.,* **47**, 2722 (1982).
248. M. Brown and L. W. Piszkiewicz, *J. Org. Chem.,* **32**, 2013 (1967).
249. P. A. Brown and D. N. Kirk, *J. Chem. Soc. C.,* 1653 (1969).

250. A. Bruggink and H. Hogeveen, *Tetrahedron Lett*., 4961 (1972).
251. S. Brunie, J. Mazan, N. Langlois and H. B. Kagan, *J. Organomet. Chem.*, **114**, 225 (1976).
252. H. Brunner, *Adv. Organomet. Chem.*, **18**, 152 (1980).
253. H. Brunner and W. Pieronczyk, *Angew. Chem.*, **91**, 655 (1979).
254. H. Brunner and W. Pieronczyk, *Ger. Pat.* 2 908 358 (1980); *C. A.* **94**, 103 556 (1981).
255. H. Brunner, W. Pieronczyk, B. Schoenhammer, K. Streng, I. Bernal and J. Korp, *Chem. Ber.*, **114**, 1137 (1981).
256. H. Brunner and M. Probster, *Inorg. Chim. Acta*, **61**, 129 (1982).
257. H. Brunner and G. Riepl, *Angew. Chem., Int. Ed. Engl.*, **21**, 377 (1982).
258. D. R. Bryant and E. Billig, *Ger. Pat.* 2 802 923 (1978); *C. A.* **89**, 163 066 (1978).
259. Z. C. Brzezinska, W. R. Cullen and G. Strukul, *Can. J. Chem.*, **58**, 750 (1980).
260. D. E. Budd, D. G. Holah, A. N. Hughes and B. C. Hui, *Can. J. Chem.*, **52**, 775 (1974).
261. N. A. Bumagin, I. O. Kalinovskii and I. P. Beletskata, *Izv. Akad. Nauk SSSR, Ser. Khim.*, 221 (1982); *C. A.* **96**, 122 341 (1982).
262. R. R. Burch, E. L. Muetterties, R. G. Teller and J. M. Williams, *J. Am. Chem. Soc.*, **104**, 4257 (1982).
263. M. G. Burnett and R. J. Morrison, *J. Chem. Soc. A*, 2325 (1971).
264. M. G. Burnett and R. J. Morrison, *J. Chem. Soc., Dalton Trans.*, 632 (1973).
265. M. G. Burnett, R. J. Morrison and C. J. Strugnell, *J. Chem. Soc., Dalton Trans.*, 701 (1973).
266. M. G. Burnett, R. J. Morrison and C. J. Strugnell, *J. Chem. Soc., Dalton Trans.*, 1663 (1974).
267. M. G. Burnett and C. J. Strugnell, *J. Chem. Res. (S)*, 250 (1977).
268. V. I. Bystrenina, *Issled. Obl. Sint. Katal. Org. Soedin.*, 36 (1975); *C. A.* **86**, 106 005 (1977).
269. G. Caccia, G. Chelucci and C. Botteghi, *Synth. Commun.*, **11**, 71 (1981).
270. H. J. Callot and C. Piechocki, *Tetrahedron Lett*., 3489 (1980).
271. R. E. Campbell, C. F. Lochow, K. P. Vora and R. G. Miller, *J. Am. Chem. Soc.*, **102**, 5824 (1980).
272. R. E. Campbell and R. G. Miller, *J. Organomet. Chem.*, **186**, C27 (1980).
273. W. H. Campbell and P. W. Jennings, *Organometallics*, **1**, 1071 (1982).
274. A. Camus, G. Mestroni and G. Zassinovich, *J. Mol. Catal.*, **6**, 231 (1979).
275. A. Camus, G. Mestroni and G. Zassinovich, *J. Organomet. Chem.*, **184**, C10 (1980).
276. A. J. Canale, W. A. Hewett, T. M. Shryne and E. A. Youngman, *Chem. Ind. (London)*, 1054 (1962).
277. J. P. Candlin and A. R. Oldham, *Faraday Discuss. Chem. Soc.*, **46**, 60 (1968).
278. M. Capka, *Collect. Czech. Chem. Commun.*, **42**, 3410 (1977).
279. M. Capka, J. Hetflejs and R. Selke, *React. Kinet. Catal. Lett.*, **10**, 225 (1979).
280. M. Capka, P. Svoboda, M. Kraus and J. Hetflejs, *Chem. Ind. (London)*, 650 (1972).
281. V. Caplar, G. Comisso and V. Sunjic, *Synthesis*, 85 (1981).
282. H. A. J. Carless, *Chem. Commun.*, 982 (1974).
283. J. T. Carlock, *U.S. Pat.* 4 173 575 (1979); *C. A.* **92**, 58 231 (1980).
284. J. T. Carlock, *U.S. Pat.* 4 178 312 (1979); *C. A.* **92**, 110 538 (1980).
285. J. T. Carlock, *U.S. Pat.* 4 178 313 (1979); *C. A.* **92**, 163 551 (1980).
286. J. T. Carlock, *U.S. Pat.* 4 178 314 (1979); *C. A.* **92**, 146 243 (1980).
287. J. T. Carlock, *U.S. Pat.* 4 183 825 (1980); *C. A.* **92**, 100 142 (1980).
288. J. T. Carlock, *U.S. Pat.* 4 198 353 (1980); *C. A.* **93**, 54 933 (1980).
289. L. Carlton and G. Read, *J. Chem. Soc., Perkin Trans. 1*, 1631 (1978).

290. L. Carlton and G. Read, *J. Mol. Catal.*, **10**, 133 (1981).
291. M. Carvalho, L. F. Wieserman and D. M. Hercules, *Appl. Spectrosc.*, **36**, 290 (1982).
292. L. Cassar, P. E. Eaton and J. Halpern, *J. Am. Chem. Soc.*, **92**, 3515 (1970).
293. T. Castrillo, H. Knoezinger, J. Lieto and M. Wolf, *Inorg. Chim. Acta*, **44**, L239 (1980).
294. M. Catellani, G. P. Chiusoli, W. Giroldini and G. Salerno, *J. Organomet. Chem.*, **199**, C21 (1980).
295. C. Cativiela, J. Fernandez, J. A. Mayoral, E. Melendez, R. Uson, L. A. Oro and M. J. Fernandez, *J. Mol. Catal.*, **16**, 19 (1982).
296. P. Cayalieri D'Oro, L. Raimondi, G. Pagani, G. Montrasi, G. Gregorio and A. Andreetta, *Chim. Ind. (Milan)*, **62**, 572 (1980).
297. J. N. Cawse and J. L. Vidal, *Ger. Pat.* 2 813 543 (1978); *C. A.* **90**, 38 522 (1979).
298. S. Cenini, R. Ugo and F. Porta, *Gazz. Chim. Ital.*, **111**, 293 (1981).
299. J. C. Chabala and M. H. Fisher, *U.S. Pat.* 4 199 569 (1980); *C. A.* **93**, 130 559 (1980).
300. P. Chabardes and Y. Querou, *Fr. Pat.* 1 582 621 (1969); *C. A.* **73**, 44 903 (1970).
301. A. J. Chalk, *J. Organomet. Chem.*, **21**, 207 (1970).
302. A. J. Chalk and J. F. Harrod, *Fr. Pat.* 1 366 279 (1964); *C. A.* **62**, 2795 (1965).
303. A. J. Chalk and J. F. Harrod, *J. Am. Chem. Soc.*, **87**, 16 (1965).
304. A. S. C. Chan and J. Halpern, *J. Am. Chem. Soc.*, **102**, 838 (1980).
305. A. S. C. Chan, J. J. Pluth and J. Halpern, *Inorg. Chim. Acta*, **37**, L477 (1979).
306. A. S. C. Chan, J. J. Pluth and J. Halpern, *J. Am. Chem. Soc.*, **102**, 5952 (1980).
307. C. Y. Chan and B. R. James, *Inorg. Nucl. Chem. Lett.*, **9**, 135 (1973).
308. G. Chandra, *Br. Pat.* 1 421 136 (1976); *C. A.* **84**, 150 754 (1976).
309. E. S. Chandrasekaran, D. A. Thompson and R. W. Rudolph, *Inorg. Chem.*, **17**, 760 (1978).
310. B. H. Chang, R. H. Grubbs and C. H. Brubaker, *J. Organomet. Chem.*, **172**, 81 (1979).
311. H. B. Charman, *J. Chem. Soc. B*, 584 (1970).
312. Y. Chauvin, D. Commereuc and R. Stern, *J. Organomet. Chem.*, **146**, 311 (1978).
313. P. S. Chekrii, M. L. Khidekel, I. V. Kalechits, O. N. Eremenko, G. I. Karyakina and A. S. Todozhokova, *Izv. Akad. Nauk SSSR, Ser. Khim.*, 1579 (1972); *C. A.* **77**, 140 605 (1972).
314. P. S. Chekrii, M. L. Khidekel, N. N. Kotova, A. M. Taber and I. V. Kalechits, *U.S.S.R. Pat.* 499 891 (1976); *C. A.* **84**, 141 310 (1976).
315. M. J. Chen and H. M. Feder, *Inorg. Chem.*, **18**, 1864 (1979).
316. C.-H. Cheng, D. E. Hendriksen and R. Eisenberg, *J. Am. Chem. Soc.*, **99**, 2791 (1977).
317. E. G. Chepaikin, M. L. Khidekel, V. Ivanova, A. Zakhariev and D. Shopov, *J. Mol. Catal.*, **10**, 115 (1980).
318. E. A. Chernyshev, G. I. Magomedov, O. V. Shkol'nik, V. G. Syrkin and Z. V. Belyakova, *Zh. Obshch. Khim.*, **48**, 1742 (1978); *C. A.* **89**, 180 087 (1978).
319. Y. Chevallier, J. P. Martinaud, F. Meiller and J. Berthoux, *Fr. Pat.* 2 243 022 (1975); *C. A.* **83**, 137 539 (1975).
320. C.-S. Chin, M. S. Sennett and L. Vaska, *J. Mol. Catal.*, **4**, 375 (1978).
321. P. Chini, S. Martinengo and G. Garlaschelli, *Chem. Commun.*, 709 (1972).
322. Chisso Corp., *Jpn. Pat.* 133 297 (1981); *C. A.* **96**, 218 018 (1982).
323. G. P. Chiusolo, W. Giroldini, L. Pallini and G. Salerno, *Eur. Pat. Appl.* 32 548 (1981); *C. A.* **96**, 68 346 (1982).

324. G. P. Chiusolo, W. Giroldini and G. Salerno, *Eur. Pat. Appl.* 15 537 (1980); *C. A.* **94**, 120 875 (1981).
325. B. R. Cho and R. M. Laine, *J. Mol. Catal.,* **15**, 383 (1982).
326. W. C. Christopfel and B. D. Vineyard, *J. Am. Chem. Soc.,* **101**, 4406 (1979).
327. P.-W. Chum and J. A. Róth, *J. Catal.,* **39**, 198 (1975).
328. F. Ciardelli, G. Braca, C. Carlini, G. Sbrana and G. Valentini, *J. Mol. Catal.,* **14**, 1 (1982).
329. D. A. Clement, J. F. Nixon and J. S. Poland, *J. Organomet. Chem.,* **76**, 117 (1974).
330. M. G. Clerici, S. Di Gioacchino, F. Maspero, E. Perrotti and A. Zanobi, *J. Organomet. Chem.,* **84**, 379 (1975).
331. R. S. Coffey, *Chem. Commun.,* 923 (1967).
332. R. S. Coffey, *Chem. Commun.,* 923 (1967).
333. R. S. Coffey; *Br. Pat.* 1 121 643 (1968); *C. A.* **69**, 60 537 (1968).
334. R. S. Coffey, *Br. Pat.* 1 130 743 (1968); *C. A.* **70**, 59 417 (1969).
335. R. S. Coffey, *Br. Pat.* 1 135 979 (1968); *C. A.* **70**, 67 840 (1969).
336. R. S. Coffey, *U.S. Pat.* 3 546 266 (1970); *C. A.* **74**, 88 123 (1971).
337. R. S. Coffey, *Br. Pat.* 1 222 216 (1971); *C. A.* **75**, 5255 (1971).
338. T. Cole, R. Ramage, K. Cann and R. Pettit, *J. Am. Chem. Soc.,* **102**, 6182 (1980).
339. J. P. Collman and L. S. Hegedus, *Principles and Applications of Organotransition Metal Chemistry*, University Science Books, CA, 1980.
340. J. P. Collman, L. S. Hegedus, M. P. Cooke, J. R. Norton, G. Dolcetti and D. N. Marquardt, *J. Am. Chem. Soc.,* **94**, 1789 (1972).
341. J. P. Collman, N. W. Hoffman and D. E. Morris, *J. Am. Chem. Soc.,* **91**, 5659 (1969).
342. J. P. Collman, M. Kubota and J. W. Hosking, *J. Am. Chem. Soc.,* **89**, 4809 (1967).
343. Compagne des Metaux Precieux, *Fr. Pat.* 2 264 801 (1975); *C. A.* **84**, 179 682 (1976).
344. N. G. Connelly, P. T. Draggett and M. Green, *J. Organomet. Chem.,* **140**, C10 (1977).
345. G. Consiglio, C. Botteghi, C. Salomon and P. Pino, *Angew. Chem., Int. Ed. Engl.,* **12**, 669 (1973).
346. G. Consiglio, D. A. Von Bezard, F. Morandini and P. Pino, *Helv. Chim. Acta,* **61**, 1703 (1978).
347. J. Cook, J. E. Hamlin, P. M. Maitlis and A. Nutton, *Brit. Pat. Appl.* 2 054 592 (1981); *C. A.* **95**, 97 075 (1981).
348. J. Cook, J. E. Hamlin, A. Nutton and P. M. Maitlis, *Chem. Commun.,* 144 (1980); *J. Chem. Soc., Dalton Trans.,* 2342 (1981).
349. J. Cook and P. M. Maitlis, *Chem. Commun.,* 924 (1981).
350. E. J. Corey and J. W. Suggs, *J. Org. Chem.,* **38**, 3224 (1973).
351. W. Cornely and B. Fell, *Chem. Ztg.,* **105**, 317 (1981).
352. B. Cornils, R. Payer and K. C. Traenckner, *Hydrocarbon Process.,* **54**, 89 (1975); *C. A.* **83**, 134 522 (1975).
353. A. J. Cornish, M. F. Lappert, G. L. Filatovs and T. A. Nile, *J. Organomet. Chem.,* **172**, 153 (1979).
354. R. J. P. Corriu and J. J. E. Moreau, *Chem. Commun.,* 812 (1971).
355. R. J. P. Corriu and J. J. E. Moreau, *J. Organomet. Chem.,* **40**, 55 (1972).
356. R. J. P. Corriu and J. J. E. Moreau, *Chem. Commun.,* 38 (1973).
357. R. J. P. Corriu and J. J. E. Moreau, *J. Organomet. Chem.,* **64**, C51 (1974); *ibid.,* **85**, 19 (1975).
358. L. A. Cosby and R. A. Fiato, *U.S. Pat.* 4 115 433 (1978); *C. A.* **90**, 103 385 (1979).
359. B. Courtis, S. P. Dent, C. Eaborn and A. Pidcock, *J. Chem. Soc., Dalton Trans.,* 2460 (1975).

360. R. H. Crabtree, *Chem. Commun.*, 647 (1975).
361. R. H. Crabtree, *Acc. Chem. Res.*, **12**, 331 (1979).
362. R. H. Crabtree, D. F. Chodosh, J. M. Quirk, H. Felkin, T. Khan-Fillebeen and G. E. Morris, *Fundam. Res. Homogeneous Catal.*, **3**, 475 (1979).
363. R. H. Crabtree and H. Felkin, *J. Mol. Catal.*, **5**, 75 (1979).
364. R. H. Crabtree, H. Felkin, T. Khan and G. B. Morris, *J. Organomet. Chem.*, **144**, C15 (1978).
365. R. H. Crabtree, H. Felkin and G. E. Morris, *Chem. Commun.*, 716 (1976); *J. Organomet. Chem.*, **141**, 205 (1977).
366. R. H. Crabtree, J. M. Mihelcic and J. M. Quirk, *J. Am. Chem. Soc.*, **101**, 7738 (1979); R. H. Crabtree, M. F. Mellea, J. M. Mihelcic and J. M. Quirk, *J. Am. Chem. Soc.*, **104**, 107 (1982).
367. J. H. Craddock, A. Hershman, F. E. Paulik and J. F. Roth, *Ind. Eng. Chem.*, *Prod. Res. Dev.*, **8**, 291 (1969).
368. J. H. Craddock, A. Hershman, F. E. Paulik and J. F. Roth, *Ger. Pat.* 1 941 501 (1970); *C. A.* **72**, 110 811 (1970).
369. R. Cramer, *J. Am. Chem. Soc.*, **87**, 4717 (1965).
370. R. Cramer, *J. Am. Chem. Soc.*, **88**, 2272 (1966).
371. R. Cramer, *J. Am. Chem. Soc.*, **89**, 1633 (1967).
372. R. Cramer, *Acc. Chem. Res.*, **1**, 186 (1968).
373. R. Cramer, *U.S. Pat.* 3 502 738 (1970); *C. A.* **73**, 14 130 (1970).
374. R. Cramer, *Trans. N. Y. Acad. Sci.*, **33**, 97 (1971).
375. R. Cramer, *Ann. N. Y. Acad. Sci.*, **172**, 507 (1971).
376. R. Cramer, *J. Am. Chem. Soc.*, **94**, 5681 (1972).
377. R. Cramer and R. V. Lindsey, *J. Am. Chem. Soc.*, **88**, 3534 (1966).
378. R. Cramer and R. V. Lindsey, *U.S. Pat.* 3 636 122 (1972); *C. A.* **76**, 71 982 (1972).
379. B. I. Cruikshank and N. R. Davies, *Aust. J. Chem.*, **26**, 1935 (1973).
380. G. Csontos, B. Heil and L. Marko, *Ann. N. Y. Acad. Sci.*, **239**, 47 (1974).
381. G. Csontos, B. Heil, L. Marko and P. Chini, *Hung. J. Ind. Chem.*, **1**, 53 (1973); *C. A.* **79**, 41 608 (1973).
382. W. R. Cullen, F. W. B. Einstein, C.-H. Huang, A. C. Willis and E. S. Yeh, *J. Am. Chem. Soc.*, **102**, 988 (1980).
383. W. R. Cullen, B. R. James, A. D. Jenkins, G. Strukel and Y. Sugi, *Inorg. Nucl. Chem. Lett.*, **13**, 577 (1977).
384. W. R. Cullen, B. R. James and G. Strukel, *Can. J. Chem.*, **56**, 1965 (1978).
385. W. R. Cullen, A. Fenster and B. R. James, *Inorg. Nucl. Chem. Lett.*, **10**, 167 (1974).
386. W. R. Cullen, D. J. Patmore, A. J. Chapman and A. D. Jenkins, *J. Organomet. Chem.*, **102**, C12 (1975).
387. W. R. Cullen and E. Shan-Yeh, *J. Organomet. Chem.*, **139**, C13 (1977).
388. W. R. Cullen and J. D. Woollins, *Can. J. Chem.*, **60**, 1793 (1982).
389. C. C. Cumbo, *Eur. Pat. Appl.* 3 753 (1979); *C. A.* **92**, 75 859 (1980).
390. C. S. Cundy, C. Eaborn and M. F. Lappert, *J. Organomet. Chem.*, **44**, 291 (1972).
391. J. A. Cusumano, R. A. Dalla Betta and R. B. Levy, *Catalysis in Coal Conversion*, Academic Press, New York, 1978.
392. M. Czakova and M. Capka, *J. Mol. Catal.*, **11**, 313 (1981).
393. J. Dahlmann, E. Hoeft and B. Giese, *East Ger. Pat.* 144 535 (1980); *C. A.* **94**, 208 370 (1981).
394. F. D'Amico, J. Von Jouanne and H. Kelm, *J. Mol. Catal.*, **6**, 327 (1979).

395. R. J. Daroda and G. Wilkinson, *Cienc. Nat.* (*St. Maria Braz.*), **2**, 33 (1980); *C. A.* **96**, 51 738 (1982).

396. J. Daub, U. Erhardt, M. Michna, J. Schmetzer and V. Trautz, *Chem. Ber.*, **111**, 2877 (1978).

397. W. G. Dauben and A. J. Kielbania, *J. Am. Chem. Soc.*, **94**, 3669 (1972).

398. W. G. Dauben, A. J. Kielbania and K. N. Raymond, *J. Am. Chem. Soc.*, **95**, 7166 (1973).

399. R. Dauby, F. Dawans and Ph. Teyssie, *J. Polym. Sci., Part C*, 1989 (1967).

400. P. J. Davidson and R. R. Hignett, *U.S. Pat.* 4 200 592 (1980); *C. A.* **93**, 113 966 (1980).

401. P. J. Davidson, M. F. Lappert and R. Pearce, *Chem. Rev.*, **76**, 219 (1976).

402. S. G. Davies, *Organotransition Metal Chemistry. Applications to Organic Synthesis*, Pergamon, Oxford, 1982.

403. J. L. Dawes, *U.S. Pat.* 4 258 215 (1981); *C. A.* **94**, 174 339 (1981).

404. J. L. Dawes and T. J. Devon, *U.S. Pat.* 4 196 096 (1980); *C. A.* **93**, 32 415 (1980).

405. V. W. Day, M. F. Fredrich, G. S. Reddy, A. J. Sivak, W. R. Pretzer and E. L. Muetterties, *J. Am. Chem. Soc.*, **99**, 809 (1977).

406. W. de Aquino, R. Bonnaire and C. Potvin, *J. Organomet. Chem.*, **154**, 159 (1978).

407. M. H. J. M. De Croon and J. W. E. Coenen, *J. Mol. Catal.*, **4**, 325 (1978).

408. M. H. J. M. De Croon and J. W. E. Coenen, *J. Mol. Catal.*, **11**, 301 (1981).

409. A. Dedieu, *Inorg. Chem.*, **19**, 375 (1980).

410. A. Dedieu, *Inorg. Chem.*, **20**, 2803 (1981).

411. A. Dedieu, A. Strich and A. Rossi, *Quantum Theory Chem. React.*, **2**, 193 (1981); *C. A.* **94**, 191 071 (1981).

412. M. Dedieu and Y. L. Pascal, *J. Mol. Catal.*, **9**, 71 (1980).

413. M. S. Delaney, C. B. Knobler and M. F. Hawthorne, *Chem. Commun.*, 849 (1980).

414. M. S. Delaney, R. G. Teller and M. F. Hawthorne, *Chem. Commun.*, 235 (1981).

415. A. Deluzarche, R. Fonseca, G. Jenner and A. Kiennemann, *Erdoel Kohle, Erdgas, Petrochem., Brennst.-Chem.*, **32**, 313 (1979); *C. A.* **91**, 157 195 (1979).

416. C. Demay, *Ger. Pat.* 3 023 025 (1981); *C. A.* **94**, 174 281 (1981).

417. G. C. Demitras and E. L. Muetterties, *J. Am. Chem. Soc.*, **99**, 2796 (1977).

418. A. Demonceau, A. F. Noels, A. J. Hubert and Ph. Teyssie, *Chem. Commun.*, 688 (1981).

419. Y. Demortier and I. De Aguirre, *Bull. Soc. Chim. Fr.*, 1614 (1974).

420. Y. Demortier and I. De Aguirre, *Bull. Soc. Chim. Fr.*, 1619 (1974).

421. N. A. De Munck, J. P. A. Notenboom, J. E. De Leur and J. J. F. Scholten, *J. Mol. Catal.*, **11**, 233 (1981).

422. N. A. De Munck and J. J. F. Scholten, *Eur. Pat. Appl.* 38 609 (1981); *C. A.* **96**, 52 173 (1982).

423. N. A. De Munck and J. J. F. Scholten, *Eur. Pat. Appl.* 40 891 (1981); *C. A.* **96**, 103 656 (1982).

424. N. A. De Munck, M. W. Verbruggen, J. E. De Leur and J. J. F. Scholten, *J. Mol. Catal.*, **11**, 331 (1981).

425. N. A. De Munck, M. W. Verbruggen and J. J. F. Scholten, *J. Mol. Catal.*, **10**, 313 (1981).

426. A. J. Dennis, *Eur. Pat. Appl.* 18 162 (1980); *C. A.* **94**, 156 338 (1981).

427. A. J. Dennis, T. F. Shevels and N. Harris, *Eur. Pat. Appl.* 16 286 (1980); *C. A.* **94**, 83 609 (1981).

428. S. P. Dent, C. Eaborn and A. Pidcock, *Chem. Commun.*, 1703 (1970).

429. G. Descotes, D. Lafont and D. Sinou, *J. Organomet. Chem.*, **150**, C14 (1978).
430. G. Descotes, D. Lafont, D. Sinou, J. M. Brown, P. A. Chaloner and D. Parker, *Nouv. J. Chim.*, **5**, 167 (1981).
431. C. Detellier, G. Gelbard and H. B. Kagan, *J. Am. Chem. Soc.*, **100**, 7556 (1978).
432. D. J. A. De Waal, T. I. A. Gerber and W. J. Louw, *Chem. Commun.*, 100 (1982).
433. K. C. Dewhirst, *U.S. Pat.* 3 366 646 (1968); *C. A.* **68**, 95 311 (1968).
434. K. C. Dewhirst, *U.S. Pat.* 3 480 659 (1969); *C. A.* **72**, 78 105 (1970).
435. K. C. Dewhirst, *U.S. Pat.* 3 489 786 (1970); *C. A.* **72**, 89 833 (1970).
436. K. C. Dewhirst, *U.S. Pat.* 3 639 439 (1972); *C. A.* **77**, 34 708 (1972).
437. K. C. Dewhirst, W. Keim and H. E. Thyret, *U.S. Pat.* 3 502 725 (1970); *C. A.* **72**, 111 023 (1970).
438. K. C. Dewhirst, W. Keim and H. E. Thyret, *U.S. Pat.* 3 652 614 (1972); *C. A.* **77**, 34 707 (1972).
439. S. E. Diamond and F. Mares, *U.S. Pat.* 4 215 218 (1980); *C. A.* **93**, 185 931 (1980).
440. S. E. Diamond, F. Mares and A. Szalkiewicz, *Fundam. Res. Homogeneous Catal.*, **3**, 345 (1979).
441. S. E. Diamond, A. Szalkiewicz and F. Mares, *J. Am. Chem. Soc.*, **101**, 490 (1979).
442. H. M. Dickers, R. N. Haszeldine, L. S. Malkin, A. P. Mather and R. V. Parish, *J. Chem. Soc., Dalton Trans.*, 308 (1980).
443. H. M. Dickers, R. N. Haszeldine, A. P. Mather and R. V. Parish, *J. Organomet. Chem.*, **161**, 91 (1978).
444. M. K. Dickson, B. P. Sudha and D. M. Roundhill, *J. Organomet. Chem.*, **190**, C43 (1980).
445. R. S. Dickson, in *Organometallic Chemistry of Rhodium and Iridium*, (P. M. Maitlis, F. G. A. Stone and R. West, Eds.), Academic Press, London, 1983.
446. I. Dietzmann, D. Tomanova and J. Hetflejs, *Collect. Czech. Chem. Commun.*, **39**, 123 (1974).
447. J. Dixon, *Br. Pat. Appl.* 2 075 857 (1981); *C. A.* **96**, 142 266 (1982).
448. C. Djerassi and J. Gutzwiller, *J. Am. Chem. Soc.*, **88**, 4537 (1966).
449. G. Dolcetti, *Inorg. Nucl. Chem. Lett.*, **9**, 705 (1973).
450. G. Dolcetti and N. W. Hoffman, *Inorg. Chim. Acta*, **9**, 269 (1974).
451. I. N. Domnin, E. F. Zhuraleva and N. V. Pronina, *Zh. Org. Khim.*, **14**, 2323 (1978); *C. A.* **90**, 137 423 (1979).
452. I. C. Douek and G. Wilkinson, *J. Chem. Soc. A*, 2604 (1969).
453. S. J. Dougherty, *U.S. Pat.* 4 263 218 (1981); *C. A.* **95**, 186 626 (1981).
454. S. J. Dougherty and R. C. Wells, *Ger. Pat.* 2 541 296 (1978); *C. A.* **89**, 46 085 (1978).
455. D. H. Doughty, M. P. Anderson, A. L. Casalnuovo, M. F. McGuiggan, C. C. Tso, H. H. Wang and L. H. Pignolet, *Adv. Chem. Ser.*, **196**, 65 (1982).
456. D. H. Doughty, M. F. McGuiggan, H. H. Wang and L. H. Pignolet, *Fundam. Res. Homogeneous Catal.*, **3**, 909 (1979).
457. D. H. Doughty and L. H. Pignolet, *J. Am. Chem. Soc.*, **100**, 7083 (1978).
458. Dow Chemical Co., *Jpn. Pat.* 109 447 (1980); *C. A.* **94**, 53 632 (1981).
459. M. P. Doyle, W. H. Tamblyn and V. Bagheri, *J. Org. Chem.*, **46**, 5094 (1981).
460. M. P. Doyle, W. H. Tamblyn, W. E. Buhro and R. L. Dorow, *Tetrahedron Lett.*, 1783 (1981).
461. M. P. Doyle and D. Van Leusen, *J. Am. Chem. Soc.*, **103**, 5917 (1981).
462. M. P. Doyle, D. Van Leusen and W. H. Tamblyn, *Synthesis*, 787 (1981).
463. R. S. Drago, E. D. Nyberg and A. G. El A'Mma, *Inorg. Chem.*, **20**, 2461 (1981).

464. R. S. Drago, E. D. Nyberg, A. G. El A'Mma and A. Zombeck, *Inorg. Chem.*, **20**, 641 (1981).
465. E. Drent, *Eur. Pat. Appl.* 42 633 (1981); *C. A.* **96**, 122 227 (1982).
466. E. Drent, *Eur. Pat. Appl.* 48 046 (1982); *C. A.* **96**, 217 268 (1982).
467. Y. Dror and J. Manassen, *J. Mol. Catal.*, **2**, 219 (1977).
468. D. J. Drury, M. J. Green, D. J. M. Ray and A. J. Stevenson, *J. Organomet. Chem.*, **236**, C23 (1982).
469. D. L. Dubois and D. W. Meek, *Inorg. Chim. Acta*, **19**, L29 (1976).
470. E. W. Duck, *Br. Pat.* 837 251 (1960); *C. A.* **54**, 20 327 (1960).
471. C. W. Dudley and G. Read, *Tetrahedron Lett.*, 5273 (1972); C. W. Dudley, G. Read and P. J. C. Walker, *J. Chem. Soc., Dalton Trans.*, 1926 (1974).
472. E. J. Dufek and G. R. List, *J. Am. Oil Chem. Soc.*, **54**, 276 (1977); *C. A.* **87**, 70 130 (1977).
473. H. Dumas, J. Levisalles and H. Rudler, *J. Organomet. Chem.*, **177**, 239 (1979).
474. W. Dumont, J. C. Poulin, T. P. Dang and H. B. Kagan, *J. Am. Chem. Soc.*, **95**, 8295 (1973).
475. N. A. Dunham and M. C. Baird, *J. Chem. Soc., Dalton Trans.*, 774 (1975).
476. D. Durand and C. Lassau, *Tetrahedron Lett.*, 2329 (1969).
477. J. Dwyer, H. S. Hilal and R. V. Parish, *J. Organomet. Chem.*, **228**, 191 (1982).
478. P. E. Eaton and S. A. Cerefice, *Chem. Commun.*, 1494 (1970).
479. P. E. Eaton and D. R. Patterson, *J. Am. Chem. Soc.*, **100**, 2573 (1978).
480. G. C. Eberhardt, M. E. Tadros and L. Vaska, *Chem. Commun.*, 290 (1972).
481. G. C. Eberhardt and L. Vaska, *J. Catal.*, **8**, 183 (1967).
482. A. Efraty and I. Feinstein, *Inorg. Chem.*, **21**, 3115 (1982).
483. R. E. Ehrenkaufer, R. R. MacGregor and A. P. Wolf, *J. Org. Chem.*, **47**, 2489 (1982).
484. R. Eisenberg and C.-H. Cheng, *U.S. Pat.* 4 107 076 (1978); *C. A.* **90**, 57 437 (1979).
485. B. Elleuch, Y. Ben Jaarit, J. M. Basset and J. Kervennal, *Angew Chem., Int. Ed. Engl.*, **21**, 687 (1982).
486. P. Ellwood, *Chem. Eng. (London)*, 148 (May 19th, 1969).
487. J. A. Elvidge, J. R. Jones, R. M. Lenk, Y. S. Tang, E. A. Evans, G. L. Guilford and D. C. Warrell, *J. Chem. Res. (S)*, 82 (1982).
488. A. Emery, A. C. Oehlschlager and A. M. Unrau, *Tetrahedron Lett.*, 4401 (1970).
489. H. Erpenbach, K. Gehrmann, H. K. Kuebbeler and K. Schmitz, *Ger. Pat.* 2 836 084 (1980); *C. A.* **93**, 45 992 (1980).
490. H. Erpenbach, K. Gehrmann, W. Lork and P. Pring, *Ger. Pat.* 3 016 900 (1981); *C. A.* **96**, 19 680 (1982).
491. L. S. Eubanks and J. L. Price, *U.S. Pat.* 4 111 982 (1978); *C. A.* **90**, 103 429 (1979).
492. D. Evans, J. A. Osborn and G. Wilkinson, *J. Chem. Soc. A*, 3133 (1968).
493. D. Evans, G. Yagupsky and G. Wilkinson, *J. Chem. Soc. A*, 2660 (1968).
494. J. Evans, *Chem. Soc. Rev.*, **10**, 159 (1981).
495. D. R. Fahey, *J. Am. Chem. Soc.*, **103**, 136 (1981).
496. J. Falbe (Ed.), *New Syntheses with Carbon Monoxide*, Springer-Verlag, Berlin, 1980.
497. J. Falbe and F. Korte, *Brennst-Chem.*, **45**, 103 (1964); *C. A.* **61**, 2953 (1964).
498. J. Falbe and J. Weber, *Fr. Pat.* 1 598 768 (1970); *C. A.* **75**, 40 965 (1971).
499. R. A. Faltynek, *Inorg. Chem.*, **20**, 1357 (1981).
500. L. W. Fannin, V. D. Phillips and T. C. Singleton, *Ger. Pat.* 2 358 410 (1974); *C. A.* **81**, 91 059 (1974).

501. J. Farrar, D. Holland and D. J. Milner, *J. Chem. Soc., Dalton Trans.*, 815 (1975).
502. M. O. Farrell, C. H. Van Dyke, L. J. Boucher and S. J. Metlin, *J. Organomet. Chem.*, **169**, 199 (1979).
503. B. Fell and H. Bahrmann, *J. Mol. Catal.*, **2**, 211 (1977).
504. B. Fell and M. Beutler, *Tetrahedron Lett.*, 3455 (1972).
505. B. Fell, W. Boll and J. Hagen, *Chem. -Ztg.*, **99**, 485 (1975).
506. B. Fell and W. Dolkemeyer, *Ger. Pat.* 2 637 262 (1978); *C. A.* **88**, 152 032 (1978).
507. B. Fell and A. Geurts: *Chem. -Ing. -Tech.*, **44**, 708 (1972); *C. A.* **77**, 66 557 (1972).
508. B. Fell, J. Hagen and W. Rupilius, *Chem. -Ztg.*, **100**, 308 (1976).
509. B. Fell and V. Hartig, *Ger. Pat.* 2 311 388 (1973); *C. A.* **82**, 3740 (1975).
510. B. Fell and E. Mueller, *Monatsh. Chem.*, **103**, 1222 (1972).
511. B. Fell and W. Rupilius; *Tetrahedron Lett.*, 2721 (1969).
512. B. Fell, W. Rupilius and F. Asinger, *Tetrahedron Lett.*, 3261 (1968).
513. A. Fenster, B. R. James and W. R. Cullen, *Inorg. Synth.*, **17**, 81 (1977).
514. D. M. Fenton, *U.S. Pat.* 3 697 600 (1972); *C. A.* **78**, 3709 (1973).
515. R. A. Fiato and R. L. Pruett, *Eur. Pat. Appl.* 26 998 (1981); *C. A.* **95**, 80 182 (1981).
516. R. A. Fiato and R. L. Pruett, *U.S. Pat.* 4 273 936 (1981); *C. A.* **95**, 97 086 (1981).
517. L. W. Fine, M. Grayson and V. H. Suggs, *J. Organomet. Chem.*, **22**, 219 (1970).
518. M. Fiorini and G. M. Giongo, *J. Mol. Catal.*, **7**, 411 (1980).
519. M. Fiorini, G. M. Giongo, F. Marcati and W. Marconi, *J. Mol. Catal.*, **1**, 451 (1975/76).
520. M. Fiorini, G. M. Giongo, F. Marcati and W. Marconi, *Ger. Pat.* 2 718 533 (1977); *C. A.* **88**, 136 978 (1978).
521. M. Fiorini, F. Marcati and G. M. Giongo, *J. Mol. Catal.*, **4**, 125 (1978).
522. R. G. Fischer, A. Zweig and S. Raghu, *U.S. Pat.* 4 168 381 (1979); *C. A.* **92**, 94 393 (1980).
523. C. Fisher and H. S. Mosher, *Tetrahedron Lett.*, 2487 (1977).
524. T. A. Foglia and P. A. Barr, *J. Am. Oil Chem. Soc.*, **53**, 737 (1976); *C. A.* **86**, 89 098 (1977).
525. D. Forster, *Inorg. Chem.*, **8**, 2556 (1969).
526. D. Forster, *J. Am. Chem. Soc.*, **98**, 846 (1976).
527. D. Forster, *Ann. N. Y. Acad. Sci.*, **295**, 79 (1977).
528. D. Forster, *Adv. Organomet. Chem.*, **17**, 255 (1979).
529. D. Forster, *J. Chem. Soc., Dalton Trans.*, 1639 (1979).
530. D. Forster and G. R. Beck, *Chem. Commun.*, 1072 (1971).
531. G. Foster, P. Johnson and M. J. Lawrenson, *Br. Pat.* 1 173 568 (1969); *C. A.* **72**, 66 388 (1970).
532. G. Foster and M. J. Lawrenson, *Ger. Pat.* 1 901 145 (1969); *C. A.* **71**, 123 572 (1969).
533. G. Foster and M. J. Lawrenson, *Ger. Pat.* 1 911 631 (1969); *C. A.* **72**, 21 322 (1970).
534. G. Foster and M. J. Lawrenson, *S. Afr. Pat.* 05 913 (1970); *C. A.* **73**, 87 430 (1970).
535. P. Fotis and J. D. McCollum, *U.S. Pat.* 3 324 018 (1967); *C. A.* **67**, 53 616 (1967).
536. R. Fowler, *Br. Pat.* 1 387 657 (1975); *C. A.* **83**, 58 122 (1975).
537. R. Fowler, H. Connor and R. A. Bachl, *Chem. Technol.*, 772 (1976).
538. C. Fragale, M. Gargano, T. Gomes and M. Rossi, *J. Am. Oil Chem. Soc.*, **56**, 498 (1979); *C. A.* **91**, 73 306 (1979).
539. C. Fragale, M. Gargano and M. Rossi, *J. Mol. Catal.*, **5**, 65 (1979).
540. N. B. Franco and M. A. Robinson, *U.S. Pat.* 3 743 664 (1973); *C. A.* **79**, 67 053 (1973).

541. E. N. Frankel, J. P. Friedrich, T. R. Bessler, W. J. Kwolek and N. L. Holy, *J. Am. Oil Chem. Soc.*, **57**, 349 (1980); *C. A.* **94**, 14 110 (1981).

542. E. N. Frankel, F. L. Thomas and W. K. Rohwedder, *Ind. Eng. Chem., Prod. Res. Dev.*, **12**, 47 (1973).

543. S. Franks and F. R. Hartley, *J. Mol. Catal.*, **12**, 121 (1981).

544. A. R. Fraser, P. H. Bird, S. A. Bezman, J. R. Shapley, R. White and J. A. Osborn, *J. Am. Chem. Soc.*, **95**, 597 (1973).

545. L. Kh. Freidlin, E. F. Litvin and L. F. Topuridze, *Zh. Org. Khim.*, **8**, 669 (1972); *C. A.* **77**, 33 741 (1972).

546. L. Kh. Freidlin, Yu. A. Kopyttsey, E. F. Litvin and N. M. Nazarova, *Zh. Org. Khim.*, **10**, 430 (1974); *C. A.* **80**, 132 707 (1974).

547. J. P. Friedrich, *U.S. Pat.* 3 899 442 (1975); *C. A.* **83**, 153 209 (1975).

548. S. J. Fritschel, J. J. H. Ackerman, T. Keyser and J. K. Stille, *J. Org. Chem.*, **44**, 3152 (1979).

549. V. M. Frolov, O. P. Parenago and L. P. Shuikina, *Kinet. Katal.*, **19**, 1608 (1978); *C. A.* **90**, 137 294 (1979).

550. M. D. Fryzuk and B. Bosnich, *J. Am. Chem. Soc.*, **99**, 6262 (1977).

551. M. D. Fryzuk and B. Bosnich, *J. Am. Chem. Soc.*, **100**, 5491 (1978).

552. T. Fuchikami and I. Ojima, *J. Am. Chem. Soc.*, **104**, 3527 (1982).

553. Fuji Chemicals Industrial Co. Ltd., *Jpn. Pat.* 129 277 (1980); *C. A.* **94**, 174 853 (1981).

554. Y. Fujikura, Y. Takaishi, Y. Inamoto and M. Nakajima, *Jpn. Pat.* 144 353 (1979); *C. A.* **92**, 214 985 (1980).

555. K. Fujimoto, S. Tanemura and T. Kunugi, *J. Chem. Soc. Jpn.*, 167 (1977).

556. H. Fujitsu, E. Matsumura, S. Shirahama, K. Takeshita and I. Mochida, *J. Chem. Soc., Perkin Trans. 1*, 855 (1982).

557. H. Fujitsu, E. Matsumura, K. Takeshita and I. Mochida, *J. Chem. Soc., Perkin Trans. 1*, 2650 (1981).

558. H. Fujitsu, E. Matsumura, K. Takeshita and I. Mochida, *J. Org. Chem.*, **46**, 5353 (1981).

559. H. Fujitsu, S. Shirahama, E. Matsumura, K. Takeshita and I. Mochida, *J. Org. Chem.*, **46**, 2287 (1981).

560. J. Furukawa, J. Kiji and S. Kadoi, *Jpn. Pat.* 06 890 (1976); *C. A.* **85**, 25 885 (1976).

561. A. Fusi, R. Ugo, F. Fox, A. Pasini and S. Cenini, *J. Organomet. Chem.*, **26**, 417 (1971).

562. D. Gagnaire and P. Vottero, *Bull. Soc. Chim. Fr.*, 164 (1970).

563. V. Yu. Gankin, L. S. Genender, L. N. Dzeniskevich, D. M. Rudkovskii, V. A. Rybokov and M. A. Khalyutina, *Gidroformilirovanie*, 66 (1972); *C. A.* **77**, 151 202 (1972).

564. V. Yu. Gankin, L. S. Genender and D. M. Rudkovskii, *Zh. Prikl. Khim. (Leningrad)*, **40**, 2029 (1967); *C. A.* **68**, 77 592 (1968).

565. V. Yu. Gankin, L. S. Genender and D. M. Rudkovskii, *Zh. Prikl. Khim. (Leningrad)*, **41**, 2275 (1968); *C. A.* **70**, 37 103 (1969).

566. M. Gargano, P. Giannoccaro and M. Rossi, *J. Organomet. Chem.*, **84**, 389 (1975).

567. M. Gargano, P. Giannoccaro and M. Rossi, *J. Organomet. Chem.*, **129**, 239 (1977).

568. J. L. Garnett, M. A. Long, A. B. McLaren and K. B. Peterson, *Chem. Commun.*, 749 (1973).

569. P. E. Garrou and G. E. Hartwell, *U.S. Pat.* 4 262 147 (1981); *C. A.* **95**, 61 616 (1981).

570. M. E. Garst and D. Lukton, *J. Org. Chem.*, **46**, 4433 (1981).

571. P. G. Gassman, *Acc. Chem. Res.*, **4**, 128 (1971).
572. P. G. Gassman and E. A. Armour, *Tetrahedron Lett.*, 1431 (1971).
573. P. G. Gassman and T. J. Atkins, *J. Am. Chem. Soc.*, **93**, 4597 (1971).
574. P. G. Gassman and T. J. Atkins, *J. Am. Chem. Soc.*, **94**, 7748 (1972).
575. P. G. Gassman, T. J. Atkins and J. T. Lumb, *J. Am. Chem. Soc.*, **94**, 7757 (1972).
576. P. G. Gassman, T. J. Atkins and F. J. Williams, *J. Am. Chem. Soc.*, **93**, 1812 (1971).
577. P. G. Gassman, G. R. Meyer and F. J. Williams, *J. Am. Chem. Soc.*, **94**, 7741 (1972).
578. P. G. Gassman and T. Nakai, *J. Am. Chem. Soc.*, **93**, 5897 (1971).
579. P. G. Gassman and T. Nakai, *J. Am. Chem. Soc.*, **94**, 5497 (1972).
580. P. G. Gassman and R. R. Reitz, *J. Am. Chem. Soc.*, **95**, 3057 (1973).
581. P. G. Gassman and F. J. Williams, *J. Am. Chem. Soc.*, **92**, 7631 (1970).
582. P. G. Gassman and F. J. Williams, *Chem. Commun.*, 80 (1972); *J. Am. Chem. Soc.*, **94**, 7733 (1972).
583. B. C. Gates, J. R. Kratzer and G. C. A. Schuitt, *Chemistry of Catalytic Processes*, McGraw-Hill, New York, 1979.
584. B. C. Gates and J. Lieto, *Chem. Technol.*, 195 (1980); *ibid.*, 248 (1980).
585. P. Gelin, Y. B. Taarit and C. Naccache, *J. Catal.*, **59**, 357 (1979).
586. P. Gelin, Y. B. Taarit and C. Naccache, *Stud. Surf. Sci. Catal.*, **7**, 898 (1981); *C. A.* **95**, 202899 (1981).
587. General Electric Co., *Br. Pat.* 1 041 237 (1966); *C. A.* **65**, 20 164 (1966).
588. L. A. Gerritsen, J. M. Herman, W. Klut and J. J. F. Scholten, *J. Mol. Catal.*, **9**, 157 (1980).
589. L. A. Gerritsen, J. M. Herman and J. J. F. Scholten, *J. Mol. Catal.* **9**, 241 (1980).
590. L. A. Gerritsen, W. Klut, M. H. Vreugdenhil and J. J. F. Scholten, *J. Mol. Catal.*, **9**, 257 (1980).
591. L. A. Gerritsen and J. J. F. Scholten, *Neth. Pat. Appl.* 02964 (1981); *C. A.* **94**, 163 360 (1981).
592. L. A. Gerritsen, A. Van Meerkerk, M. H. Vreugdenhill and J. J. F. Scholten, *J. Mol. Catal.*, **9**, 139 (1980).
593. A. Giarrusso, P. Gronchi, G. Ingrosso and L. Porri, *Makromol. Chem.*, **178**, 1375 (1977).
594. D. S. Gill, C. White and P. M. Maitlis, *J. Chem. Soc., Dalton Trans.*, 617 (1978).
595. R. D. Gillard, J. A. Osborn, P. B. Stockwell and G. Wilkinson, *Proc. Chem. Soc. (London)*, 284 (1964).
596. B. Giovannitti, M. Ghedini, G. Dolcetti and G. Denti, *J. Organomet. Chem.*, **157**, 457 (1978).
597. R. M. Gipson, *Ger. Pat.* 1 802 895 (1969); *C. A.* **71**, 70 067 (1969).
598. M. Giustiniani, G. Dolcetti, M. Nicolini and U. Belluco, *J. Chem. Soc. A*, 1961 (1969).
599. R. Glaser, *Tetrahedron Lett.*, 2127 (1975).
600. R. Glaser, J. Blumenfeld and M. Twaik, *Tetrahedron Lett.*, 4639 (1977).
601. R. Glaser and S. Geresh, *Tetrahedron Lett.*, 2527 (1977).
602. R. Glaser and S. Geresh, *Tetrahedron*, **35**, 2381 (1979).
603. R. Glaser, S. Geresh and J. Blumenfeld, *J. Organomet. Chem.*, **112**, 355 (1976).
604. R. Glaser, S. Geresh, J. Blumenfeld and M. Twaik, *Tetrahedron*, **34**, 2405 (1978).
605. R. Glaser, S. Geresh and M. Twaik, *Isr. J. Chem.*, **20**, 102 (1980).
606. R. W. Goetz, *U.S. Pat.* 4 200 765 (1980); *C. A.* **93**, 113 951 (1980).
607. A. Goldup and M. T. Westaway, *Ger. Pat.* 2 029 625 (1971); *C. A.* **75**, 22 008 (1971).

608. A. Goldup, M. T. Westaway and G. Walker, *Ger. Pat.* 1 953 641 (1971); *C. A.* **74**, 80 298 (1971).
609. N. M. Goncharova and G. S. Grinenko, *Khim. -Farm. Zh.*, **14**, 61 (1980); *C. A.* **93**, 72 076 (1980).
610. L. V. Gorbunova, I. 'L. Knyazeva, B. K. Nefedov, Kh. O. Khoshdurdyev and V. I. Manov-Yuvenskii, *Izv. Akad. Nauk SSSR, Ser. Khim.*, 1644 (1981); *C. A.* **95**, 115 000 (1981).
611. C. Graillat, H. Jacobelli, M. Bartholin and A. Guyot, *Rev. Port. Quim.*, **19**, 279 (1977); *C. A.* **93**, 94 735 (1980).
612. H. B. Gray, A. Rembaum and A. Gupta, *U.S. Pat.* 4 127 506 (1978); *C. A.* **90**, 86 744 (1979).
613. R. T. Gray and A. J. De Jong, *Ger. Pat.* 2 753 644 (1978); *C. A.* **89**, 108 350 (1978).
614. W. F. Graydon and M. D. Langan, *J. Catal.*, **69**, 180 (1981).
615. M. Green and T. A. Kuc, *J. Chem. Soc., Dalton Trans.*, 832 (1972).
616. M. Green and T. A. Kuc, *Ger. Pat.* 2 153 314 (1972); *C. A.* **77**, 37 216 (1972).
617. M. Green and T. A. Kuc, *Ger. Pat.* 2 153 332 (1972); *C. A.* **77**, 37 215 (1972).
618. G. Gregorio, G. Montrasi, M. Tampieri, P. Cavalieri D'Oro, G. Pagani and A. Andreetta, *Chim. Ind.* (*Milan*), **62**, 389 (1980).
619. G. Gregorio, G. Pregaglia and R. Ugo, *Inorg. Chim. Acta*, **3**, 89 (1969).
620. P. A. Grieco and N. Marinovic, *Tetrahedron Lett.*, 2545 (1978).
621. P. A. Grieco, M. Nishizawa, N. Marinovic and W. H. Ehmann, *J. Am. Chem. Soc.*, **98**, 7102 (1976).
622. R. E. Grigg, *Eur. Pat. Appl.* 34 480 (1981); *C. A.* **96**, 5744 (1982).
623. R. Grigg, R. Hayes and A. Sweeney, *Chem. Commun.*, 1248 (1971).
624. R. Grigg, T. R. B. Mitchell and A. Ramasubbu, *Chem. Commun.*, 27 (1980).
625. R. Grigg, T. R. B. Mitchell and S. Sutthivaiyakit, *Tetrahedron Lett.*, 739 (1979).
626. R. Grigg, T. R. B. Mitchell and S. Sutthivaiyakit, *Tetrahedron*, **37**, 4313 (1981).
627. R. Grigg, T. R. B. Mitchell, S. Sutthivaiyakit and N. Tongpenyai, *Chem. Commun.*, 611 (1981).
628. R. Grigg, T. R. B. Mitchell, S. Sutthivaiyakit and N. Tongpenyai, *Tetrahedron Lett.*, 4107 (1981).
629. R. Grigg, T. R. B. Mitchell and N. Tongpenyai, *Synthesis*, 442 (1981).
630. R. Grigg and G. Shelton, *Chem. Commun.*, 1247 (1971).
631. J. Grimblot, J. P. Bonnelle, A. Mortreux and F. Petit, *Inorg. Chim. Acta*, **34**, 29 (1979).
623. J. Grimblot, J. P. Bonnelle, C. Vaccher, A. Mortreux, F. Petit and G. Peiffer, *J. Mol. Catal.*, **9**, 357 (1980).
633. A. A. Grinberg, B. D. Babitskii, Yu. S. Varshavskii, M. I. Gel'fman, N. V. Kiseleva, V. A. Kormer, D. B. Smolenskaya, T. G. Cherkasova and N. Y. Chesnokova, *Dokl. Adad. Nauk SSSR*, **170**, 1334 (1966); *Dokl. Chem.* (*Engl. Trans.*), **170**, 1008 (1966).
634. R. H. Grubbs and R. A. Devries, *Tetrahedron Lett.*, 1879 (1977).
635. R. H. Grubbs, C. Gibbons, L. C. Kroll, W. D. Bonds and C. H. Brubaker, *J. Am. Chem. Soc.*, **95**, 2373 (1973).
636. R. H. Grubbs and L. C. Kroll, *J. Am. Chem. Soc.*, **93**, 3062 (1971).
637. R. H. Grubbs, L. C. Kroll and E. M. Sweet, *J. Macromol. Sci., Chem.*, **7**, 1047 (1973).
638. R. H. Grubbs and H. Su Shiu-Chin, *J. Organomet. Chem.*, **122**, 151 (1976).
639. R. H. Grubbs and E. M. Sweet, *Macromolecules*, **8**, 241 (1975).
640. R. H. Grubbs and E. M. Sweet, *J. Mol. Catal.*, **3**, 259 (1978).

641. M. Gullotti, R. Ugo and S. Colonna, *J. Chem. Soc. C*, 2652 (1971).
642. A. Guyot, C. Graillat and M. Bartholin, *J. Mol. Catal.*, **3**, 39 (1977).
643. W. O. Haag and D. D. Whitehurst, *U.S. Pat.* 4 098 727 (1978); *C. A.* **91**, 19 901 (1979).
644. Y. M. Y. Haddad, H. B. Henbest, J. Husbands and T. R. B. Mitchell, *Proc. Chem. Soc. (London)*, 361 (1964); Y. M. Y. Haddad, H. B. Henbest, J. Husbands, T. R. B. Mitchell and J. Trocha-Grimshaw, *J. Chem. Soc., Perkin Trans. 2*, 596 (1974).
645. M. Haga, K. Kawakami and T. Tanaka, *Inorg. Chem.*, **15**, 1946 (1976).
646. J. Hagen and K. Bruns, *Ger. Pat.* 2 849 742 (1980); *C. A.* **93**, 186 613 (1980).
647. J. Hagen, R. Lehmann and K. Bansemir, *Ger. Pat.* 2 914 187 (1980); *C. A.* **94**, 71 565 (1981).
648. J. P. Hagenbuch and P. Vogel, *Tetrahedron Lett.*, 561 (1979).
649. J. Halpern, in *Organic Synthesis via Metal Carbonyls*, (I. Wender and P. Pino, Eds.), Wiley, New York, 1977.
650. J. Halpern, *Trans. Am. Crystallogr. Assoc.*, **14**, 59 (1978).
651. J. Halpern, T. Okamoto and A. Zakhariev, *J. Mol. Catal.*, **2**, 65 (1977).
652. J. Halpern, D. P. Riley, A. S. C. Chan and J. J. Pluth, *J. Am. Chem. Soc.*, **99**, 8055 (1977).
653. J. Halpern and C. S. Wong, *Chem. Commun.*, 629 (1973).
654. J. E. Hamlin, K. Hirai, V. C. Gibson and P. M. Maitlis, *J. Mol. Catal.*, **15**, 337 (1982).
655. K. Hanaki, K. Kashiwabara and J. Fujita, *Chem. Lett.*, 489 (1978).
656. R. D. Hancock, I. V. Howell and R. C. Pitkethly, *Br. Pat.* 1 426 881 (1976); *C. A.* **84**, 179 674 (1976).
657. R. E. Harmon, J. L. Parsons, D. W. Cooke, S. K. Gupta and J. Schoolenberg, *J. Org. Chem.*, **34**, 3684 (1969).
658. R. E. Harmon, J. L. Parsons and S. K. Gupta, *Chem. Commun.*, 1365 (1969).
659. R. E. Harmon, J. L. Parsons and S. K. Gupta, *Org. Prep. Proced. Int.*, **2**, 25 (1970).
660. N. Harris, A. J. Dennis and G. E. Harrison, *Eur. Pat. Appl.* 18 161 (1980); *C. A.* **94**, 156 315 (1981).
661. N. Harris, A. J. Dennis and G. E. Harrison, *Eur. Pat. Appl.* 18 163 (1980); *C. A.* **94**, 156 295 (1981).
662. N. Harris and T. F. Shevels, *Eur. Pat. Appl.* 7768 (1980); *C. A.* **93**, 71 021 (1980).
663. J. F. Harrod and A. J. Chalk, *J. Am. Chem. Soc.*, **86**, 1776 (1964).
664. J. F. Harrod and A. J. Chalk, *J. Am. Chem. Soc.*, **88**, 3491 (1966).
665. J. F. Harrod and A. J. Chalk, in *Organic Syntheses via Metal Carbonyls*, (I. Wender and P. Pino, Eds.), Vol. 2, Wiley-Interscience, New York, 1977, p. 673.
666. D. W. Hart and J. Schwartz, *J. Organomet. Chem.*, **87**, C11 (1975).
667. P. W. Hart, *U.S. Pat.* 4 302 547 (1981); *C. A.* **96**, 34 549 (1982).
668. F. R. Hartley, S. G. Murray and P. N. Nicholson, *J. Mol. Catal.*, **16**, 363 (1982).
669. F. R. Hartley, S. G. Murray and P. N. Nicholson, *J. Organomet. Chem.*, **231**, 369 (1982).
670. F. R. Hartley and P. N. Vezey, *Adv. Organomet. Chem.*, **15**, 189 (1976).
671. G. E. Hartwell and P. E. Garrou, *U.S. Pat.* 4 144 191 (1979); *C. A.* **90**, 203 478 (1979).
672. G. E. Hartwell and P. E. Garrou, *Braz. Pat.* 01 172 (1980); *C. A.* **94**, 139 208 (1981).
673. R. N. Haszeldine, R. V. Parrish and D. J. Parry, *J. Chem. Soc. A*, 683 (1969).
674. R. N. Haszeldine, R. V. Parrish and R. J. Taylor, *J. Chem. Soc., Dalton Trans.*, 2311 (1974).
675. S. Hattori, N. Morita and T. Imaki, *Jpn. Pat.* 38 533 (1977); *C. A.* **88**, 104 692 (1978).

676. T. Hayashi, K. Kanehira and M. Kumada, *Tetrahedron Lett.*, 4417 (1981).
677. T. Hayashi, A. Katsumura, M. Konishi and M. Kumada, *Tetrahedron Lett.*, 425 (1979).
678. T. Hayashi and M. Kumada, *Fundam. Res. Homogeneous Catal.*, **2**, 159 (1978).
679. T. Hayashi, T. Mise, M. Fukushima, M. Kagotani, N. Nagashima, Y. Hamada *et al.*, *Bull. Chem. Soc. Jpn.*, **53**, 1138 (1980).
680. T. Hayashi, T. Mise, S. Mitachi and W. Marconi, *J. Mol. Catal.*, **1**, 451 (1976).
681. T. Hayashi, M. Tanaka, Y. Ikeda and I. Ogata, *Bull. Chem. Soc. Jpn.*, **52**, 2605 (1979).
682. T. Hayashi, M. Tanaka and I. Ogata, *Tetrahedron Lett.*, 295 (1977).
683. T. Hayashi, M. Tanaka and I. Ogata, *J. Mol. Catal.*, **6**, 1 (1979).
684. T. Hayashi, M. Tanaka and I. Ogata, *J. Mol. Catal.*, **13**, 323 (1981).
685. T. Hayashi, K. Yamamoto and M. Kumda, *Tetrahedron Lett.*, 331 (1974).
686. B. L. Haymore and J. A. Ibers, *J. Am. Chem. Soc.*, **96**, 3325 (1974).
687. P. Haynes, L. H. Slaugh and J. F. Kohnle, *Tetrahedron Lett.*, 365 (1970).
688. C. H. Heathcock and S. R. Poulter, *Tetrahedron Lett.*, 2755 (1969).
689. R. F. Heck, *Adv. Catal.*, **27**, 323 (1977).
690. L. S. Hegedus, P. M. Kendall, S. M. Lo and J. R. Sheats, *J. Am. Chem. Soc.*, **97**, 5448 (1975).
691. B. Heil and L. Marko, *Chem. Ber.*, **99**, 1086 (1966).
692. B. Heil and L. Marko, *Acta Chim. Acad. Sci. Hung.*, **55**, 107 (1968).
693. B. Heil and L. Marko, *Chem. Ber.*, **101**, 2209 (1968).
694. B. Heil and L. Marko, *Chem. Ber.*, **102**, 2238 (1969).
695. B. Heil, L. Marko and G. Bor, *Chem. Ber.*, **104**, 3418 (1971).
696. B. Heil, S. Toros, J. Bakos and L. Marko, *J. Organomet. Chem.*, **175**, 229 (1979).
697. B. Heil, S. Toros, S. Vastag and L. Marko, *J. Organomet. Chem.*, **94**, C47 (1975).
698. W. Heitz, K. Arlt and W. Mehnert, *Makromol. Chem.*, **177**, 1625 (1976).
699. R. F. Heldeweg and H. Hogeveen, *J. Am. Chem. Soc.*, **98**, 6040 (1976).
700. H. B. Henbest and T. R. B. Mitchell, *J. Chem. Soc. C*, 785 (1970).
701. H. B. Henbest and J. Trocha-Grimshaw, *J. Chem. Soc., Perkin Trans. 1*, 601 (1974).
702. H. B. Henbest and J. Trocha-Grimshaw, *J. Chem. Soc., Perkin Trans. 1*, 607 (1974).
703. W. A. Henderson, *U.S. Pat.* 4 166 824 (1979); *C. A.* **92**, 6530 (1980).
704. W. A. Henderson, R. G. Fischer, A. Zweig and S. Raghu, *Ger. Pat.* 2 824 861 (1979); *C. A.* **90**, 204 251 (1979).
705. D. E. Hendriksen and R. Eisenberg, *J. Am. Chem. Soc.*, **98**, 4662 (1976).
706. D. E. Hendriksen, C. D. Meyer and R. Eisenberg, *Inorg. Chem.*, **16**, 970 (1977).
707. G. Henrici-Olivé and S. Olivé, *Top. Curr. Chem.*, **67**, 107 (1976).
708. G. Henrici-Olivé and S. Olivé, *Angew. Chem.*, **91**, 83 (1979).
709. P. M. Henry, *Adv. Organomet. Chem.*, **13**, 363 (1975).
710. P. M. Henry, *Palladium-catalyzed Oxidation of Hydrocarbons*, Reidel, Dordrecht, 1980.
711. J. L. Herde and C. V. Senoff, *Inorg. Nucl. Chem. Lett.*, **7**, 1029 (1971).
712. W. A. Herrmann, *Angew. Chem., Int. Ed. Engl.*, **21**, 117 (1982).
713. H. Hershcovitz and G. Schmuckler, *J. Inorg. Nucl. Chem.*, **41**, 687 (1979).
714. A. Hershman, K. K. Robinson, J. H. Craddock and J. F. Roth, *Ind. Eng. Chem., Prod. Res. Dev.*, **8**, 372 (1969).
715. M. Hidai, M. Orishku and Y. Uchida, *Chem. Lett.*, 753 (1980).
716. R. R. Hignett and P. J. Davidson, *Belg. Pat.* 868 278 (1978); *C. A.* **90**, 103 417 (1979).

717. J. E. Hill and T. A. Nile, *J. Organomet. Chem.*, **137**, 293 (1977).
718. W. Himmele and W. Aquila, *Ger. Pat.* 2 064 279 (1972); *C. A.* **77**, 114 126 (1972).
719. W. Himmele and W. Aquila, *Ger. Pat.* 2 111 116 (1972); *C. A.* **77**, 152 153 (1972).
720. W. Himmele, W. Aquila and F. J. Mueller, *Ger. Pat.* 1 957 300 (1971); *C. A.* **75**, 36 336 (1971).
721. W. Himmele, W. Aquila and R. Prinz, *Ger. Pat.* 1 918 694 (1970); *C. A.* **73**, 130 643 (1970).
722. W. Himmele, F. J. Mueller and W. Aquila, *Ger. Pat.* 2 039 078 (1972); *C. A.* **76**, 112 700 (1972).
723. W. Himmele and H. Siegel, *Ger. Pat.* 2 235 466 (1974); *C. A.* **81**, 3593 (1974).
724. W. Himmele, H. Siegel and W. Aquila, *Ger. Pat.* 2 219 168 (1973); *C. A.* **80**, 14 740 (1974).
725. H. Hirai, S. Komatsuzaki, S. Hamasaki and N. Toshima, *J. Chem. Soc. Jpn.*, 316 (1982).
726. K. Hirai, A. Nutton and P. M. Maitlis, *J. Mol. Catal.*, **10**, 203 (1981).
727. E. Hitzel, *Z. Naturforsch., Teil B*, **33**, 997 (1978).
728. J. Hjortkjaer, *J. Mol. Catal.*, **5**, 377 (1979).
729. J. Hjortkjaer and V. W. Jensen, *Ind. Eng. Chem., Prod. Res. Dev.*, **15**, 46 (1976).
730. J. Hjortkjaer and J. C. A. Joergensen, *J. Mol. Catal.*, **4**, 199 (1978).
731. J. Hjortkjaer and J. C. Joergensen, *J. Chem. Soc., Perkin Trans. 2*, 763 (1978).
732. J. Hjortkjaer, M. S. Scurrell and P. Simonsen, *J. Mol. Catal.*, **6**, 405 (1979).
733. J. Hjortkjaer, M. S. Scurrell and P. Simonsen, *J. Mol. Catal.*, **10**, 127 (1980).
734. C. F. Hobbs and W. S. Knowles, *J. Org. Chem.*, **46**, 4422 (1981).
735. R. V. Hoffman, *Tetrahedron Lett.*, 2415 (1974).
736. H. Hogeveen and T. B. Middlekoop, *Tetrahedron Lett.*, 3671 (1973).
737. H. Hogeveen and T. B. Middlekoop, *Tetrahedron Lett.*, 4325 (1973).
738. H. Hogeveen and B. J. Nusse, *Tetrahedron Lett.*, 3667 (1973).
739. H. Hogeveen and B. J. Nusse, *J. Am. Chem. Soc.*, **100**, 3110 (1978).
740. H. Hogeveen and H. C. Volger, *J. Am. Chem. Soc.*, **89**, 2486 (1967).
741. J. Hojo, S. Yuasa, N. Yamazoe, I. Mochida and T. Seiyama, *J. Catal.*, **36**, 93 (1975).
742. D. G. Holah, I. M. Hoodless, A. N. Hughes, B. C. Hu and D. Martin, *Can. J. Chem.*, **52**, 3758 (1974).
743. D. Holland and D. J. Milner, *J. Chem. Soc., Dalton Trans.*, 2440 (1975).
744. D. Holland and D. J. Milner, *J. Chem. Res. (S)*, 317 (1979).
745. J. M. Holovka and E. Hurley, *U.S. Pat.* 4 169 110 (1979); *C. A.* **92**, 75 858 (1980).
746. N. L. Holy, *J. Org. Chem.*, **43**, 4686 (1978).
747. N. L. Holy, *J. Org. Chem.*, **44**, 239 (1979).
748. N. L. Holy, W. A. Logan and K. D. Stein, *Ger. Pat.* 2 816 231 (1978); *C. A.* **90**, 71 239 (1979).
749. N. L. Holy, W. A. Logan and K. D. Stein, *U.S. Pat.* 4 313 018 (1982); *C. A.* **96**, 141 870 (1982).
750. E. H. Homeier, A. R. Dodds and T. Imai, *U.S. Pat.* 4 292 196 (1981); *C. A.* **95**, 193, 193 092 (1981).
751. E. H. Homeier and T. Imai, *U.S. Pat.* 4 204 066 (1980); *C. A.* **93**, 168 142 (1980).
752. P. Hong, B.-R. Cho and H. Yamazaki, *Chem. Lett.*, 339 (1979).
753. P. Hong, T. Mise and H. Yamazaki, *Chem. Lett.*, 989 (1981).
754. P. Hong, T. Mise and H. Yamazaki, *Chem. Lett.*, 361 (1982).
755. P. Hong, K. Sonogashira and N. Hagihara, *J. Chem. Soc. Jpn.*, 74 (1968).
756. P. Hong and H. Yamazaki, *Chem. Lett.*, 1335 (1979).

757. P. Hong, H. Yamazaki, K. Sonogashira and N. Hagihara, *Chem. Lett.*, 535 (1978).
758. L. Horner, H. Büthe and H. Siegel, *Tetrahedron Lett.*, 4023 (1968).
759. L. Horner and M. Jordan, *Phosphorus Sulfur*, 225 (1980).
760. L. Horner and B. Schlotthauer, *Phosphorus Sulfur*, 155 (1978).
761. L. Horner and F. Schumacher, *Justus Liebigs Ann. Chem.*, 633 (1976).
762. L. Horner and H. Siegel, *Phosphorus*, 1, 199 (1972).
763. L. Horner and H. Siegel, *Phosphorus*, 1, 209 (1972).
764. L. Horner, H. Siegel and H. Büthe, *Angew Chem., Int. Ed. Engl.*, 7, 942 (1968).
765. A. B. Hornfeldt, J. S. Gronowitz and S. Gronowitz, *Acta Chem. Scand.*, 22, 2725 (1968).
766. R. P. Houghton, *Metal Complexes in Organic Chemistry*, Cambridge University Press, Cambridge, 1979.
767. R. P. Houghton, M. Voyle and R. Price, *Chem. Commun.*, 884 (1980).
768. J. P. Howe, K. Lung and T. Nile, *J. Organomet. Chem.*, 208, 401 (1981).
769. R. F. Howe, *J. Catal.*, 50, 196 (1977).
770. B. W. Howk and J. C. Sauer, *U.S. Pat.* 3 055 949 (1962); *C. A.* 58, 7870 (1963).
771. J. K. Hoyano and W. A. G. Graham, *J. Am. Chem. Soc.*, 104, 3722 (1982).
772. F. Hrabak and J. Roda, *Czech. Pat.* 162 162 (1976); *C. A.* 85, 94 941 (1976).
773. I.-D. Huang, A. A. Westner, A. A. Oswald and T. G. Jermansen, *PCT Int. Pat. Appl.* 01 690 (1980); *C. A.* 94, 156 314 (1981).
774. D. A. Hucul and A. Brenner, *J. Am. Chem. Soc.*, 103, 217 (1981).
775. D. A. Hucul and A. Brenner, *J. Phys. Chem.*, 85, 496 (1981).
776. B. Hudson, P. C. Taylor, D. E. Webster and P. B. Wells, *Faraday Discuss. Chem. Soc.*, 46, 37 (1968).
777. B. Hudson, D. E. Webster and P. B. Wells, *J. Chem. Soc., Dalton Trans.*, 1204 (1972).
778. O. R. Hughes, *U.S. Pat.* 4 066 705 (1978); *C. A.* 88, 136 138 (1978).
779. O. R. Hughes, *U.S. Pat.* 4 169 861 (1979); *C. A.* 92, 6066 (1980).
780. O. R. Hughes, *U.S. Pat.* 4 201 728 (1980); *C. A.* 93, 149 796s (1980).
781. O. R. Hughes and J. D. Unruh, *J. Mol. Catal.*, 12, 71 (1981).
782. V. L. Hughes and R. S. Brodkey, *U.S. Pat.* 2 839 580 (1958); *C. A.* 52, P17 109 (1958).
783. W. B. Hughes, *U.S. Pat.* 3 511 885 (1970); *C. A.* 73, 120 186 (1970).
784. W. B. Hughes, *U.S. Pat.* 3 514 497 (1970); *C. A.* 73, 25 016 (1970).
785. W. B. Hughes, *U.S. Pat.* 3 697 615 (1972); *C. A.* 78, 15 474 (1973).
786. B. C. Y. Hui and G. L. Rempel, *Chem. Commun.*, 1195 (1970).
787. B. C. Y. Hui, W. K. Teo and G. L. Rempel, *Inorg. Chem.*, 12, 757 (1973).
788. A. S. Hussey and Y. Takeuchi, *J. Am. Chem. Soc.*, 91, 672 (1969).
789. A. S. Hussey and Y. Takeuchi, *J. Org. Chem.*, 35, 643 (1970).
790. T. G. Hyde, *Br. Pat.* 1 143 065 (1969); *C. A.* 70, 106 004 (1969).
791. M. Ichikawa, *Bull. Chem. Soc. Jpn.*, 51, 2268 (1978).
792. M. Ichikawa, *Bull. Chem. Soc. Jpn.*, 51, 2273 (1978).
793. M. Ichikawa, *Chem. Commun.*, 566 (1978).
794. M. Ichikawa, *J. Catal.*, 56, 127 (1979).
795. M. Ichikawa, *J. Catal.*, 59, 67 (1979).
796. M. Ichikawa, *Jpn. Pat.* 41 291 (1979); *C. A.* 91, 10 046 (1979).
797. M. Ichikawa, *Chem. Technol.*, 674 (1982).
798. M. Ichikawa and Y. Kido, *Jpn. Pat.* 41 292 (1979); *C. A.* 91, 157 273 (1979).
799. M. Ichikawa and Y. Kido, *Jpn. Pat.* 41 293 (1979); *C. A.* 91, 56 355 (1979).

800. M. Ichikawa, K. Sekizawa, K. Shikakura and M. Kawai, *J. Mol. Catal.*, **11**, 167 (1981).
801. M. Ichikawa and K. Shikakura, *Stud. Surf. Sci. Catal.*, **7**, 925 (1981); *C. A.* **95**, 203 250 (1981).
802. M. Ichikawa, A. Uda and K. Tamaru, *Jpn. Pat.* 18 217 (1973); *C. A.* **78**, 158 953 (1973).
803. F. Igersheim and H. Mimoun, *Nouv. J. Chim.*, **4**, 161 (1980).
804. T. Iizuka and J. H. Lunsford, *J. Mol. Catal.*, **8**, 391 (1980).
805. D. E. Iley and B. Fraser-Reid, *J. Am. Chem. Soc.*, **97**, 2563 (1975).
806. T. Imai, *U.S. Pat.* 4 220 764 (1978); *C. A.* **93**, 239 429 (1980).
807. T. Imai, *U.S. Pat.* 4 207 260 (1980); *C. A.* **94**, 102 809 (1980).
808. T. Imai, *U.S. Pat.* 4 219 684 (1980); *C. A.* **93**, 204 056 (1980).
809. T. Imai, *Ger. Pat.* 3 012 729 (1980); *C. A.* **94**, 120 838 (1981).
810. T. Imai, *U.S. Pat.* 4 250 115 (1981); *C. A.* **95**, 42 335 (1981).
811. T. Imai, E. H. Homeier and D. E. Mackowiak, *U.S. Pat.* 4 275 252 (1981); *C. A.* **95**, 149 943 (1981).
812. Imperial Chemical Industries Ltd., *Neth. Pat. Appl.* 6 501 822 (1965); *C. A.* **64**, 4944 (1966).
813. Imperial Chemical Industries Ltd., *Neth. Pat.* 6 502 601 (1965); *C. A.* **64**, 11 126 (1966).
814. Imperial Chemical Industries Ltd., *Neth. Pat. Appl.* 6 602 062 (1966); *C. A.* **66**, 10 556 (1967).
815. Imperial Chemical Industries Ltd., *Neth. Pat.* 6 603 612 (1966); *C. A.* **66**, 28 511 (1967).
816. Imperial Chemical Industries Ltd., *Neth. Pat. Appl.* 6 605 627 (1966); *C. A.* **66**, 55 581 (1967).
817. Imperial Chemical Industries Ltd., *Neth. Pat. Appl.* 6 608 122 (1966); *C. A.* **67**, 26 248 (1967).
818. Imperial Chemical Industries Ltd., *Neth. Pat.* 6 612 227 (1967); *C. A.* **67**, 64 522 (1967).
819. Imperial Chemical Industries Ltd., *Fr. Pat.* 1 521 991 (1968); *C. A.* **70**, 114 572 (1969).
820. N. S. Imyanitov, *Hung. J. Ind. Chem.*, **3**, 331 (1975); *C. A.* **83**, 137 395 (1975).
821. N. S. Imyanitov and D. M. Rudkovskii, *Neftekhimiya*, **3**, 198 (1963); *C. A.* **59**, 7396 (1963).
822. N. S. Imyanitov and D. M. Rudkovskii, *Zh. Prikl. Khim.*, **39**, 2811 (1966); *C. A.* **66**, 75 465 (1967).
823. N. S. Imyanitov and D. M. Rudkovskii, *Zh. Prikl. Khim.*, **40**, 2020 (1967); *C. A.* **68**, 95 367 (1968).
824. G. Innorta, A. Modelli, F. Scagnolari and A. Foffani, *J. Organomet. Chem.*, **185**, 403 (1980).
825. Institut Francais du Petrole, des Carburants et Lubrifiants, *Fr. Pat.* 1 475 624 (1967); *C. A.* **67**, 82 544 (1967).
826. G. M. Intille, *U.S. Pat.* 3 796 739 (1974); *C. A.* **80**, 121 117 (1974).
827. Ionics Inc., *Br. Pat.* 1 034 298 (1966); *C. A.* **65**, 13 551 (1966).
828. A. F. M. Iqbal, *Tetrahedron Lett.*, 3381 (1971).
829. A. F. M. Iqbal, *Tetrahedron Lett.*, 3385 (1971).
830. A. F. M. Iqbal, *Helv. Chim. Acta*, **55**, 798 (1972).
831. A. F. M. Iqbal, *Helv. Chim. Acta*, **55**, 2637 (1972).

832. M. Iriuchijima and T. Anezaki, *Jpn. Pat.* 26 762 (1972); *C. A.* **77**, 131 259 (1972).
833. H. Irngartinger, A. Goldmann, R. Schappert, P. Garner and P. Dowd, *Chem. Commun.*, 455 (1981).
834. K. Ishimi, M. Iwase, E. Tanaka, M. Hidai and Y. Uchida, *Yakugaku Zasshi,* **23**, 408 (1974); *C. A.* **81**, 169 878 (1974).
835. J. Ishiyama, Y. Senda, I. Shinoda and S. Imaizumi, *Bull. Chem. Soc. Jpn.,* **52**, 2353 (1979).
836. T. Isshiki, Y. Kijima and Y. Miyauchi, *Ger. Pat.* 2 846 709 (1979); *C. A.* **91**, 20 108 (1979).
837. T. Isshiki, Y. Kijima and Y. Miyauchi, *Ger. Pat.* 2 847 241 (1979); *C. A.* **91**, 19 920 (1979).
838. T. Isshiki, Y. Kijima and Y. Miyauchi, *U.S. Pat.* 4 212 989 (1980); *C. A.* **93**, 238 850 (1980).
839. K. J. Ivin, G. Lapienis and J. J. Rooney, *Makromol. Chem.*, 183 (1982).
840. E. Iwamoto, T. Saito and S. Arai, *Jpn. Pat.* 105 590 (1977); *C. A.* **87**, 207 278 (1977).
841. Y. Iwasawa, T. Hayasaka and S. Ogasawara, *Chem. Lett.*, 131 (1982).
842. Y. Iwashita and M. Sakuraba, *Tetrahedron Lett.*, 2409 (1971).
843. Y. Iwashita and F. Tamura, *Bull. Chem. Soc. Jpn.,* **43**, 1517 (1970).
844. F. Jachimowicz, *Belg. Pat.* 887 627 (1981); *C. A.* **95**, 133 642 (1981).
845. F. Jachimowicz, *Belg. Pat.* 887 630 (1981); *C. A.* **95**, 152 491 (1981).
846. F. Jachimowicz and J. W. Raksis, *J. Org. Chem.*, **47**, 445 (1982).
847. R. Jackson and D. J. Thompson, *J. Organomet. Chem.*, **159**, C29 (1978).
848. S. E. Jacobson, W. Clements, H. Hiramoto and C. U. Pittman, *J. Mol. Catal.*, **1**, 73 (1975).
849. S. E. Jacobson and C. U. Pittman, *Chem. Commun.*, 187 (1975).
850. M. Jakoubkova and M. Capka, *Collect. Czech. Chem. Commun.*, **45**, 2219 (1980).
851. B. R. James, *Homogeneous Hydrogenation*, Wiley, New York, 1973.
852. B. R. James, *Adv. Organomet. Chem.*, **17**, 319 (1979).
853. B. R. James and M. Kastner, *Can. J. Chem.*, **50**, 1698 (1972).
854. B. R. James and M. Kastner, *Can. J. Chem.*, **50**, 1708 (1972).
855. B. R. James, R. S. McMillan, R. H. Morris and D. K. W. Wang, *Adv. Chem. Ser.*, **167**, 122 (1978).
856. B. R. James and M. A. Memon, *Can. J. Chem.*, **46**, 217 (1968).
857. B. R. James and R. H. Morris, *Chem. Commun.*, 929 (1978).
858. B. R. James and F. T. T. Ng, *Chem. Commun.*, 908 (1970); *Can. J. Chem.*, **53**, 797 (1975).
859. B. R. James, F. T. T. Ng and G. L. Rempel, *Inorg. Nucl. Chem. Lett.*, **4**, 197 (1968); B. R. James and F. T. T. Ng, *J. Chem. Soc., Dalton Trans.*, 1321 (1972).
860. B. R. James and E. Ochiai, *Can. J. Chem.*, **49**, 975 (1971).
861. B. R. James and G. L. Rempel, *J. Am. Chem. Soc.*, **91**, 863 (1969).
862. A. H. Janowicz and R. G. Bergman, *J. Am. Chem. Soc.*, **104**, 352 (1982).
863. Japan Synthetic Rubber Co., Ltd., *Jpn. Pat.* 26 706 (1967); *C. A.* **68**, 79 398 (1968).
864. Japan Synthetic Rubber Co., Ltd., *Br. Pat.* 1 184 751 (1970); *C. A.* **72**, 112 035 (1970).
865. F. H. Jardine, *Prog. Inorg. Chem.*, **28**, 64 (1981).
866. F. H. Jardine, J. A. Osborne and G. Wilkinson, *J. Chem. Soc. A*, 1574 (1967).
867. F. H. Jardine, J. A. Osborne, G. Wilkinson and F. J. Young, *Chem. Ind. (London)*, 560 (1965).

868. F. H. Jardine and G. Wilkinson, *J. Chem. Soc. C*, 270 (1967).
869. I. Jardine and F. J. McQuillan, *Chem. Commun.*, 477 (1969).
870. I. Jardine and F. J. McQuillan, *Chem. Commun.*, 626 (1970).
871. M. S. Jarrell and B. C. Gates, *J. Catal.*, **40**, 255 (1975).
872. M. S. Jarrell and B. C. Gates, *J. Catal.*, **54**, 81 (1978).
873. M. S. Jarrell, B. C. Gates and E. D. Nicholson, *J. Am. Chem. Soc.*, **100**, 5727 (1978).
874. J. Jenck, *Fr. Pat.* 2 478 078 (1981); *C. A.* **96**, 85 054 (1982).
875. J. A. Joensson and H. Pscheidl, *Z. Chem.*, **22**, 105 (1982).
876. P. Johnson and M. J. Lawrenson, *U.S. Pat.* 3 660 493 (1972); *C. A.* **77**, 61 247 (1972).
877. T. H. Johnson and T. F. Baldwin, *J. Org. Chem.*, **45**, 140 (1980).
878. T. H. Johnson, K. C. Klein and S. Thomen, *J. Mol. Catal.*, **12**, 37 (1981).
879. T. H. Johnson and G. Rangarajan, *J. Org. Chem.*, **45**, 62 (1980).
880. Johnson Matthey and Co. Ltd., *Neth. Pat.* 6 510 941 (1966); *C. A.* **64**, 19 408 (1966).
881. Johnson Matthey and Co. Ltd., *Jpn. Pat.* 53 293 (1975); *C. A.* **85**, 176 838 (1976).
882. Johnson Matthey and Co. Ltd., *Neth. Pat. Appl.* 00 875 (1981); *C. A.* **96**, 7277 (1982).
883. F. N. Jones, *Fr. Pat.* 1 545 065 (1968); *C. A.* **72**, 3056 (1970).
884. G. Jones and B. R. Ramachandran, *J. Org. Chem.*, **41**, 798 (1976).
885. W. D. Jones and F. J. Feher, *J. Am. Chem. Soc.*, **104**, 4240 (1982).
886. F. Joó and E. Trócsányi, *J. Organomet. Chem.*, **231**, 63 (1982).
887. L. Joosten, M. J. Mirbach and A. Saus, *J. Mol. Catal.*, **6**, 441 (1979).
888. A. T. Jurewicz, L. D. Rollmann and D. D. Whitehurst, *Adv. Chem. Ser.*, **132**, 240 (1974).
889. J. A. Kaduk and J. A. Ibers, *Inorg. Chem.*, **14**, 3070 (1975).
890. J. A. Kaduk, T. H. Tulip, J. R. Budge and J. A. Ibers, *J. Mol. Catal.*, **12**, 239 (1981).
891. H. B. Kagan and T. P. Dang, *Chem. Commun.*, 481 (1971).
892. H. B. Kagan and T. P. Dang, *J. Am. Chem. Soc.*, **94**, 6429 (1972).
893. H. B. Kagan and T. P. Dang, *Ger. Pat.* 2 161 200 (1972); *C. A.* **77**, 114 567 (1972).
894. H. B. Kagan and J. C. Frian, *Top. Stereochem.*, **10**, 175 (1977).
895. H. B. Kagan, J. F. Peyronel and T. Yamagishi, *Adv. Chem. Ser.*, **173**, 50 (1979).
896. I. V. Kalechits, G. A. Balaev, A. D. Shebaldova, L. N. Ryzhenko, P. S. Chekrii, M. L. Khidekel, T. V. Fokina, Z. V. Arkhipova, A. A. Taber, *et al.*, *U.S.S.R. Pat.* 394 385 (1973); *C. A.* **81**, 50 312 (1974).
897. I. V. Kalechits, T. V. Fokina, P. S. Chekrii, O. M. Eremenko, G. J. Karyakina, M. L. Khidekel and A. M. Taber, *U.S.S.R. Pat.* 326 973 (1972); *C. A.* **77**, 6100 (1972).
898. M. I. Kalinkin, S. M. Markosyan, G. D. Kolomnikova, Z. N. Parnes and D. N. Kursanov, *Dokl. Akad. Nauk SSSR*, **253**, 1137 (1980); *C. A.* **94**, 29 871 (1981).
899. M. I. Kalinkin, S. M. Markosyan, D. N. Kursanov and Z. N. Parnes, *Izv. Akad. Nauk SSSR, Ser. Khim.*, 675 (1981); *C. A.* **95**, 41 917 (1981).
900. P. Kalk, R. Poilblanc and A. Gaset, *Belg. Pat.* 871 814 (1979); *C. A.* **91**, 129 602 (1979).
901. P. Kalk, R. Poilblanc and A. Gaset, *Fr. Pat.* 2 408 388 (1979); *C. A.* **92**, 58 229 (1980).
902. P. Kalk, R. Poilblanc, R. P. Martin, A. Rovera and A. Gaset, *J. Organomet. Chem.*, **195**, C9 (1980).
903. N. Kameda, *Makromol. Chem., Rapid Commun.*, **2**, 461 (1981).
904. N. Kameda and E. Ishii, *Kobunshi Ronbunshu*, **36**, 347 (1979); *C. A.* **91**, 39 979 (1979).

905. N. Kameda and E. Ishii, *J. Chem. Soc. Jpn.*, 122 (1979).
906. N. Kameda and N. Itagaki, *Bull. Chem. Soc. Jpn.*, **46**, 2597 (1973).
907. T. Kametani, T. Honda, J. Sasaki, H. Terasawa and K. Fukumoto, *J. Chem. Soc., Perkin Trans. 1*, 1884 (1981).
908. Y. Kamiya, K. Kawato and H. Ota, *Chem. Lett.*, 1549 (1980).
909. J. A. Kampmeier, S. H. Harris and D. K. Wedegaertner, *J. Org. Chem.*, **45**, 315 (1980).
910. J. A. Kampmeier, R. M. Rodehorst and J. B. Phillip, *J. Am. Chem. Soc.*, **103**, 1847 (1981).
911. K. Kaneda, H. Azuma, M. Wayaku and S. Teranishi, *Chem. Lett.*, 215 (1974).
912. K. Kaneda, M. Hiraki, T. Imanaka and S. Teranishi, *J. Mol. Catal.*, **12**, 385 (1981).
913. K. Kaneda, M. Hiraki, K. Sano, T. Imanaka and S. Teranishi, *J. Mol. Catal.*, **9**, 227 (1980).
914. K. Kaneda, T. Itoh, Y. Fujiwara and S. Teranishi, *Bull. Chem. Soc. Jpn.*, **46**, 3810 (1973).
915. K. Kaneda, M. Yasumura, T. Imanaka and S. Teranishi, *Chem. Commun.*, 935 (1982).
916. K. Kaneda, M. Yasumura, M. Hiraki, T. Imanaka and S. Teranishi, *Chem. Lett.*, 1763 (1981).
917. H. C. Kang, C. H. Mauldin, R. Cole, W. Slegeir, K. Cann and R. Petit, *J. Am. Chem. Soc.*, **99**, 8323 (1977).
918. J. W. Kang, *U.S. Pat.* 3 993 855 (1976); *C. A.* **86**, 56 558 (1977).
919. L. Kaplan, *Ger. Pat.* 2 643 193 (1977); *C. A.* **87**, 41 652 (1977).
920. L. Kaplan, *Ger. Pat.* 2 643 897 (1977); *C. A.* **87**, 41 733 (1977).
921. L. Kaplan, *Ger. Pat.* 2 643 913 (1977); *C. A.* **89**, 214 898 (1978).
922. L. Kaplan, *Ger. Pat.* 2 743 630 (1978); *C. A.* **89**, 77 593 (1978).
923. L. Kaplan, *Ger. Pat.* 2 823 127 (1978); *C. A.* **90**, 186 350 (1979).
924. L. Kaplan, *U.S. Pat.* 4 190 598 (1980); *C. A.* **93**, 25 899 (1980).
925. L. Kaplan, *Eur. Pat. Appl.* 6634 (1980); *C. A.* **93**, 113 942 (1980).
926. L. Kaplan and F. A. Cotton, *U.S. Pat.* 4 188 335 (1980); *C. A.* **93**, 45 956 (1980).
927. L. Kaplan and R. L. Pruett, *U.S. Pat.* 4 199 521 (1980); *C. A.* **93**, 132 071 (1980).
928. L. Kaplan, W. E. Walker and G. L. O'Conner, *U.S. Pat.* 4 153 623 (1979); *C. A.* **91**, 107 657 (1979).
929. K. Kashiwabara, K. Hanaki and J. Fujita, *Bull. Chem. Soc. Jpn.*, **53**, 2275 (1980).
930. J. Kaspar, R. Spogliarich and M. Graziani, *J. Organomet. Chem.*, **231**, 71 (1982).
931. J. Kaspar, R. Spogliarich, G. Mestroni and M. Graziani, *J. Organomet. Chem.*, **208**, C15 (1981).
932. R. V. Kastrup, J. S. Merola and A. A. Oswald, *Adv. Chem. Ser.*, **196**, 43 (1982).
933. N. Katsin and R. Ikan, *Synth. Commun.*, 185 (1977).
934. T. J. Katz and S. A. Cerefice, *J. Am. Chem. Soc.*, **91**, 2405 (1969); *ibid.*, **91**, 6519 (1969).
935. T. J. Katz and S. A. Cerefice, *Tetrahedron Lett.*, 2561 (1969).
936. Y. Kawabata, C. U. Pittman and R. Kobayashi, *J. Mol. Catal.*, **12**, 113 (1981).
937. H. Kawazura and T. Okmori, *Bull. Chem. Soc. Jpn.*, **45**, 2213 (1972).
938. V. Kavan and M. Capka, *Collect. Czech. Chem. Commun.*, **45**, 2100 (1980).
939. K. A. Keblys, *U.S. Pat.* 3 859 359 (1975); *C. A.* **82**, 98 152 (1975).
940. K. A. Keblys, *U.S. Pat.* 3 907 847 (1975); *C. A.* **84**, 58 635 (1976).
941. W. Keim (Ed.), *Catalysis in C_1 Chemistry*, Reidel, Dordrecht, 1983.
942. K. Kellner, A. Tzschach, Z. Nagy-Magos and L. Marko, *J. Organomet. Chem.*, **193**, 307 (1980).

943. R. J. Kern, *Chem. Commun.*, 706 (1968).
944. R. J. Kern, *J. Polym. Sci., Part A-1*, 7, 621 (1969).
945. A. D. Ketley, L. P. Fisher, A. J. Berlin, C. R. Morgan, E. H. Gorman and T. R. Steadman, *Inorg. Chem.*, 6, 657 (1967).
946. L. A. Kheifits, A. E. Gol'dovskii, I. S. Kolomnikov and M. E. Volpin, *Izv. Akad. Nauk SSSR, Ser. Khim.*, 2078 (1970); *Bull. Acad. Sci. USSR (Engl. Transl.)*, 9, 1951 (1970).
947. N. Kihara and K. Saeki, *Jpn. Pat.* 15 411 (1980); *C. A.* 93, 168 127 (1980).
948. J. Kiji, S. Kadoi and J. Furukawa, *Angew. Makromol. Chem.*, 46, 163 (1975); *C. A.* 83, 180 548 (1975).
949. J. Kiji, S. Yoshikawa and J. Furukawa, *Bull. Chem. Soc. Jpn.*, 47, 490 (1974).
950. L. Kim and K. C. Dewhirst, *J. Org. Chem.*, 38, 2722 (1973).
951. L. Kim, T. E. Paxson and S. C. Tang, *Eur. Pat. Appl.* 2557 (1979); *C. A.* 92, 94 574 (1980).
952. L. Kim, T. E. Paxson and S. C. Tang, *U.S. Pat.* 4 306 085 (1981); *C. A.* 96, 103 652 (1982).
953. D. L. King, J. A. Cusumano and R. L. Garten, *Catal. Rev.*, 23, 233 (1981).
954. R. B. King, J. Bakos, C. D. Hoff and L. Marko, *J. Org. Chem.*, 44, 1729 (1979).
955. R. B. King, J. Bakos, C. D. Hoff and L. Marko, *J. Org. Chem.*, 44, 3095 (1979).
956. R. B. King, A. D. King and M. Z. Iqbal, *J. Am. Chem. Soc.*, 101, 4893 (1979).
957. T. Kitamura, T. Joh and N. Hagihara, *Chem. Lett.*, 203 (1975).
958. T. Kitamura, M. Matsumoto and M. Tamura, *Fr. Pat.* 2 477 140 (1981); *C. A.* 96, 34 600 (1982).
959. T. Kitamura, N. Sakamoto and T. Joh, *Chem. Lett.*, 379 (1973).
960. U. Klabunde and G. W. Parshall, *J. Am. Chem. Soc.*, 94, 9081 (1972).
961. A. Kleeman, J. Martens, M. Samson and W. Bergstein, *Synthesis*, 740 (1981).
962. M. V. Klyuev and M. L. Khidekel, *Transition Met. Chem.*, 5, 134 (1980).
963. J. F. Knifton, *J. Organomet. Chem.*, 188, 223 (1980).
964. J. F. Knifton, *Hydrocarbon Process., Int. Ed.*, 60, 113 (1981); *C. A.* 96, 54 256 (1982).
965. J. F. Knifton, *U.S. Pat.* 4 315 994 (1982); *C. A.* 96, 162 122 (1982).
966. H. Knoezinger, *Inorg. Chim. Acta*, 37, L537 (1979).
967. H. Knoezinger, E. W. Thornton and M. Wolf, *J. Chem. Soc., Faraday Trans. 1*, 75, 1888 (1979).
968. W. S. Knowles, *Acc. Chem. Res.*, 16, 106 (1983).
969. W. S. Knowles and M. J. Sabacky, *Chem. Commun.*, 1445 (1968).
970. W. S. Knowles and M. J. Sabacky, *Ger. Pat.* 2 123 063 (1971); *C. A.* 76, 60 074 (1972).
971. W. S. Knowles and M. J. Sabacky, *U.S. Pat.* 3 849 480 (1974); *C. A.* 82, 86 086 (1975).
972. W. S. Knowles, M. J. Sabacky and B. D. Vineyard, *Chem. Commun.*, 10 (1972).
973. W. S. Knowles, M. J. Sabacky and B. D. Vineyard, *Ger. Pat.* 2 210 938 (1972); *C. A.* 77, 165 073 (1972).
974. W. S. Knowles, M. J. Sabacky and B. D. Vineyard, *Ger. Pat.* 2 456 937 (1975); *C. A.* 83, 164 367 (1975).
975. W. S. Knowles, M. J. Sabacky and B. D. Vineyard, *U.S. Pat.* 4 005 127 (1977); *C. A.* 86, 190 463 (1977).
976. W. S. Knowles, M. J. Sabacky, B. D. Vineyard and D. J. Weinkauff, *J. Am. Chem. Soc.*, 97, 2567 (1975).

977.   H. Knözinger and E. Rumpf, *Inorg. Chim. Acta,* **30**, 51 (1978).
978.   H. Knözinger, E. W. Thornton and M. Wolf, *Fundam. Res. Homogeneous Catal.,* **3**, 461 (1979).
979.   K. Ko and H. Yamasaki, *Jpn. Pat.* 117433 (1979); *C. A.* **92**, 110699 (1980).
980.   G. K. Koch and J. W. Dalenberg, *J. Labelled Comp. Radiopharm.,* **6**, 395 (1970).
981.   J. K. Kochi, *Organometallic Mechanisms and Catalysis,* Academic Press, New York, 1978.
982.   K. E. Koenig, G. L. Bachman and B. D. Vineyard, *J. Org. Chem.,* **45**, 2362 (1980).
983.   K. E. Koenig and W. S. Knowles, *J. Am. Chem. Soc.,* **100**, 7561 (1978).
984.   K. E. Koenig, M. J. Sabacky, G. L. Bachman, W. C. Christopfel, H. D. Barnstorff, R. B. Friedman, W. S. Knowles, B. R. Stults, B. D. Vineyard and D. J. Weinkauff, *Ann. N. Y. Acad. Sci.,* **333**, 16 (1980).
985.   J. Koettner and G. Greber, *Chem. Ber.,* **113**, 2323 (1980).
986.   I. Koga, Y. Terui and M. Ohgushi, *U.S. Pat.* 4309558 (1982); *C. A.* **96**, 181427 (1982).
987.   I. Koga, Y. Terui, M. Ohgushi and T. Kitahara, *Ger. Pat.* 2814750 (1978); *C. A.* **90**, 23241 (1979).
988.   I. Koga, Y. Terui, M. Ohgushi and N. Otake, *Jpn. Pat.* 43051 (1980); *C. A.* **93**, 95394 (1980).
989.   I. Kolb and J. Hetflejs, *Collect. Czech. Chem. Commun.,* **45**, 2224 (1980).
990.   I. Kolb and J. Hetflejs, *Collect. Czech. Chem. Commun.,* **45**, 2808 (1980).
991.   L. Kollar, S. Toros, B. Heil and L. Marko, *J. Organomet. Chem.,* **192**, 253 (1980).
992.   I. S. Kolomnikov, I. L. Nakhshunova, F. Prukhnik, N. A. Belikova and M. E. Vol'pin, *Izv. Akad. Nauk SSSR, Ser. Khim.,* 1180 (1972); *Bull, Acad. Sci. USSR (Engl. Transl.),* **21**, 1130 (1972).
993.   H. Kono, I. Ojima, M. Matsumoto and Y. Nagai, *Org. Prep. Proced. Int.,* **5**, 135 (1973).
994.   H. Kono, M. Wakao, I. Ojima and Y. Nagai, *Chem. Lett.,* 189 (1975).
995.   L. I. Kopylova, V. B. Pukhnarevich, I. I. Tsykhanskaya, E. N. Satsuk, B. V. Timokhin, V. I. Dmitriev, V. Chvalovsky, M. Capka, A. V. Kalabina and M. G. Voronkov, *Zh. Obsch. Khim.,* **51**, 1851 (1981); *C. A.* **95**, 204040 (1981).
996.   L. I. Kopylova, M. V. Sigalov, E. N. Satsuk, M. Capka, V. Chvalovsky, V. B. Pukhnarevich, E. Lukevics and M. G. Voronkov, *Zh. Obshch. Khim.,* **51**, 385 (1981); *C. A.* **95**, 62298 (1981).
997.   L. G. Korableva and I. P. Lavrent'ev, *Izv. Akad. Nauk SSSR, Ser. Khim.,* 137 (1981); *C. A.* **94**, 157365 (1981).
998.   V. A. Kormer, B. D. Babitskii, M. I. Lobach and N. N. Chesnokova, *J. Polym. Sci., Part C,* **16**, 4351 (1965).
999.   L. M. Koroleva, V. K. Latov, M. B. Saporovskaya and V. M. Belikov, *Izv. Akad. Nauk SSSR, Ser. Khim.,* 2390 (1979); *C. A.* **92**, 129284 (1980).
1000. M. N. Kotova, P. S. Cherkii, A. M. Taber and I. V. Kalechits, *U.S.S.R. Pat.* 461931 (1975); *C. A.* **83**, 10380 (1975).
1001. W. Kotowski, *Chem. Tech. (Leipzig),* **30**, 360 (1978); *C. A.* **89**, 146362 (1978).
1002. W. Kotowski and Z. Pokorska, *Chem. Tech. (Leipzig),* **32**, 637 (1980); *C. A.* **94**, 191612 (1981).
1003. Z. Kozak and M. Capka, *Collect. Czech. Chem. Commun.,* **44**, 2624 (1979).
1004. H. Krause, *East Ger. Pat.* 133230 (1978); *C. A.* **91**, 158101 (1979).
1005. H. Krause, *React. Kinet. Catal. Lett.,* **10**, 243 (1979).
1006. H. Krause and H. Mise, *East Ger. Pat.* 133199 (1978); *C. A.* **91**, 63336 (1979).

1007. J. Krepelka and J. Zachoval, *Collect. Czech. Chem. Commun.*, **35**, 3800 (1970); *Chem. Prum.*, **22**, 624 (1972).
1008. J. Krepelka, J. Zachoval and P. Strop, *Sb. Vys. Sk. Chem.-Technol. Praze, Org. Chem. Technol.*, **C18**, 5 (1973); *C. A.* **80**, 3888 (1974).
1009. H. J. Kreuzfeld and C. Doebler, *React. Kinet. Catal. Lett.*, **16**, 229 (1981).
1010. A. Krzywicki and M. Marczewski, *J. Mol. Catal.*, **6**, 431 (1979).
1011. A. Krzywicki and G. Pannetier, *Bull. Soc. Chim. Fr.*, 1093 (1975).
1012. C. P. Kubiak and R. Eisenberg, *J. Am. Chem. Soc.*, **102**, 3637 (1980).
1013. C. P. Kubiak, C. Woodcock and R. Eisenberg, *Inorg. Chem.*, **21**, 2119 (1982).
1014. H. K. Kuebbeler, H. Erpenbach, K. Gehrmann and G. Kehl, *Ger. Pat.* 2 941 232 (1981); *C. A.* **95**, 61 530 (1981).
1015. E. J. Kuhlmann and J. J. Alexander, *Inorg. Chim. Acta*, **34**, 197 (1979).
1016. E. J. Kuhlmann and J. J. Alexander, *J. Organomet. Chem.*, **174**, 81 (1979).
1017. R. Kummer and H. W. Schneider, *Ger. Pat.* 2 614 799 (1977); *C. A.* **87**, 207 275 (1977).
1018. R. Kummer, K. Schwirten and H. D. Schindler, *Ger. Pat.* 2 448 005 (1976); *C. A.* **85**, 77 622 (1976).
1019. R. Kummer, K. Schwirten and H. D. Schindler, *Can. Pat.* 1 055 512 (1979); *C. A.* **91**, 140 340 (1979).
1020. G. Kuncova and V. Chalovsky, *Collect. Czech. Chem. Commun.*, **45**, 2085 (1980).
1021. G. Kuncova and V. Chalovsky, *Collect. Czech. Chem. Commun.*, **45**, 2240 (1980).
1022. V. P. Kurkov, S. J. Lapporte and W. G. Toland, *U.S. Pat.* 3 751 453 (1973); *C. A.* **79**, 91 598 (1971).
1023. V. P. Kurkov, J. Z. Pasky and J. B. Lavigne, *J. Am. Chem. Soc.*, **90**, 4 743 (1968).
1024. K. Kurtev, D. Ribola, R. A. Jones, D. J. Cole-Hamilton and G. Wilkinson, *J. Chem. Soc., Dalton Trans.*, 55 (1980).
1025. P. Kvintovics, B. Heil and L. Marko, *Adv. Chem. Ser.*, **173**, 26 (1979).
1026. P. Kvintovics, B. Heil, J. Palágyi and L. Marko, *J. Organomet. Chem.*, **148**, 311 (1978).
1027. E. Kwaskowska-Chec and J. J. Ziolkowski, *Chem. Stosow.*, **24**, 433 (1980); *C. A.* **95**, 131 749 (1981).
1028. E. P. Kyba, R. E. Davis, P. N. Juri and K. R. Shirley, *Inorg. Chem.*, **20**, 3616 (1981).
1029. D. Lafont, D. Sinou and G. Descotes, *J. Organomet. Chem.*, **169**, 87 (1979).
1030. D. Lafont, D. Sinou and G. Descotes, *J. Chem. Res.*, 117 (1982).
1031. D. Lafont, D. Sinou, G. Descotes, R. Glaser and S. Geresh, *J. Mol. Catal.*, **10**, 305 (1981).
1032. R. Lai and E. Ucciani, *C.R. Acad. Sci., Ser. C*, **276**, 425 (1973).
1033. R. Lai and E. Ucciani, *J. Mol. Catal.*, **4**, 401 (1978).
1034. R. M. Laine, *J. Am. Chem. Soc.*, **100**, 6451 (1978).
1035. R. M. Laine, *J. Org. Chem.*, **45**, 3370 (1980).
1036. R. M. Laine, D. W. Thomas and L. W. Cary, *J. Org. Chem.*, **44**, 4964 (1979).
1037. R. M. Laine, D. W. Thomas and L. W. Cary, *J. Am. Chem. Soc.*, **104**, 1763 (1982).
1038. R. M. Laine, D. W. Thomas, L. W. Cary and S. E. Buttrill, *J. Am. Chem. Soc.*, **100**, 6527 (1978).
1039. S. Lamalle, A. Mortreux, M. Evrard, F. Petit, J. Grimblot and J. P. Bonnelle, *J. Mol. Catal.*, **6**, 11 (1979).
1040. G. La Monica and S. Cenini, *J. Organomet. Chem.*, **216**, C35 (1981).
1041. W. H. Lang, A. T. Jurewicz, W. O. Haag, D. D. Whitehurst and L. D. Rollmann, *J. Organomet. Chem.*, **134**, 85 (1977).

1042.   N. Langlois, T. P. Dang and H. B. Kagan, *Tetrahedron Lett.*, 4865 (1973).
1043.   A. L. Lapidus, S. D. Pirozhkov and A. A. Koryakin, *Izv. Akad. Nauk SSSR, Ser. Khim.*, 2814 (1978); *C. A.* **90**, 103 345 (1979).
1044.   A. L. Lapidus and M. M. Savel'ev, *Izv. Akad. Nauk SSSR, Ser. Khim.*, 335 (1980); *C. A.* **93**, 7585 (1980).
1045.   A. L. Lapidus, M. M. Savel'ev and L. T. Kondrat'ev, *Khim. Tverd. Topl. (Moscow)*, **1**, 23 (1981); *C. A.* **94**, 211 106 (1981).
1046.   A. L. Lapidus, M. M. Savel'ev, L. T. Kondrat'ev and E. V. Yastrebova, *Izv. Akad. Nauk SSSR, Ser. Khim.*, 474 (1981); *C. A.* **94**, 128 092 (1981).
1047.   A. L. Lapidus, M. M. Savel'ev, L. T. Kondrat'ev and E. V. Yastrebova, *Izv. Akad. Nauk SSSR, Ser. Khim.*, 1564 (1981); *C. A.* **95**, 222 615 (1981).
1048.   A. L. Lapidus and Y. B. Yan, *U.S.S.R. Pat.* 859 349 (1981); *C. A.* **96**, 122 205 (1982).
1049.   A. L. Lapidus, Y. B. Yan, N. I. Alpatov and Y. N. Ogibin, *U.S.S.R. Pat.* 859 350 (1981); *C. A.* **96**, 122 206 (1982).
1050.   S. J. Lapporte and W. G. Toland, *Ger. Pat.* 2 302 568 (1973); *C. A.* **79**, 115 158 (1973).
1051.   R. C. Larock, *U.S. Pat.* 4 105 705 (1978); *C. A.* **90**, 71 240 (1979).
1052.   R. C. Larock, *U.S. Pat.* 4 288 613 (1981); *C. A.* **96**, 6262 (1982).
1053.   R. C. Larock and J. C. Bernhardt, *J. Org. Chem.*, **42**, 1680 (1977).
1054.   R. C. Larock and S. A. Hershberger, *J. Org. Chem.*, **45**, 3840 (1980).
1055.   R. C. Larock and S. A. Hershberger, *Tetrahedron Lett.*, 2443 (1981).
1056.   R. C. Larock, K. Oertle and G. F. Potter, *J. Am. Chem. Soc.*, **102**, 190 (1980).
1057.   S. Lars, T. Anderson and M. S. Scurrel, *J. Catal.*, **59**, 340 (1979).
1058.   C. Lassau, Y. Chauvin and G. Lefebvro, *Ger. Pat.* 1 902 560 (1969); *C. A.* **72**, 21 358 (1970).
1059.   C. Lassau, R. Stern and L. Sajus, *Fr. Pat.* 2 041 776 (1971); *C. A.* **75**, 129 945 (1971).
1060.   K. S. Y. Lau, Y. Becker, F. Huang, N. Baenziger and J. K. Stille, *J. Am. Chem. Soc.*, **99**, 5664 (1977).
1061.   M. Lauer, O. Samuel and H. B. Kagan, *J. Organomet. Chem.*, **177**, 309 (1979).
1062.   P. M. Lausarot, G. A. Vaglio and M. Valle, *J. Organomet. Chem.*, **204**, 249 (1981).
1063.   P. M. Lausarot, G. A. Vaglio and M. Valle, *J. Organomet. Chem.*, **215**, 111 (1981).
1064.   M. J. Lawrenson, *Ger. Pat.* 2 031 380 (1971); *C. A.* **74**, 87 397 (1971).
1065.   M. J. Lawrenson, *Ger. Pat.* 2 058 814 (1971); *C. A.* **75**, 109 845 (1971).
1066.   M. J. Lawrenson, *Br. Pat.* 1 243 190 (1971); *C. A.* **75**, 109 847 (1971).
1067.   M. J. Lawrenson, *Br. Pat.* 1 254 222 (1971); *C. A.* **76**, 33 787 (1972).
1068.   M. J. Lawrenson, *Br. Pat.* 1 284 615 (1972); *C. A.* **77**, 125 982 (1972).
1069.   M. J. Lawrenson, *Br. Pat.* 1 296 435 (1972); *C. A.* **78**, 83 824 (1973).
1070.   M. J. Lawrenson and G. Foster, *Ger. Pat.* 1 806 293 (1969); *C. A.* **71**, 70 109 (1969).
1071.   M. J. Lawrenson and G. Foster, *Ger. Pat.* 1 812 504 (1969); *C. A.* **71**, 101 313 (1969).
1072.   H. S. Leach, T. C. Singleton and Y. W. Wei, *U.S. Pat.* 4 007 130 (1977); *C. A.* **86**, 111 643 (1977).
1073.   H. S. Leach, T. C. Singleton and Y. W. Wei, *Ger. Pat.* 2 659 173 (1977); *C. A.* **87**, 91 407 (1977).
1074.   P. Legzdins, G. L. Rempel and G. Wilkinson, *Chem. Commun.*, 825 (1969); P. Legzdins, R. W. Mitchell, G. L. Rempel, J. D. Ruddick and G. Wilkinson, *J. Chem. Soc. A*, 3322 (1970).

1075. A. Levi, G. Modena and G. Scorrano, *Chem. Commun.*, 6 (1975).
1076. I. Ya. Levitin, L. G. Volkova, T. M. Ushakova, A. L. Sigan and M. E. Volpin, *Izv. Akad. Nauk SSSR, Ser. Khim.*, 1188 (1973); *Bull. Acad. Sci. USSR (Engl. Transl.)*, 22, 1156 (1973).
1077. J. Lieto, *Chem. Technol.*, 46 (1983).
1078. J. Lieto, J. J. Rafalko and B. C. Gates, *J. Catal.*, 62, 149 (1980).
1079. J. Lieto, J. J. Rafalko, J. V. Minkiewicz, P. W. Rafalko and B. C. Gates, *Fundam. Res. Homogeneous Catal.*, 3, 637 (1979).
1080. J. R. C. Light and H. H. Zeiss, *J. Organomet. Chem.*, 21, 517 (1970).
1081. E. Lindner and A. Thasitis, *Chem. Ber.*, 107, 2418 (1974).
1082. D. C. Lini, K. C. Ramey and W. B. Wise, *U.S. Pat.* 3 465 043 (1969); *C. A.* 72, 12 106 (1970).
1083. E. F. Litvin, L. K. Freidlin, L. F. Krokhmaleva, L. M. Kozlova and N. M. Nazarova, *Izv. Akad. Nauk SSSR, Ser. Khim.*, 811 (1981); *C. A.* 95, 61 629 (1981).
1084. A. T. C. Liu, *Br. Pat.* 1 119 728 (1968); *C. A.* 69, 86 412 (1968).
1085. C. F. Lochow and R. G. Miller, *J. Am. Chem. Soc.*, 98, 1281 (1976).
1086. C. J. Love and F. J. McQuillin, *J. Chem. Soc., Perkin Trans. 1*, 2509 (1973).
1087. A. Luchetti and D. M. Hercules, *J. Mol. Catal.*, 16, 95 (1982).
1088. E. Lukevics, *Russ. Chem. Rev. (Engl. Transl.)*, 46, 264 (1977).
1089. E. Lukevics, Z. V. Belyakova, M. G. Pomerantseva and M. G. Voronkov, *Organomet. Chem. Rev., J. Organomet. Chem. Libr.*, 5, 1 (1977).
1090. J. E. Lyons, *Chem. Commun.*, 564 (1969).
1091. J. E. Lyons, *Ger. Pat.* 2 002 596 (1970); *C. A.* 73, 120 191 (1970).
1092. J. E. Lyons, *J. Org. Chem.*, 36, 2497 (1971).
1093. J. E. Lyons, *J. Catal.*, 30, 490 (1973).
1094. J. E. Lyons, *Aspects Homogeneous Catal.*, 3, 1 (1977).
1095. J. E. Lyons, *U.S. Pat.* 4 123 465 (1978); *C. A.* 90, 168 426 (1979).
1096. J. E. Lyons and J. O. Turner, *J. Org. Chem.*, 37, 2881 (1972).
1097. J. E. Lyons and J. O. Turner, *Tetrahedron Lett.*, 2903 (1972).
1098. V. Macho, L. Jurecek and M. Polievka, *Chem. Prum.*, 24, 237 (1974); *C. A.* 81, 36 940 (1974).
1099. V. Macho and M. Minarska, *Petrochemia*, 13, 149 (1973); *C. A.* 82, 30 581 (1975).
1100. V. Macho and M. Minarska, *Hung. J. Ind. Chem.*, 2 (1, Suppl.), 147 (1974); *C. A.* 83, 113 362 (1975).
1101. P. A. MacNeil, N. K. Roberts and B. Bosnich, *J. Am. Chem. Soc.*, 103, 2273 (1981).
1102. T. D. Madden, W. E. Peel, P. J. Quinn and D. Chapman, *J. Biochem. Biophys. Methods*, 2, 19 (1980).
1103. T. D. Madden and P. J. Quinn, *Biochem. Soc. Trans.*, 6, 1345 (1978).
1104. G. K. I. Magomedov, K. A. Andrianov, O. V. Shkolnik, B. A. Izmailov and V. N. Kalinin, *J. Organomet. Chem.*, 149, 29 (1978).
1105. G. K. I. Magomedov, G. V. Druzhkova, V. G. Syrkin and O. V. Shkol'nik, *Koord. Khim.*, 6, 767 (1980); *C. A.* 93, 167 148 (1980).
1106. G. K. I. Magomedov, L. V. Morozova, V. N. Kalinin and L. I. Zakharkin, *J. Organomet. Chem.*, 149, 23 (1978).
1107. G. K. I. Magomedov and O. V. Shkol'nik, *Zh. Obshch. Khim.*, 50, 1103 (1980); *C. A.* 93, 167 153 (1980).
1108. G. K. I. Magomedov, O. V. Shkol'nik, V. G. Syrkin and S. A. Sigachev, *Zh. Obshch. Khim.*, 48, 2257 (1978); *C. A.* 90, 86 314 (1979).
1109. J. T. Mague, M. O. Nutt and E. H. Gausse, *J. Chem. Soc., Dalton Trans.*, 2578 (1974).

1110. J. T. Mague and G. Wilkinson, *J. Chem. Soc. A*, 1736 (1966).
1111. P. M. Maitlis, *Acc. Chem. Res.*, **11**, 301 (1978).
1112. P. M. Maitlis, *Adv. Chem. Ser.*, **173**, 31 (1979).
1113. P. M. Maitlis, C. White, J. W. Kang and D. S. Gill, *Can. Pat.* 915 697 (1972); *C. A.* **78**, 76 413 (1973).
1114. F. D. Mango, *Coord. Chem. Rev.*, **15**, 109 (1975).
1115. V. I. Manov-Yuvenskii and B. K. Nefedov, *Izv. Akad. Nauk SSSR, Ser. Khim.*, 1055 (1981); *C. A.* **95**, 97 236 (1981).
1116. V. I. Manov-Yuvenskii, A. V. Smetanin and B. K. Nefedov, *Izv. Akad. Nauk SSSR, Ser. Khim.*, 2556 (1980); *C. A.* **94**, 83 245 (1981).
1117. V. I. Manov-Yuvenskii, A. V. Smetanin, B. K. Nefedov and A. L. Chimishkyan, *Kinet. Katal.*, **21**, 1335 (1980); *C. A.* **94**, 29 883 (1981).
1118. E. Mantovani, N. Palladino and A. Zanobi, *J. Mol. Catal.*, **3**, 285 (1978).
1119. E. Mantovani, N. Palladino and A. Zanobi, *Ger. Pat.* 2 804 307 (1978); *C. A.* **89**, 136 452 (1978).
1120. B. Marciniec, Z. W. Kornetka and W. Urbaniak, *J. Mol. Catal.*, **12**, 221 (1981).
1121. B. Marciniec, W. Urbaniak and P. Pawlak, *J. Mol. Catal.*, **14**, 323 (1982).
1122. F. Mares and R. Tang, *J. Org. Chem.*, **43**, 4631 (1978).
1123. L. Marko, *Aspects Homogeneous Catal.*, **2**, 1 (1974).
1124. L. Marko and J. Bakos, *J. Organomet. Chem.*, **81**, 411 (1974).
1125. L. Marko and J. Bakos, *Aspects Homogeneous Catal.*, **4**, 145 (1981).
1126. S. M. Markosyan, M. I. Kalinkin, Z. N. Parnes and D. N. Kursanog, *Dokl. Akad. Nauk SSSR*, **255**, 599 (1980); *C. A.* **94**, 138 922 (1981).
1127. F. Martinelli, G. Mestroni, A. Camus and G. Zassinovich, *J. Organomet. Chem.*, **220**, 383 (1981).
1128. K. Maruyama, K. Terada and Y. Yamamoto, *Chem. Lett.*, 839 (1981); *J. Org. Chem.*, **46**, 5294 (1981).
1129. G. Marx, *J. Org. Chem.*, **36**, 1725 (1971).
1130. R. Mason, D. M. P. Mingos, G. Rucci and J. A. Connor, *J. Chem. Soc., Dalton Trans.*, 1729 (1972).
1131. F. Maspero and E. Perrotti, *Ger. Pat.* 2 149 934 (1972); *C. A.* **77**, 61 251 (1972).
1132. F. Maspero, E. Perrotti and F. Simonetti, *Ger. Pat.* 2 263 882 (1973); *C. A.* **79**, 92 386 (1973).
1133. C. Masters, *Adv. Organomet. Chem.*, **17**, 61 (1979).
1134. C. Masters, *Homogeneous Transition Metal Catalysis – A Gentle Art*, Chapman and Hall, London, 1981.
1135. C. Masters, A. A. Kiffen and J. P. Visser, *J. Am. Chem. Soc.*, **98**, 1357 (1976).
1136. A. Masuda, H. Mitani, K. Oku and Y. Yamazaki, *J. Chem. Soc. Jpn.*, 249 (1982).
1137. T. Masuda and J. K. Stille, *J. Am. Chem. Soc.*, **100**, 268 (1978).
1138. N. Matsuhira and U. Ono, *Jpn. Pat.* 44 446 (1978); *C. A.* **90**, 121 023 (1979).
1139. E. Matsui and T. Tsuruta, *Polym. J.*, **10**, 133 (1978).
1140. Y. Matsui, *Tetrahedron Lett.*, 1107 (1976).
1141. M. Matsumoto and M. Tamura, *Ger. Pat.* 2 904 782 (1979); *C. A.* **91**, 174 826 (1979).
1142. M. Matsumoto and M. Tamura, *Ger. Pat.* 2 922 757 (1979); *C. A.* **92**, 214 886 (1980).
1143. M. Matsumoto and M. Tamura, *Jpn. Pat.* 45 643 (1980); *C. A.* **93**, 113 962 (1980).
1144. M. Matsumoto and M. Tamura, *Jpn. Pat.* 45 642 (1980); *C. A.* **93**, 239 228 (1980).
1145. M. Matsumoto and M. Tamura, *Ger. Pat.* 3 030 108 (1981); *C. A.* **95**, 6510 (1981).

1146. M. Matsumoto and M. Tamura, *J. Mol. Catal.*, **16**, 195 (1982).

1147. M. Matsumoto and M. Tamura, *J. Mol. Catal.*, **16**, 209 (1982).

1148. T. Matsumoto, T. Mizoroki and A. Osaki, *J. Catal.*, **51**, 96 (1978).

1149. T. Matsumoto, K. Mori, T. Mizoroki and A. Ozaki, *Bull. Chem. Soc. Jpn.*, **50**, 2337 (1977).

1150. Y. Matsushima, T. Sakakibara, Y. Nagashima, S. Kaneko, N. Okamoto, Y. Ishii and S. Wada, *Jpn. Pat.* 49 334 (1980); *C. A.* **93**, 238 837 (1980).

1151. M. Mazzei, W. Marconi and M. Riocci, *J. Mol. Catal.*, **9**, 381 (1980).

1152. M. Mazzei, M. Riocci and W. Marconi, *Belg. Pat.* 871 320 (1979); *C. A.* **91**, 74 198 (1979).

1153. J. J. McCoy and J. S. Yoo, *U.S. Pat.* 3 714 185 (1973); *C. A.* **78**, 147 785 (1973).

1154. F. J. McQuillin, *Homogeneous Hydrogenation in Organic Chemistry*, Reidel, Dordrecht, 1976.

1155. F. J. McQuillin, *Br. Pat.* 1 555 331 (1979); *C. A.* **93**, 12 948 (1980).

1156. F. J. McQuillin and P. Abley, *J. Chem. Soc. C*, 844 (1971).

1157. F. J. McQuillin and D. G. Parker, *J. Chem. Soc., Perkin Trans. 1*, 2092 (1975).

1158. G. B. McVicker, *Ger. Pat.* 2 424 526 (1975); *C. A.* **84**, 89 616 (1976).

1159. G. B. McVicker, *U.S. Pat.* 3 939 188 (1976); *C. A.* **84**, 164 166 (1976).

1160. G. B. McVicker, *U.S. Pat.* 4 217 249 (1980); *C. A.* **94**, 106 198 (1981).

1161. G. B. McVicker and M. A. Vannice, *J. Catal.*, **63**, 25 (1980).

1162. P. Meakin, J. P. Jesson and C. A. Tolman, *J. Am. Chem. Soc.*, **94**, 3240 (1972).

1163. W. Mehnert, *Ger. Pat.* 2 645 650 (1978); *C. A.* **89**, 25 098 (1978).

1164. E. B. Meier, R. R. Burch, E. L. Muetterties and V. W. Dat, *J. Am. Chem. Soc.*, **104**, 2661 (1982).

1165. G. D. Mercer, W. B. Beaulieu and D. M. Roundhill, *J. Am. Chem. Soc.*, **99**, 6551 (1977).

1166. G. D. Mercer, J. S. Shu, T. B. Rauchfuss and D. M. Roundhill, *J. Am. Chem. Soc.*, **97**, 1967 (1975).

1167. G. Mestroni, R. Spogliarich, A. Camus, F. Martinelli and G. Zassinovich, *J. Organomet. Chem.*, **157**, 345 (1978).

1168. G. Mestroni, G. Zassinovich and A. Camus, *J. Organomet. Chem.*, **140**, 63 (1977).

1169. G. Mestroni, G. Zassinovich and A. Camus, *Ger. Pat.* 3 008 671 (1980); *C. A.* **93**, 238 931 (1980).

1170. G. Mestroni, G. Zassinovich and A. Camus, *Fr. Pat.* 2 474 026 (1981); *C. A.* **96**, 34 866 (1982).

1171. G. Mestroni, G. Zassinovich, A. Camus and F. Martinelli, *J. Organomet. Chem.*, **198**, 87 (1980).

1172. A. Y. Meyer and J. Schlesinger, *Isr. J. Chem.*, **8**, 671 (1970).

1173. C. D. Meyer and R. Eisenberg, *J. Am. Chem. Soc.*, **98**, 1364 (1976).

1174. D. Meyer, J. C. Poulin, H. B. Kagan, H. Levine-Pinto, J. L. Morgat and P. Fromageot, *J. Org. Chem.*, **45**, 4680 (1980).

1175. Z. Michalska, *Zesz. Nauk. Politech Lodz., Chem.*, **32**, 135 (1976); *C. A.* **86**, 140 232 (1977).

1176. Z. Michalska, *J. Mol. Catal.*, **3**, 125 (1977).

1177. Z. M. Michalska, *Transition Met. Chem.*, **5**, 125 (1980).

1178. Z. M. Michalska, M. Capka and J. Stoch, *J. Mol. Catal.*, **11**, 323 (1981).

1179. Z. M. Michalska and D. E. Webster, *Chem. Technol.*, 117 (1975).

1180. M. Michman and M. Balog, *J. Organomet. Chem.*, **31**, 395 (1971).

1181. F. Mikes and J. Kalal, *Chem. Prum.*, **20**, 17 (1970).

1182. F. Mikes and J. Kalal, *Chem. Prum.,* **20**, 218 (1970).
1183. A. Millan, E. Towns and P. M. Maitlis, *Chem. Commun.,* 673 (1981).
1184. D. Milstein, O. Buchman and J. Blum, *Tetrahedron Lett.,* 2257 (1974).
1185. H. Mimoun, *Pure Appl. Chem.,* **53**, 2389 (1981).
1186. H. Mimoun, R. Charpentier and I. S. DeRoch, *Fr. Pat.* 2 395 244 (1979); *C. A.* **91**, 174 899 (1979).
1187. H. Mimoun, M. M. P. Machirant and I. S. DeRoch, *J. Am. Chem. Soc.,* **100**, 5437 (1978).
1188. M. J. Mirbach, N. Topalsavoglu, N. P. Tuyet, M. F. Mirbach and A. Saus, *Angew. Chem., Int. Ed. Engl.,* **20**, 381 (1981).
1189. T. Mise, P. Hong and H. Yamazaki, *Chem. Lett.,* 439 (1980).
1190. T. Mise, P. Hong and H. Yamazaki, *Chem. Lett.,* 993 (1981).
1191. T. Mise, P. Hong and H. Yamazaki, *Chem. Lett.,* 401 (1982).
1192. R. W. Mitchell, J. D. Ruddick and G. Wilkinson, *J. Chem. Soc. A*, 3224 (1971).
1193. T. R. B. Mitchell, *J. Chem. Soc. B*, 823 (1979).
1194. Mitsubishi Chemical Industries Co., Ltd., *Jpn. Pat.* 02 994 (1981); *C. A.* **96**, 68 297 (1982).
1195. Mitsubishi Chemical Industries Co., Ltd., *Jpn. Pat.* 38 577 (1981); *C. A.* **96**, 6178 (1982).
1196. Mitsubishi Chemical Industries Co., Ltd. *Jpn. Pat.* 65 946 (1981); *C. A.* **95**, 154 521 (1981); *Jpn. Pat.* 65 947 (1981); *C. A.* **95**, 173 237 (1981); *Jpn. Pat.* 65 948 (1981); *C. A.* **95**, 154 522 (1981).
1197. Mitsubishi Gas Chemical Co., Inc., *Jpn. Pat.* 30 930 (1981); *C. A.* **95**, 97 012 (1981).
1198. Mitsubishi Gas Chemical Co., Inc., *Jpn. Pat.* 150 031 (1981); *C. A.* **96**, 103 674 (1982).
1199. Mitsubishi Petrochemical Co., Ltd., *Jpn. Pat.* 16 482 (1981); *C. A.* **95**, 81 276 (1981).
1200. Mitsubishi Petrochemical Co., Ltd., *Jpn. Pat.* 30 938 (1981); *C. A.* **95**, 114 918 (1981).
1201. Mitsubishi Petrochemical Co., Ltd., *Jpn. Pat.* 87 549 (1981); *C. A.* **95**, 203 367 (1981).
1202. Mitsui Toatsu Chemicals, Inc., *Jpn. Pat.* 160 727 (1980); *C. A.* **95**, 24 575 (1981).
1203. S. Miyano, A. Mori, K. Kato, Y. Kawashima and H. Hashimoto, *Chem. Lett.,* 1379 (1982).
1204. A. Miyashita, A. Yasuda, H. Takaya, K. Toriumi, T. Ito, T. Souchi and R. Noyari, *J. Am. Chem. Soc.,* **102**, 7932 (1980).
1205. K. Miyata, Y. Hirai and H. Yoshida, *U.S. Pat.* 4 134 880 (1979); *C. A.* **90**, 137 537 (1979).
1206. T. Mizoroki, K. Seki, S. Meguro and A. Ozaki, *Bull. Chem. Soc. Jpn.,* **50**, 2148 (1977).
1207. I. Mochida, S. Shirahama, H. Fujitsu and K. Takeshita, *Chem. Lett.,* 1025 (1975).
1208. I. Mochida, S. Shirahama, H. Fujitsu and K. Takeshita, *Chem. Lett.,* 421 (1977).
1209. Monsanto Co., *Belg. Pat.* 713 296 (1968).
1210. Monsanto Co., *Br. Pat.* 1 185 453 (1970); *C. A.* **72**, 132 059 (1970).
1211. Monsanto Co., *Br. Pat.* 1 278 354 (1972); *C. A.* **77**, 127 218 (1972).
1212. Monsanto Co., *Neth. Pat.* 00 745 (1975); *C. A.* **85**, 20 619 (1976).
1213. Monsanto Co., *Br. Pat.* 1 507 376 (1978); *C. A.* **90**, 23 260 (1979).

1214. Montecatini Edison S.p.A., *Ital. Pat.* 792 570 (1967); *C. A.* **70**, 20 504 (1967).
1215. S. Montelatici, A. Van der Ent, J. A. Osborne and G. Wilkinson, *J. Chem. Soc. A*, 1054 (1968).
1216. G. Montrasi, G. Pagani, G. Gregorio, P. C. D'Oro and A. Andreetta, *Chim. Ind. (Milan)*, **62**, 737 (1980).
1217. J. R. Morandi and H. B. Jensen, *J. Org. Chem.*, **34**, 1889 (1969).
1218. F. Morandini, B. Longato and S. Bresadola, *J. Organomet. Chem.*, **239**, 377 (1982).
1219. D. Morel, *Eur. Pat. Appl.* 9429 (1980); *C. A.* **93**, 113 947 (1980).
1220. D. Morel, *Eur. Pat. Appl.* 15 845 (1980); *C. A.* **94**, 102 846 (1981).
1221. D. Morel, *Eur. Pat. Appl.* 44 771 (1982); *C. A.* **96**, 199 115 (1982).
1222. Y. Mori and T. Akitani, *Jpn. Pat.* 79 252 (1979); *C. A.* **91**, 210 971 (1979).
1223. D. G. Morrell and P. D. Sherman, *Ger. Pat.* 2 802 922 (1978); *C. A.* **89**, 179 540 (1978).
1224. D. E. Morris, *Ger. Pat.* 2 310 808 (1973); *C. A.* **79**, 136 534 (1973).
1225. D. E. Morris, *U.S. Pat.* 3 944 603 (1976); *C. A.* **84**, 179 686 (1976).
1226. D. E. Morris and H. B. Tinker, *Ger. Pat.* 2 359 377 (1974); *C. A.* **81**, 91 053 (1974).
1227. D. E. Morris and H. B. Tinker, *U.S. Pat.* 3 948 962 (1976); *C. A.* **85**, 33 191 (1976).
1228. J. D. Morrison, W. F. Masler and S. Hathaway, in *Catalysis in Organic Synthesis*, (R. N. Rylander and H. Greenfield, Eds.), Academic Press, New York, 1976.
1229. K. Moseley, J. W. Kang and P. M. Maitlis, *Chem. Commun.*, 1155 (1969).
1230. K. Moseley and P. M. Maitlis, *Chem. Commun.*, 1156 (1969); *J. Chem. Soc. A*, 2884 (1970).
1231. D. Moy, *U.S. Pat.* 4 266 070 (1981); *C. A.* **95**, 80 526 (1981).
1232. E. Müller, W. Hoffmann, W. Aguila and H. Siegel, *Ger. Pat.* 2 226 212 (1973); *C. A.* **80**, 70 326 (1974); *Ger. Pat.* 2 227 400 (1974); *C. A.* **80**, 145 414 (1974).
1233. E. Müller and A. Segnitz, *Justus Liebigs Ann. Chem.*, **759**, 1583 (1973); *C. A.* **80**, 26 868 (1974).
1234. E. Müller, A. Segnitz and E. Langer, *Tetrahedron Lett.*, 1129 (1969).
1235. P. Mueller and C. Bobillier, *Tetrahedron Lett.*, 5157 (1981).
1236. E. L. Muetterties, *Chem. Soc. Rev.*, **11**, 283 (1982).
1237. E. L. Muetterties, A. J. Sivak, R. K. Brown, J. W. Williams, M. F. Fredrich and V. W. Day, *Fundam. Res. Homogeneous Catal.*, **3**, 487 (1979).
1238. E. L. Muetterties and J. Stein, *Chem. Rev.*, **79**, 479 (1979).
1239. S. Murai, Y. Seki, A. Hidaka and N. Sonoda, *Jpn. Pat.* 119 830 (1978); *C. A.* **90**, 72 341 (1979).
1240. S. Murai, R. Sugise and N. Sonoda, *Angew. Chem.*, **93**, 481 (1981).
1241. I. Murata and T. Tatsuoka, *Tetrahedron Lett.*, 2697 (1975).
1242. I. Murata, T. Tatsuoka and Y. Sugihara, *Angew. Chem., Int. Ed. Engl.*, **13**, 142 (1974).
1243. L. L. Murrell, in *Advanced Materials in Catalysis*, (J. J. Burton and R. L. Garten, Eds.), Academic Press, New York, 1977, pp. 235–265.
1244. A. J. Naatktgeboren, R. J. M. Nolte and W. Drenth, *J. Mol. Catal.*, **11**, 343 (1981).
1245. Y. Nagai, I. Ojima and S. Inaba, *Ger. Pat.* 2 409 010 (1974); *C. A.* **81**, 169 625 (1974).
1246. Y. Nagai, I. Ojima, I. Ono, Y. Hiraga and Y. Suzuki, *Jpn. Pat.* 88 923 (1976); *C. A.* **86**, 106 766 (1977).
1247. U. Nagel, H. Menzel, P. W. Lednor, W. Beck, A Guyot and M. Bartholin, *Z. Naturforsch., Teil B*, **36**, 578 (1981).

1248. Z. Nagy-Magos, B. Heil and L. Marko, *Transition Met. Chem.*, **1**, 215 (1976).
1249. Z. Nagy-Magos, S. Vastag, B. Heil and L. Marko, *Transition Met. Chem.*, **3**, 123 (1978).
1250. Z. Nagy-Magos, S. Vastag, B. Heil and L. Marko, *J. Organomet. Chem.*, **171**, 97 (1979).
1251. A. Nakamara and M. Tsutsui, *Principles and Applications of Homogeneous Catalysis*, Wiley, New York 1980.
1252. Y. Nakamura, *Jpn. Pat.* 44 608 (1979); *C. A.* **91**, 123 409 (1979).
1253. Y. Nakamura and H. Hirai, *Chem. Lett.*, 823 (1975).
1254. Y. Nakamura and Y. Sado, *Jpn. Pat.* 119 814 (1978); *C. A.* **90**, 71 767 (1979).
1255. Y. Nakamura, S. Saito and Y. Morita, *Chem. Lett.*, 7 (1980).
1256. G. Natta, G. Dall'asta and G. Motroni, *J. Polym. Sci., Part B*, **2**, 349 (1964).
1257. K. M. Nicholas, *J. Organomet. Chem.*, **188**, C10 (1980).
1258. J. K. Nicholson and B. L. Shaw, *Tetrahedron Lett.*, 3533 (1965).
1259. J. Niewahner and D. W. Meek, *Adv. Chem. Ser.*, **196**, 257 (1982).
1260. J. Niewahner and D. W. Meek, *Inorg. Chim. Acta*, **64**, L123 (1982).
1261. E. E. Nifant'ev, A. T. Teleshev, M. P. Koroteev, S. A. Ermishkina and E. M. Abbasov, *Koord. Khim.*, **7**, 311 (1981); *C. A.* **95**, 133 267 (1981).
1262. T. Nishiguchi, K. Tachi and K. Fukuzumi, *J. Am. Chem. Soc.*, **94**, 8916 (1972).
1263. T. Nishiguchi, K. Tanaka and K. Fukuzumi, *J. Org. Chem.*, **43**, 2968 (1978).
1264. T. Nishiguchi, K. Tanaka and K. Fukuzumi, *J. Organomet. Chem.*, **193**, 37 (1980).
1265. Nissan Chemical Industries Ltd., *Jpn. Pat.* 55 323 (1981); *C. A.* **95**, 168 556 (1981).
1266. J. F. Nixon and J. R. Swain, *J. Organomet. Chem.*, **72**, C15 (1974).
1267. J. F. Nixon and B. Wilkins, *J. Organomet. Chem.*, **44**, C25 (1972).
1268. J. F. Nixon and B. Wilkins, *J. Organomet. Chem.*, **87**, 341 (1975).
1269. A. F. Noels, A. J. Hubert and Ph. Teyssie, *J. Organomet. Chem.*, **166**, 79 (1979).
1270. S. S. Novikov, V. I. Manov-Yuvenskii, A. V. Smetanin and B. K. Nefedov, *Dokl. Akad. Nauk SSSR*, **251**, 371 (1980); *C. A.* **93**, 131 686 (1980).
1271. M. Novotny and L. R. Anderson, *U.S. Pat.* 4 223 001 (1980); *C. A.* **93**, 222 609 (1980).
1272. G. Nowlin and H. D. Lyons, *U.S. Pat.* 2 918 459 (1959); *C. A.* **54**, 7230 (1959).
1273. H. R. Null, H. L. Epstein, L. S. Eubanks, R. T. Eby, C. M. Cruse and F. E. Rosenberger, *Ger. Pat.* 2 211 203 (1973); *C. A.* **79**, 18 101 (1971).
1274. R. G. Nuzzo, D. Feitler and G. M. Whitesides, *J. Am. Chem. Soc.*, **101**, 3683 (1979).
1275. E. D. Nyberg and R. S. Drago, *J. Am. Chem. Soc.*, **103**, 4966 (1981).
1276. C. O'Connor and G. Wilkinson, *J. Chem. Soc. A*, 2665 (1968).
1277. C. O'Connor and G. Wilkinson, *Tetrahedron Lett.*, 1375 (1969).
1278. C. O'Connor, G. Yagupski, D. Evans and G. Wilkinson, *Chem. Commun.*, 420 (1968).
1279. A. L. Odell, J. B. Richardson and W. R. Roper, *J. Catal.*, **8**, 393 (1967).
1280. H. C. Odom and A. R. Pinder, *J. Chem. Soc., Perkin Trans. 1*, 2193 (1972).
1281. I. Ogata and Y. Ikeda, *Chem. Lett.*, 487 (1972).
1282. I. Ogata and Y. Ikeda, *Tokyo Kogyo Shikensho Hokoku*, **67**, 340 (1972); *C. A.* **78**, 136 702 (1973).
1283. I. Ogata, Y. Ikeda and T. Asakawa, *Kogyo Kagaku Zasshi*, **74**, 1839 (1971); *C. A.* **75**, 140 422 (1971).
1284. I. Ogata, F. Mizukami, Y. Ikeda and M. Tanaka, *Jpn. Pat.* 39 589 (1976); *C. A.* **85**, 99 914 (1976).

1285. T. Ohara, T. Kondo, N. Takaya and K. Kubota, *Jpn. Pat.* 115 309 (1979); *C. A.* 92, 58 230 (1980).
1286. K. Ohkubo, K. Fujimori and K. Yoshinaga, *Inorg. Nucl. Chem. Lett.,* 15, 231 (1979).
1287. K. Ohkubo, M. Haga, K. Yoshinaga and Y. Motozato, *Inorg. Nucl. Chem. Lett.,* 16, 155 (1980).
1288. K. Ohkubo, M. Haga, K. Yoshinaga and Y. Motozato, *Inorg. Nucl. Chem. Lett.,* 17, 215 (1981).
1289. K. Ohkubo, K. Hirata, T. Ohgushi and K. Yoshinaga, *J. Coord. Chem.,* 6, 185 (1977).
1290. K. Ohkubo, T. Ohgushi and K. Yoshinaga, *Chem. Lett.,* 775 (1976).
1291. K. Ohkubo, T. Ohgushi and K. Toshinaga, *J. Coord. Chem.,* 8, 195 (1979).
1292. K. Ohkubo, M. Setogushi and K. Yoshinaga, *Inorg. Nucl. Chem. Lett.,* 15, 235 (1979).
1293. K. Ohno and J. Tsuji, *J. Am. Chem. Soc.,* 90, 99 (1968).
1294. Y. Ohtani, A. Yamagishi and M Fujimoto, *Bull. Chem. Soc. Jpn.,* 52, 69 (1979).
1295. I. Ojima, *Fund. Research Homogeneous Catal.,* 2, 181 (1978).
1296. I. Ojima, *Jpn. Pat.* 105 429 (1978); *C. A.* 90, 39 030 (1979).
1297. I. Ojima and T. Fuchikami, *Fr. Pat.* 2 477 533 (1981); *C. A.* 96, 85 253 (1982).
1298. I. Ojima, S. Inaba, T. Kogure, M. Matsumoto, H. Matsumoto, H. Watanabe and Y. Nagai, *J. Organomet. Chem.,* 55, C4 (1973).
1299. I. Ojima and T. Kogure, *Chem. Lett.,* 1145 (1978).
1300. I. Ojima and T. Kogure, *Chem. Lett.,* 641 (1979).
1301. I. Ojima and T. Kogure, *Jpn. Pat.* 128 511 (1979); *C. A.* 92, 163 581 (1980).
1302. I. Ojima and T. Kogure, *J. Organomet. Chem.,* 195, 239 (1980).
1303. I. Ojima, T. Kogure and K. Achiwa, *Chem. Commun.,* 428 (1977).
1304. I. Ojima, T. Kogure and K. Achiwa, *Chem. Lett.,* 567 (1978).
1305. I. Ojima, T. Kogure and M. Kumagai, *J. Org. Chem.,* 42, 1671 (1977).
1306. I. Ojima, T. Kogure, M. Kumagai, S. Horiuchi and T. Sato, *J. Organomet. Chem.,* 122, 83 (1976).
1307. I. Ojima, T. Kogure and Y. Nagai, *Tetrahedron Lett.,* 5035 (1972).
1308. I. Ojima, T. Kogure and Y. Nagai, *Chem. Lett.,* 541 (1973).
1309. I. Ojima, T. Kogure, M. Nihonyanagi, H. Kono, S. Inaba and Y. Nagai, *Chem. Lett.,* 501 (1973).
1310. I. Ojima, T. Kogure, M. Nihonyanagi and Y. Nagai, *Bull. Chem. Soc. Jpn.,* 45, 3506 (1972).
1311. I. Ojima, T. Kogure, T. Terasaki and K. Achiwa, *J. Org. Chem.,* 43, 3444 (1978).
1312. I. Ojima, T. Kogure and N. Yoda, *Chem. Lett.,* 495 (1979); *J. Org. Chem.,* 45, 4728 (1980).
1313. I. Ojima and M. Kumagai, *J. Organomet. Chem.,* 134, C6 (1977).
1314. I. Ojima and M. Kumagai, *J. Organomet. Chem.,* 157, 359 (1978).
1315. I. Ojima, M. Kumagai and Y. Nagai, *J. Organomet. Chem.,* 66, C14 (1974).
1316. I. Ojima and Y. Nagai, *Chem. Lett.,* 223 (1974).
1317. I. Ojima, M. Nihonyanagi, T. Kogure, M. Kumagai, S. Horiuchi and K. Nakatsugawa, *J. Organomet. Chem.,* 94, 449 (1975).
1318. I. Ojima, M. Nihonyanagi and Y. Nagai, *Chem. Commun.,* 938 (1972); *Bull. Chem. Soc. Jpn.,* 45, 3722 (1972).
1319. I. Ojima and T. Suzuki, *Tetrahedron Lett.,* 1239 (1980).
1320. I. Ojima, T. Tanaka and T. Kogure, *Chem. Lett.,* 823 (1981).
1321. I. Ojima, K. Yamamoto and M. Kumada, *Aspects Homogeneous Catal.,* 3, 1 (1977).

1322. I. Ojima and N. Yoda, *Tetrahedron Lett.*, 1051 (1980).
1323. T. Okano, T. Kobayashi, H. Konishi and J. Kiji, *Bull. Chem. Soc. Jpn.,* **54**, 3799 (1981).
1324. T. Okano, K. Tsukinyama, H. Konishi and J. Kiji, *Chem. Lett.*, 603 (1982).
1325. I. Okura and T. Keii, *J. Chem. Soc. Jpn.*, 257 (1972).
1326. I. Okura, N. Takahashi and T. Keii, *J. Mol. Catal.,* **4**, 237 (1978).
1327. H. Okushima, I. Nitta and S. Hayakawa, *Jpn. Pat.* 130 649 (1978); *C. A.* **90**, 87 741 (1979).
1328. G. A. Olah and P. Kreienbuehl, *J. Org. Chem.,* **32**, 1614 (1967).
1329. K. L. Olivier, *U.S. Pat.* 3 530 190 (1970); *C. A.* **73**, 132 733 (1970).
1330. J. J. Oltyoort, C. A. A. Van Borckel, J. H. De Koning and J. H. Van Boom, *Synthesis*, 305 (1981).
1331. T. Onoda, Y. Tsunoda, Y. Koyama, T. Kawatsu and K. Tano, *Jpn. Pat.* 24 928 (1978); *C. A.* **89**, 163 068 (1978).
1332. T. Onoda, Y. Tsunoda, T. Nomura, T. Nonaka, O. Kurashiki and T. Masuyama, *Ger. Pat.* 2 438 847 (1975); *C. A.* **82**, 174 055 (1975).
1333. K. Onuma, T. Ito and A. Nakamura, *Jpn. Pat.* 57 490 (1979); *C. A.* **91**, 97 421 (1979).
1334. K. Onuma, T. Ito and A. Nakamura, *Chem. Lett.*, 905 (1979).
1335. K. Onuma, T. Ito and A. Nakamura, *Chem. Lett.*, 481 (1980).
1336. K. Onuma, T. Ito and A. Nakamura, *Bull. Chem. Soc. Jpn.,* **53**, 2012 (1980).
1337. K. Onuma, T. Ito and A. Nakamura, *Bull. Chem. Soc. Jpn.,* **53**, 2016 (1980).
1338. I. A. Oreshkin, I. Ya. Ostrovskaya, V. A. Yakovlev, E. I. Tinyakova and B. A. Dolgoplosk, *Dokl. Adad. Nauk SSSR*, **173**, 1349 (1967), *Dokl. Chem. (Engl. Transl.)*, **173**, 402 (1967).
1339. L. A. Oro, A. Manrique and M. Royo, *Transition Met. Chem.,* **3**, 383 (1978).
1340. J. C. Orr, M. Mersereau and A. Sanford, *Chem. Commun.*, 162 (1970).
1341. K. Osakada, T. Ikariya, M. Saburi and S. Yoshikawa, *Chem. Lett.*, 1691 (1981).
1342. J. A. Osborn, F. H. Jardine, J. F. Young and G. Wilkinson, *J. Chem. Soc. A*, 1711 (1966).
1343. J. A. Osborn and G. Wilkinson, *Inorg. Synth.,* **10**, 67 (1967).
1344. J. A. Osborn, G. Wilkinson and J. F. Young, *Chem. Commun.*, 17 (1965).
1345. Y. Osumi and M. Yamaguchi, *Jpn. Pat.* 21 604 (1971); *C. A.* **75**, 129 312 (1971).
1346. Y. Osumi and M. Yamaguchi, *Jpn. Pat.* 21 605 (1971); *C. A.* **75**, 129 313 (1971).
1347. A. A. Oswald, *PCT Int. Pat. Appl.* 01 692 (1980); *C. A.* **94**, 157 091 (1981).
1348. A. A. Oswald and L. L. Murrell, *Ger. Pat.* 2 332 167 (1974); *C. A.* **80**, 83 252 (1974).
1349. A. A. Oswald and L. L. Murrell, *U.S. Pat.* 4 134 906 (1979); *C. A.* **90**, 142 652 (1979).
1350. Z. Otero-Schipper, J. Lieto, J. J. Rafalko and B. C. Gates, *Stud. Surf. Sci. Catal.,* **4**, 535 (1980); *C. A.* **93**, 238 307 (1980).
1351. S. Otsuka, A. Nakamura and H. Minamida, *Chem. Commun.*, 191 (1969).
1352. S. Otsuka, A. Nakamura and K. Tani, *J. Chem. Soc. Jpn.,* **70**, 2007 (1967).
1353. S. Otsuka, A. Nakamura, S. Ueda and H. Minamida, *J. Chem. Soc. Jpn.,* **72**, 1809 (1969).
1354. W. Otte, W. H. E. Mueller and M. Z. Hausen, *Ger. Pat.* 2 912 230 (1980); *C. A.* **94**, 21 133 (1981).

1355. G. F. Ottman, E. H. Kober and D. F. Gavin, *Fr. Pat.* 1 567 321 (1969); *C. A.* **72**, 111 015 (1970).
1356. G. Paiaro and L. Pandolfo, *Gazz. Chim. Ital.,* **107**, 467 (1977).
1357. G. Paiaro and L. Pandolfo, *Chim. Ind. (Milan),* **60**, 709 (1978).
1358. S. A. Panichev, G. V. Kudryavstev and G. V. Lisichkin, *Koord. Khim.,* **5**, 1141 (1979); *C. A.* **91**, 163 567 (1979).
1359. V. B. Panov, M. L. Khidekel and S. A. Shchepinov, *Izv. Akad. Nauk SSSR, Ser. Khim.,* 2397 (1968); *Bull. Acad. Sci. USSR (Engl. Transl.),* **17**, 2272 (1968).
1360. L. A. Paquette, R. A. Boggs, W. B. Farnham and R. S. Beckley, *J. Am. Chem. Soc.,* **97**, 1112 (1975).
1361. L. A. Paquette, R. A. Boggs and J. S. Ward, *J. Am. Chem. Soc.,* **7**, 1118 (1975).
1362. L. A. Paquette and M. R. Delty, *Tetrahedron Lett.,* 713 (1978).
1363. L. A. Paquette and R. Grée, *J. Organomet. Chem.,* **146**, 319 (1978).
1364. S. H. Park and G. S. Chin, *Taehan Hwahakhoe Chi,* **25**, 57 (1981); *C. A.* **95**, 25 711 (1981).
1365. D. G. Parker, R. Pearce and D. W. Prest, *Chem. Commun.,* 1193 (1982).
1366. G. W. Parshall, *Catalysis,* **1**, 335 (1977).
1367. G. W. Parshall, *Homogeneous Catalysis,* Wiley-Interscience, New York 1980.
1368. G. W. Parshall, D. L. Thorn and T. H. Tulip, *Chem. Technol.,* 571 (1982).
1369. H. Pasternak, T. Glowiak and F. Pruchnik, *Inorg. Chim. Acta,* **19**, 11 (1976).
1370. J. L. Paul, W. L. Pieper and L. E. Wade, *Ger. Pat.* 2 721 792 (1977); *C. A.* **88**, 50 287 (1978).
1371. F. E. Paulik, A. Hershman, W. R. Knox and J. F. Roth, *Ger. Pat.* 1 941 448 (1970); *C. A.* **73**, 34 811 (1970).
1372. F. E. Paulik, A. Hershman, W. R. Knox and J. F. Roth, *Ger. Pat.* 1 941 449 (1970); *C. A.* **72**, 110 807 (1970).
1373. F. E. Paulik, A. Hershman, J. F. Roth, J. H. Craddock and D. Forster, *Fr. Pat.* 1 573 130 (1969); *C. A.* **72**, 100 054 (1970).
1374. F. E. Paulik, K. K. Robinson and J. F. Roth, *U.S. Pat.* 3 487 112 (1969); *C. A.* **72**, 68 984 (1970).
1375. F. E. Paulik and J. F. Roth, *Chem. Commun.,* 1578 (1968).
1376. F. E. Paulik, J. F. Roth and K. K. Robinson, *Fr. Pat.* 1 560 961 (1969); *C. A.* **72**, 54 762 (1970).
1377. N. C. Payne and D. W. Stephan, *Inorg. Chem.,* **21**, 182 (1982).
1378. R. G. Pearson, *Symmetry Rules for Chemical Reactions,* Wiley, New York, 1976.
1379. R. Pellicciari, R. Fringuelli, P. Ceccherelli and E. Sisani, *Chem. Commun.,* 959 (1979).
1380. V. N. Perchenko, I. S. Mirskova and N. S. Nametkin, *Dokl. Akad. Nauk SSSR,* **251**, 1437 (1980); *C. A.* **93**, 150 853 (1980).
1381. J. R. Peterson, D. W. Bennett and L. D. Spicer, *J. Catal.,* **71**, 223 (1981).
1382. N. Petiniot, A. J. Anciaux, A. F. Noels, A. J. Hubert and Ph. Teyssie, *Tetrahedron Lett.,* 1239 (1978).
1383. N. Petiniot, A. F. Noels, A. J. Anciaux, A. J. Hubert and Ph. Teyssie, *Fundam. Res. Homogeneous Catal.,* **3**, 421 (1979).
1384. R. Pettit, J. Wristers and L. Brener, *J. Am. Chem. Soc.,* **92**, 7499 (1970).
1385. J. F. Peyronel, J. C. Fiaud and H. B. Kagan, *J. Chem. Res. (S),* 320 (1980).
1386. J. F. Peyronel and H. B. Kagan, *Nouv. J. Chim.,* **2**, 211 (1978).

1387. P. N. Phung and G. Le Febvre, *C.R. Acad. Sci., Ser. C,* **265**, 519 (1967).
1388. Y. Pickholtz, Y. Sasson and J. Blum, *Tetrahedron Lett.*, 1263 (1974).
1389. E. Piers, R. W. Britton and W. De Waal, *Can. J. Chem.*, **49**, 12 (1971).
1390. E. Piers and K. F. Cheng, *Can. J. Chem.*, **46**, 377 (1968).
1391. F. Pinna, C. Candilira, G. Strukul, M. Bonivento and M. Graziani, *J. Organomet. Chem.*, **159**, 91 (1978).
1392. T. J. Pinnavaia, R. Raythata, J. G. Lee, L. J. Halloran and J. F. Hoffman, *J. Am. Chem. Soc.*, **101**, 6891 (1979).
1393. T. J. Pinnavaia and P. K. Welty, *J. Am. Chem. Soc.*, **97**, 3819 (1975).
1394. P. Pino, *J. Organomet. Chem.*, **200**, 223 (1980).
1395. P. Pino, C. Botteghi, G. Consiglio and C. Salomon, *Ger. Pat.* 2 359 101 (1974); *C. A.* **81**, 90 644 (1979).
1396. P. Pino, F. Piacenti and M. Bianchi, in *Organic Syntheses via Metal Carbonyls*, (I. Wender and P. Pino, Eds.), Vol. 2, Wiley-Interscience, New York, 177, p. 43.
1397. P. Pino and D. Von Bezard, *Ger. Pat.* 2 807 251 (1978); *C. A.* **90**, 6235 (1979).
1398. C. U. Pittman and R. M. Hanes, *J. Am. Chem. Soc.*, **98**, 5402 (1976).
1399. C. U. Pittman and R. M. Hanes, *J. Org. Chem.*, **42**, 1194 (1977).
1400. C. U. Pittman and A. Hirao, *J. Org. Chem.*, **43**, 640 (1978).
1401. C. U. Pittman, A. Hirao, C. Jones, R. M. Hanes and Q. Ng, *Ann. N.Y. Acad. Sci.*, **295**, 15 (1977).
1402. C. U. Pittman and W. D. Honnick, *J. Org. Chem.*, **45**, 2132 (1980).
1403. C. U. Pittman, W. D. Honnick and J. J. Yang, *J. Org. Chem.*, **45**, 684 (1980).
1404. C. U. Pittman, S. E. Jacobsen and H. Hiramoto, *J. Am. Chem. Soc.*, **97**, 4774 (1975).
1405. C. U. Pittman and C.-C. Lin, *J. Org. Chem.*, **43**, 4928 (1978).
1406. C. U. Pittman and L. R. Smith, *J. Am. Chem. Soc.*, **97**, 1749 (1975).
1407. C. U. Pittman, L. R. Smith and R. M. Hanes, *J. Am. Chem. Soc.*, **97**, 1742 (1975).
1408. M. Polievka, V. Macho and L. Uhlar, *Petrochemia*, **19**, 5 (1979); *C. A.* **91**, 91 084 (1979).
1409. M. Polievka and L. Uhlar, *Petrochemia*, **19**, 21 (1979); *C. A.* **91**, 107 648 (1979).
1410. M. Polievka, L. Uhlar and V. Macho, *Petrochemia*, **18**, 9 (1978); *C. A.* **89**, 146 328 (1978).
1411. M. Polievka, L. Uhlar and V. Macho, *Petrochemia*, **20**, 33 (1980); *C. A.* **93**, 238 160 (1980).
1412. R. V. Porcelli and V. S. Bhise, *Ger. Pat.* 3 024 353 (1981); *C. A.* **94**, 120 902 (1981).
1413. L. Porri, P. Diversi, A. Lucherini and R. Rossi, *Makromol. Chem.*, **176**, 3121 (1975).
1414. L. Porri, R. Rossi, P. Diversi and A. Lucherini, *Makromol. Chem.*, **175**, 3097 (1974).
1415. J. C. Poulin, T.-P. Dang and H. B. Kagan, *J. Organomet. Chem.*, **84**, 87 (1975).
1416. J. C. Poulin, D. Meyer and H. B. Kagan, *C.R. Acad. Sci., Ser. C,* **291**, 69 (1980).
1417. K. G. Powell and F. J. McQuillin, *Chem. Commun.*, 931 (1971).
1418. G. F. Pregaglia, F. G. Ferrari, A. Andreetta, G. Capparella, F. Genoni and R. Ugo, *J. Organomet. Chem.*, **70**, 89 (1974).
1419. M. Primet, J. C. Vedrine and C. Naccache, *J. Mol. Catal.*, **4**, 411 (1978).
1420. R. H. Prince and K. A. Raspin, *Chem. Commun.*, 156 (1966).
1421. R. H. Prince and K. A. Raspin, *J. Chem. Soc. A*, 612 (1969).
1422. L. G. Privalova, L. D. Tyutchenkova, Z. K. Maizus and N. M. Emanuel, *Teor.*

Prakt., Zhidkofazn. Okisleniya, (Mater. Vses Konf. Okisleniyu Org. Soedin. Zhidk. Faze), 127 (1974); *C. A.* **83**, 57 764 (1975).

1423. R. V. Procelli, V. S. Bhise and A. J. Shapiro, *Ger. Pat.* 2 940 752 (1980); *C. A.* **93**, 71 046 (1980).

1424. J. Protiva, L. Lepsa, E. Klinotova, J. Klinot, V. Krecek and A. Vystrcil, *Collect. Czech. Chem. Commun.*, **46**, 2734 (1981).

1425. F. Pruchnik, *Inorg. Nucl. Chem. Lett.*, **9**, 1229 (1973).

1426. F. Pruchnik, *Inorg. Nucl. Chem. Lett.*, **10**, 661 (1974).

1427. F. Pruchnik, *Pol. Pat.* 92 925 (1977); *C. A.* **90**, 104 656 (1979).

1428. F. Pruchnik, *Pol. Pat.* 99 111 (1978); *C. A.* **92**, 41 313 (1980).

1429. F. Pruchnik, H. Pasternak, M. Zuber and G. Kluczewska-Patrzalek, *Pol. Pat.* 101 169 (1979); *C. A.* **92**, 93 874 (1980).

1430. F. Pruchnik, M. Zuber and S. Krzystofik, *Rocz. Chem.*, **51**, 1177 (1977); *C. A.* **87**, 207 127 (1977).

1431. R. L. Pruett, *Ann. N. Y. Acad. Sci.*, **295**, 239 (1977).

1432. R. L. Pruett, *Adv. Organomet. Chem.*, **17**, 1 (1979).

1433. R. L. Pruett and K. O. Groves, *U.S. Pat.* 3 499 932 (1970); *C. A.* **72**, 110 900 (1970).

1434. R. L. Pruett and K. P. Groves, *U.S. Pat.* 3 499 933 (1970); *C. A.* **73**, 14 291 (1970).

1435. R. L. Pruett and J. A. Smith, *S. Afr. Pat.* 04 937 (1968); *C. A.* **71**, 90 819 (1969).

1436. R. L. Pruett and J. A. Smith, *J. Org. Chem.*, **34**, 327 (1969).

1437. R. L. Pruett and J. A. Smith, *Ger. Pat.* 2 062 703 (1971); *C. A.* **75**, 109 844 (1971).

1438. R. L. Pruett and J. A. Smith, *U.S. Pat.* 3 917 661 (1975); *C. A.* **84**, 30 433 (1976).

1439. R. Psaro, A. Fusi, R. Ugo, J. M. Basset, A. R. Smith and F. Hughes, *J. Mol. Catal.*, **7**, 511 (1980).

1440. H. Pscheidl, K. Bethke and D. Haberland, *Z. Chem.*, **18**, 393 (1978).

1441. H. Pscheidl, E. Moeller, H. U. Juergens and D. Haberland, *East Ger. Pat.* 138 153 (1979); *C. A.* **92**, 65 370 (1980).

1442. H. Pscheidl, E. Moeller, H. U. Juergens and D. Haberland, *East Ger. Pat.* 138 735 (1979); *C. A.* **92**, 186 550 (1980).

1443. J. Pugach, *U.S. Pat.* 4 251 458 (1981); *C. A.* **95**, 61 524 (1981).

1444. J. Pugach, *Belg. Pat.* 886 853 (1981); *C. A.* **95**, 168 568 (1981).

1445. J. Pugach, *Belg. Pat.* 886 854 (1981); *C. A.* **95**, 149 974 (1981).

1446. J. Pugiach, *Fr. Pat.* 2 472 556 (1981); *C. A.* **96**, 6205 (1982).

1447. O. Puglisi, F. A. Bottino, A. Recca and J. K. Stille, *J. Chem. Res. (S)*, 216 (1980).

1448. V. B. Pukhnarevich, L. I. Kopylova, M. Capka, J. Hetflejs, E. N. Satsuk, M. V. Sigalov, V. Chvalovsky and M. G. Voronkov, *Zh. Obshch. Khim.*, **50**, 1554 (1980); *C. A.* **93**, 239 508 (1980).

1449. V. B. Pukhnarevich, L. I. Kopylova, E. O. Tsetlina, V. A. Pestunovich, V. Chvalovsky, J. Hetflejs and M. G. Voronkov, *Dokl. Akad. Nauk SSSR*, **231**, 1366 (1976); *Dokl. Chem. (Engl. Transl.)*, **231**, 764 (1976).

1450. M. H. Quick, *U.S. Pat.* 4 312 779 (1981); *C. A.* **96**, 110 982 (1982).

1451. P. J. Quinn and C. E. Taylor, *J. Mol. Catal.*, **13**, 389 (1981).

1452. P. J. Quinn and G. Wilkinson, *Br. Pat.* 1 594 603 (1981); *C. A.* **96**, 31 229 (1982).

1453. S. Quinn and A. Shaver, *Inorg. Chim. Acta*, **38**, 243 (1980).

1454. J. J. Rafalko, J. Lieto, B. C. Gates and G. L. Schrader, *Chem. Commun.*, 540 (1978).

1455. S. Raghu, A. Zweig and W. A. Henderson, *Fundam. Res. Homogeneous Catal.*, **3**, 565 (1979).

1456. J. W. Raksis and F. Jachimowicz, *Belg. Pat.* 887 628 (1981); *C. A.* **95**, 151 486 (1981).
1457. T. B. Rauchfaus, *Fundam. Res. Homogeneous Catal.,* **3**, 1021 (1979).
1458. T. B. Rauchfaus, J. L. Clements, S. F. Agnew and D. M. Roundhill, *Inorg. Chem.,* **16**, 775 (1977).
1459. B. K. Ravi Shankar and H. Shechter, *Tetrahedron Lett.,* 2277 (1982).
1460. R. Raythatha and T. J. Pinnavaia, *J. Organomet. Chem.,* **218**, 115 (1981).
1461. G. Read and P. J. C. Walker, *J. Chem. Soc., Dalton Trans.,* 883 (1977).
1462. A. Recca, J. Garapon and J. K. Stille, *Macromolecules,* **10**, 1344 (1977).
1463. J. Reed, P. Eisenberg. B.-K. Teo and B. M. Kincaid, *J. Am. Chem. Soc.,* **100**, 2375 (1978).
1464. S. L. Regen and D. P. Lee, *Isr. J. Chem.,* **17**, 284 (1979).
1465. W. Reimann, W. Abhoud, J. M. Basset, R. Mutin, G. L. Rempel and A. K. Smith, *J. Mol. Catal.,* **9**, 349 (1980).
1466. J. Rejhon and J. Hetflejs, *Collect. Czech. Chem. Commun.,* **40**, 3190 (1975).
1467. J. Rejhon and J. Hetflejs, *Collect. Czech. Chem. Commun.,* **40**, 3680 (1975).
1468. C. J. Restall, P. Williams, M. P. Percival, P. J. Quinn and D. Chapman, *Biochim. Biophys. Acta,* **555**, 119 (1979).
1469. J. M. Reuter, A. Sinha and R. G. Salomon, *J. Org. Chem.,* **43**, 2438 (1978).
1470. Rhone-Poulenc Industries S. A., *Jpn. Pat.* 14 911 (1979); *C. A.* **91**, 39 681 (1979).
1471. W. Richter, R. Kummer and K. Schwirten, *Ger. Pat.* 2 840 168 (1980); *C. A.* **93**, 71 017 (1980).
1472. D. P. Riley, *Eur. Pat. Appl.* 36 741 (1981); *C. A.* **96**, 69 221 (1982).
1473. D. P. Riley, *J. Organomet. Chem.,* **234**, 85 (1982).
1474. D. P. Riley and R. E. Shumate, *J. Org. Chem.,* **45**, 5187 (1980).
1475. R. E. Rinehart, *J. Polym. Sci., Part C,* **27**, 7 (1966).
1476. R. E. Rinehart, *U.S. Pat.* 3 433 808 (1969); *C. A.* **70**, 106 368 (1969).
1477. R. E. Rinehart, *Ger. Pat.* 1 943 624 (1970); *C. A.* **72**, 101 242 (1970).
1478. R. E. Rinehart and H. P. Smith, *J. Polym. Sci., Part B,* **3**, 1049 (1965).
1479. R. E. Rinehart, H. P. Smith, H. S. Witt and H. Romeyn, *J. Am. Chem. Soc.,* **83**, 4864 (1961).
1480. N. Rizkalla and C. N. Winnick, *Ger. Pat.* 2 610 035 (1976); *C. A.* **85**, 176 870 (1976).
1481. K. K. Robinson, A. Hershman, J. H. Craddock and J. F. Roth, *J. Catal.,* **27**, 389 (1972).
1482. K. K. Robinson, F. E. Paulik, A. Hershman and J. F. Roth, *J. Catal.,* **15**, 245 (1969).
1483. F. Roehrscheid, *Ger. Pat.* 2 140 644 (1973); *C. A.* **78**, 124 047 (1973).
1484. C. K. Rofer-Depoorter, *Chem. Rev.,* **81**, 447 (1981).
1485. B. G. Rogachev and M. L. Khidekel, *Izv. Akad. Nauk SSSR, Ser. Khim.,* 141 (1969); *Bull. Acad. Sci. USSR (Engl. Transl.),* **18**, 127 (1969).
1486. L. D. Rollman and D. D. Whitehurst, *Br. Pat.* 1 448 255 (1973); *C. A.* **86**, 71 909 (1977).
1487. P. R. Rony, *J. Catal.,* **14**, 142 (1969).
1488. P. R. Rony and J. F. Roth, *J. Mol. Catal.,* **1**, 13 (1975).
1489. R. Rossi, P. Diversi, A. Lucherini and L. Porri, *Tetrahedron Lett.,* 879 (1974).
1490. J. F. Roth, *Platinum Met. Rev.,* **19**, 12 (1975).
1491. J. F. Roth, J. H. Craddock, A. Hershman and F. E. Paulik, *Chem. Technol.,* 600 (1971).

1492. J. F. Roth and T. J. Katz, *Tetrahedron Lett.*, 2503 (1972).
1493. R. Roulet, J. Wenger, M. Hardy and P. Vogel, *Tetrahedron Lett.*, 1479 (1974).
1494. D. M. Roundhill, R. A. Bechtold and S. G. N. Roundhill, *Inorg. Chem.*, **19**, 284 (1980).
1495. D. M. Roundhill, M. K. Dickson, N. S. Dixit and B. P. Sudha-Dixit, *J. Am. Chem. Soc.*, **102**, 5538 (1980).
1496. D. M. Roundhill, M. K. Dickson, N. S. Dixit and B. P. Sudha-Dixit, *Adv. Chem. Ser.*, **196**, 291 (1982).
1497. C. Rousseau, M. Evrard and F. Pettit, *J. Mol. Catal.*, **3**, 309 (1978).
1498. C. Rousseau, M. Evrard and F. Pettit, *J. Mol. Catal.*, **5**, 163 (1979).
1499. M. Royo, F. Melo, A. Manrique and L. Oro, *Transition Met. Chem.*, **7**, 44 (1982).
1500. C. Rüger, A. Mehlhorn and K. Schwetlick, *J. Prakt. Chem.*, **317**, 583 (1975).
1501. M. J. H. Russell, C. White and P. M. Maitlis, *Chem. Commun.*, 427 (1977).
1502. R. C. Ryan, G. M. Wilemon, M. P. Dalsanto and C. U. Pittman, *J. Mol. Catal.*, **5**, 319 (1979).
1503. P. N. Rylander, *Catalytic Hydrogenation in Organic Synthesis*, Academic Press, New York, 1979.
1504. Y. Sado and K. Tajima, *Noguchi Kenkyusho Jiho*, **23**, 52 (1980); *C. A.* **94**, 174 245 (1981).
1505. Y. Sado and M. Tajima, *Jpn. Pat.* 51 037 (1980); *C. A.* **93**, 149 816 (1980).
1506. Sagami Chemical Research Center, *Jpn. Pat.* 81 549 (1981); *C. A.* **96**, 52 670 (1982).
1507. Sagami Chemical Research Center, *Jpn. Pat.* 161 340 (1981); *C. A.* **96**, 180 787 (1982).
1508. N. Saito, S. Arai and Y. Tsutsumi, *Jpn. Pat.* 45 646 (1980); *C. A.* **93**, 204 061 (1980).
1509. T. Saito, Y. Tsutsumi and S. Arai, *Ger. Pat.* 3 017 682 (1980); *C. A.* **95**, 42 342 (1981).
1510. K. Sakai and O. Oda, *Tetrahedron Lett.*, 4375 (1972).
1511. M. Sakai, M. Takahashi, Y. Sakakibara and N. Uchino, *J. Chem. Soc. Jpn.*, 1283 (1981).
1512. M. Sakai, H. Yamaguchi and S. Masamune, *Chem. Commun.*, 486 (1971).
1513. T. Sakakibara and H. Alper, *Chem. Commun.*, 458 (1979).
1514. T. Sakakibara, Y. Matsushima, Y. Nagashima, N. Okamoto, K. Kaneko, Y. Ishii and S. Wada, *Eur. Pat. Appl.* 14 796; *C. A.* **94**, 175 261 (1981).
1515. M. F. Salomon and R. G. Salomon, *Chem. Commun.*, 89 (1976).
1516. R. G. Salomon and N. El Sanadi, *J. Am. Chem. Soc.*, **97**, 6214 (1975).
1517. R. G. Salomon, M. F. Salomon and J. L. C. Kachinski, *J. Am. Chem. Soc.*, **99**, 1043 (1977).
1518. P. Salvadori, R. Lazzaroni, A. Raffaelli, S. Pucci, S. Bertozzi, D. Pini and G. Fatti, *Chim. Ind. (Milan)*, **63**, 492 (1981).
1519. O. Samuel, R. Couffignal, M. Lauer, S. Y. Zhang and H. B. Kagan, *Nouv. J. Chem.*, **5**, 15 (1981).
1520. A. R. Sanger, *J. Mol. Catal.*, **3**, 221 (1978).
1521. A. R. Sanger and L. R. Schallig, *J. Mol. Catal.*, **3**, 101 (1977).
1522. A. R. Sanger and K. G. Tan, *Inorg. Chim. Acta*, **31**, L439 (1978).
1523. K. Sanui, W. J. MacKnight and R. W. Lenz, *Macromolecules*, **7**, 952 (1974).
1524. S. Sarel, *Acc. Chem. Res.*, **11**, 204 (1978).

1525. Y. Sasson, A. Zoran and J. Blum., *J. Mol. Catal.,* **11**, 293 (1981).
1526. S. Sata, M. Takesada and H. Wakamatsu, *J. Chem. Soc. Jpn.,* 579 (1969).
1527. J. C. Sauer, *U.S. Pat.* 3 097 237 (1963); *C. A.* **60**, 2788 (1972).
1528. B. M. Savehenko, V. Z. Sharf, V. N. Krutii and L. K. Freidlin, *Izv. Akad. Nauk SSSR, Ser. Khim.,* 2632 (1979); *C. A.* **92**, 128 494 (1980).
1529. H. F. Schaeffer, *Acc. Chem. Res.,* **10**, 287 (1977).
1530. L. Schmerling and E. H. Homeier, *U.S. Pat.* 4 100 359 (1978); *C. A.* **90**, 22 340 (1979).
1531. H. J. Schmidt, F. A. Wunder, H. J. Arpe and E. Leupold, *Ger. Pat.* 2 814 365 (1979); *C. A.* **92**, 41 349 (1980).
1532. D. L. Schmitt and H. B. Jonassen, *J. Organomet. Chem.,* **49**, 469 (1973).
1533. H. J. Schmitt and H. Singer, *J. Organomet. Chem.,* **153**, 165 (1978).
1534. E. Schmitz, R. Urban, G. Zimmermann, E. Gruendemann and G. Benndorf, *Z. Chem.,* **12**, 100 (1972).
1535. A. Schneider, H. K. Myers and G. Suld, *U.S. Pat.* 4 275 254 (1981); *C. A.* **95**, 117 899 (1981).
1536. R. C. Schoenig, J. L. Vidal and R. A. Fiato, *J. Mol. Catal.,* **13**, 83 (1981).
1537. J. P. Scholten and H. J. van der Ploeg, *J. Polym. Sci., Part A-1,* **10**, 3067 (1972).
1538. J. P. Scholten and H. J. van der Ploeg, *J. Polym. Sci., Polym. Chem. Ed.,* **11**, 3205 (1973).
1539. G. N. Schrauzer, R. K. Y. Ho and G. Schlesinger, *Tetrahedron Lett.,* 543 (1970).
1540. H. J. Schrepfer, H. Siegel and H. Theobald, *Ger. Pat.* 2 715 923 (1978); *C. A.* **90**, 23 253 (1979).
1541. R. R. Schrock and J. A. Osborn, *Chem. Commun.,* 567 (1970).
1542. R. R. Schrock and J. A. Osborn, *J. Am. Chem. Soc.,* **93**, 3089 (1971).
1543. R. R. Schrock and J. A. Osborn, *J. Am. Chem. Soc.,* **98**, 2134 (1976).
1544. R. R. Schrock and J. A. Osborn, *J. Am. Chem. Soc.,* **98**, 2143 (1976).
1545. R. R. Schrock and J. A. Osborn, *J. Am. Chem. Soc.,* **98**, 4450 (1976).
1546. R. G. Schultz and P. D. Montgomery, *J. Catal.,* **13**, 105 (1969).
1547. R. A. Schunn, *Inorg. Chem.,* **9**, 2567 (1970).
1548. R. A. Schunn, G. C. Demitras, H. W. Choi and E. L. Muetterties, *Inorg. Chem.,* **20**, 4023 (1981).
1549. V. Schurig, *J. Mol. Catal.,* **6**, 75 (1979).
1550. V. Schurig and E. Bayer, *Chem. Technol.,* 212 (1976).
1551. K. Schwetlick, K. Unverferth, R. Hoentsch and M. Pfeifer, *East Ger. Pat.* 121 509 (1976); *C. A.* **87**, 39 143 (1977).
1552. K. Schwetlick, K. Unverferth, R. Teitz and J. Pels, *J. Prakt. Chem.,* **320**, 955 (1978).
1553. K. Schwirten, R. Kummer and W. Richter, *Ger. Pat.* 2 833 469 (1980); *C. A.* **93**, 7649 (1980).
1554. J. W. Scott, D. D. Keith, G. Nix, D. R. Parrish, S. Remington, G. R. Roth, J. M. Townsend, D. Valentine and R. Yang, *J. Org. Chem.,* **46**, 5086 (1981).
1555. M. S. Scurrell, *Platinum Met. Rev.,* **21**, 92 (1977).
1556. M. S. Scurrell, *Chim. Ind. (Milan),* **61**, 652 (1979).
1557. M. S. Scurrell, *J. Mol. Catal.,* **10**, 57 (1980).
1558. M. S. Scurrell and T. Hauberg, *Appl. Catal.,* **2**, 225 (1982).
1559. M. S. Scurrell and R. F. Howe, *J. Mol. Catal.,* **7**, 535 (1980).
1560. Y. Seki, S. Murai, A. Hidaka and N. Sonoda, *Angew. Chem.,* **89**, 919 (1977).
1561. Y. Seki and T. Suzukamo, *Jpn. Pat.* 154 712 (1979); *C. A.* **92**, 198 782 (1980).

1562. A. Sekiya and J. K. Stille, *J. Am. Chem. Soc.*, **103**, 5096 (1981).
1563. R. Selke, *React. Kinet. Catal. Lett.*, **10**, 135 (1979).
1564. R. Selke and H. Pracejus, *East Ger. Pat.* 140 036 (1980); *C. A.* **94**, 66 066 (1981).
1565. H. H. Seltzman, S. D. Wyrick and C. G. Pitt, *J. Labelled Comp. Radiopharm.*, **18**, 1365 (1981).
1566. M. F. Semmelhack and R. N. Misra, *J. Org. Chem.*, **47**, 2469 (1982).
1567. M. F. Semmelhack and L. Ryono, *Tetrahedron Lett.*, 2967 (1973).
1568. Y. Senda, T. Iwasaki and S. Mitsui, *Tetrahedron Lett.*, 4059 (1972).
1569. Y. Senda, S. Mitsui, H. Sugiyama and S. Seto, *Bull. Chem. Soc. Jpn.*, **45**, 3498 (1972).
1570. J. R. Shapley, R. R. Schrock and J. A. Osborn, *J. Am. Chem. Soc.*, **91**, 2816 (1969).
1571. V. Z. Sharf, L. K. Freidlin, V. N. Krutii and I. S. Shekoyan, *Izv. Akad. Nauk SSSR, Ser. Khim.*, 1330 (1974); *Bull. Acad. Sci. USSR (Engl. Transl.)*, **23**, 1251 (1974).
1572. V. Z. Sharf, L. K. Freidlin, I. S. Portyahova and V. N. Krutii, *Izv. Akad. Nauk SSSR, Ser. Khim.*, 1414 (1979); *C. A.* **91**, 107 466 (1979).
1573. V. Z. Sharf, L. K. Freidlin, I. S. Shekoyan and V. N. Krutii, *Bull. Acad. Sci. USSR (Engl. Transl.)*, **27**, 919 (1978).
1574. V. Z. Sharf, L. K. Freidlin, B. M. Savchenko and V. N. Krutii, *Izv. Akad. Nauk SSSR, Ser. Khim.*, 1134 (1979); *C. A.* **91**, 91 268 (1979).
1575. V. Z. Sharf, L. K. Freidlin, B. M. Savchenko and V. N. Krutii, *Izv. Akad. Nauk SSSR, Ser. Khim.*, 1393 (1979); *C. A.* **91**, 107 737 (1979).
1576. V. Z. Sharf, E. A. Mistryukov, L. K. Freidlin, I. S. Portyakova and V. N. Krutii, *Izv. Akad. Nauk SSSR, Ser. Khim.*, 1411 (1979); *C. A.* **91**, 123 152 (1979).
1577. V. Z. Sharf, B. M. Savchenko, V. N. Krutii and L. K. Freidlin, *Izv. Akad. Nauk SSSR, Ser. Khim.*, 456 (1980); *C. A.* **93**, 25 569 (1980).
1578. S. A. Shchepinov, M. L. Khidekel and G. V. Lagodzinskaya, *Izv. Akad. Nauk SSSR, Ser. Khim.*, 2165 (1968); *Bull. Acad. Sci. USSR (Engl. Transl.)*, **17**, 2061 (1968).
1579. A. D. Shebaldova, V. I. Bystrenina, V. N. Kravtsova and M. L. Khidekel, *Izv. Akad. Nauk SSSR, Ser. Khim.*, 2101 (1975); *Bull. Acad. Sci. USSR (Engl. Transl.)*, **24**, 1986 (1975).
1580. R. A. Sheldon, *Chemicals from Synthesis Gas*, Reidel, Dordrecht, 1983.
1581. R. A. Sheldon and J. K. Kochi, *Metal Catalyzed Oxidations of Organic Compounds*, Academic Press, New York 1981.
1582. Shell Intermationale Research Maatschappij b.V., *Br. Pat.* 916 092 (1963); *C. A.* **58**, 9312 (1963).
1583. Shell Internationale Research Maatschappij b.V., *Belg. Pat.* 621 662 (1963); *C. A.* **59**, P11 268 (1963).
1584. Shell Internationale Research Maatschappij b.V., *Neth. Pat.* 6 601 668 (1966); *C. A.* **66**, 11 791 (1967).
1585. Shell International Research Maatschappij, *Res. Discl.*, **208**, 327 (1981); *C. A.* **95**, 149 909 (1981).
1586. P. D. Sherman, S. C. Winans and D. G. Morrell, *U.S. Pat.* 4 262 142 (1981); *C. A.* **95**, 42 368 (1981).
1587. V. S. Shestakova, T. N. Omel'chenko, S. M. Brailovskii and O. N. Temkin, *Zh. Org. Khim.*, **14**, 2346 (1978); *C. A.* **90**, 137 184 (1979).
1588. T. F. Shevels and N. Harris, *Eur. Pat. Appl.* 16 285 (1980); *C. A.* **94**, 83 610 (1981).

1589.   A. E. Shilov, *Activation of Saturated Hydrocarbons*, Reidel, Dordrecht, 1983.
1590.   A. E. Shilov and A. A. Shteinman, *Coord. Chem. Rev.*, **24**, 97 (1977).
1591.   T. Shimiza, S. Moriya, T. Tsurumaru and M. Tamura, *Jpn. Pat.* 68 709 (1978);
        C. A. **89**, 146 415 (1978).
1592.   S. Shinoda, T. Kojima and Y. Saito, *Stud. Surf. Sci. Catal.*, **7**, 1504 (1981).
1593.   S. Shinoda, K. Nakamura and Y. Saito, *J. Mol. Catal.*, **17**, 77 (1982).
1594.   N. V. Shulyakovskaya, L. V. Vlasova, M. L. Khidekel and I. A. Markushina, *Izv.*
        Akad. Nauk SSSR, Ser. Khim., 1799 (1971); *Bull. Acad. Sci. USSR (Engl. Transl.)*,
        20, 1689 (1971).
1595.   Y. Shvo, D. W. Thomas and J. Laine, *J. Am. Chem. Soc.*, **103**, 2461 (1981).
1596.   J. W. Sibert, *U.S. Pat.* 3 515 757 (1970); *C. A.* **73**, 130 630 (1970).
1597.   H. Siegel and W. Himmelle, *Angew. Chem., Int. Ed. Engl.*, **19**, 178 (1980).
1598.   S. Siegel, *J. Catal.*, **30**, 139 (1973).
1599.   S. Siegel and J. J. Davis, *Stud. Surf. Sci. Catal.*, **7**, 1506 (1981).
1600.   S. Siegel and D. W. Ohrt, *Inorg. Nucl. Chem. Lett.*, **8**, 15 (1972).
1601.   S. Siegel and D. W. Ohrt, *Tetrahedron Lett.*, 5155 (1972).
1602.   H. Simon and O. Berngruber, *Tetrahedron*, **26**, 1401 (1970).
1603.   J. J. Sims, V. K. Honwad and L. H. Selman, *Tetrahedron Lett.*, 87 (1969).
1604.   J. H. Sinfelt, in *Catalysis – Science and Technology*, (J. R. Anderson and M.
        Boudart, Eds.), Vol. 1, Chap. 5, Springer-Verlag, Berlin, 1981.
1605.   H. Singer and K. Duckart, *Fette, Seifen, Anstrichm.*, **81**, 348 (1979); *C. A.* **91**,
        194 933 (1979).
1606.   H. Singer and W. Stein, *Ger. Pat.* 2 120 201 (1972); *C. A.* **78**, 45 175 (1973).
1607.   H. Singer, W. Stein and H. Lepper, *Fette, Seifen, Anstrichm.*, **74**, 193 (1972);
        C. A. **77**, 87 769 (1972).
1608.   H. Singer, W. Stein and H. Lepper, *Ger. Pat.* 2 049 937 (1972); *C. A.* **77**, 100 826
        (1972).
1609.   H. Singer and G. Wilkinson, *J. Chem. Soc. A*, 849 (1968).
1610.   B. Singh, *Ger. Pat.* 2 800 461 (1978); *C. A.* **89**, 180 371 (1978).
1611.   D. M. Singleton, *Tetrahedron Lett.*, 1245 (1973).
1612.   T. C. Singleton, *U.S. Pat.* 4 132 734 (1979); *C. A.* **90**, 86 758 (1979).
1613.   T. C. Singleton, L. J. Park, J. L. Price and D. Forster, *Am. Chem. Soc., Div. Pet.*
        Chem., Prep., **24**, 329 (1979); *C. A.* **94**, 163 276 (1981).
1614.   D. Sinou, *Tetrahedron Lett.*, 2987 (1981).
1615.   D. Sinou and G. Descotes, *React. Kinet. Catal. Lett.*, **14**, 463 (1980).
1616.   D. Sinou and H. B. Kagan, *J. Organomet. Chem.*, **114**, 325 (1976).
1617.   D. Sinou, D. Lafont, G. Descotes and A. G. Kent, *J. Organomet. Chem.*, **217**,
        119 (1981).
1618.   A. J. Sivac and E. L. Muetterties, *J. Am. Chem. Soc.*, **101**, 4878 (1979).
1619.   D. A. Slack and M. C. Baird, *J. Organomet. Chem.*, **142**, C69 (1977).
1620.   L. H. Slaugh and R. D. Mullineaux, *U.S. Pat.* 3 239 566 (1966); *C. A.* **64**, 15 745
        (1966).
1621.   L. H. Slaugh and R. D. Mullineaux, *U.S. Pat.* 3 239 571 (1966); *C. A.* **65**, 618
        (1966).
1622.   A. K. Smith, W. Abboud, J. M. Basset, W. Reimann, G. L. Rempel, J. L. Bilhou,
        V. Bilhou-Bougnol, W. F. Graydon, J. Dunogues *et al.*, *Fundam. Res. Homogeneous*
        Catal., **3**, 621 (1979).
1623.   A. K. Smith, F. Hughes, A. Theolier, J. M. Basset, R. Ugo, G. M. Zanderighi, J. L.
        Bilhou and W. F. Graydon, *Inorg. Chem.*, **18**, 3104 (1979).

1624.   G. V. Smith and R. J. Shuford, *Tetrahedron Lett.*, 525 (1970).
1625.   G. V. Smith and R. J. Shuford, *Am. Chem. Soc., Div. Pet. Chem., Prep.*, **16**, A62 (1971); *C. A.* **78**, 15 280 (1973).
1626.   H. P. Smith, *U.S. Pat.* 3 667 850 (1971); *C. A.* **76**, 35 057 (1972).
1627.   H. P. Smith and G. Wilkinson, *U.S. Pat.* 3 025 286 (1962); *C. A.* **57**, P1016 (1962).
1628.   T. W. Smith, *U.S. Pat.* 4 252 678 (1981); *C. A.* **94**, 181 537 (1981).
1629.   W. E. Smith, *Ger. Pat.* 2 758 473 (1978); *C. A.* **89**, 163 050 (1978).
1630.   W. E. Smith, *U.S. Pat.* 4 091 041 (1978); *C. A.* **90**, 5896 (1979).
1631.   W. E. Smith, *U.S. Pat.* 4 123 444 (1978); *C. A.* **90**, 87 245 (1979).
1632.   W. E. Smith, *U.S. Pat.* 4 139 542 (1979); *C. A.* **90**, 151 967 (1979).
1633.   W. E. Smith, *PCT Int. Pat. Appl.* 00 081 (1980); *C. A.* **93**, 70 978 (1980).
1634.   J. A. Sofranko, R. Eisenberg and J. A. Kampmeier, *J. Am. Chem. Soc.*, **102**, 1163 (1980).
1635.   M. Sohn, J. Blum and J. Halpern, *J. Am. Chem. Soc.*, **101**, 2694 (1979).
1636.   V. A. Sokolenko, *Izv. Akad. Nauk SSSR, Ser. Khim.*, 479 (1982); *Bull. Acad. Sci. USSR (Engl. Transl.)*, **31**, 435 (1982).
1637.   V. N. Sokolov, B. D. Babitskii, V. A. Kormer, I. Ya. Poddubnyi and N. N. Chesnokova, *J. Polym. Sci., Part C*, **16**, 4345 (1965).
1638.   A. J. Solodar, *Ger. Pat.* 2 306 222 (1973); *C. A.* **79**, 146 179 (1973).
1639.   A. J. Solodar, *Ger. Pat.* 2 312 924 (1973); *C. A.* **80**, 3672 (1974).
1640.   J. Solodar, *J. Org. Chem.*, **37**, 1840 (1972).
1641.   J. Solodar, *Chem. Technol.*, 421 (1975).
1642.   L. H. Sommer, J. E. Lyons and H. Fujimoto, *J. Am. Chem. Soc.*, **91**, 7051 (1969).
1643.   J. L. Speier, *Adv. Organomet. Chem.*, **17**, 407 (1979).
1644.   J. L. Speier, J. A. Webster and G. H. Barnes, *J. Am. Chem. Soc.*, **79**, 974 (1957).
1645.   A. Spencer, *J. Organomet. Chem.*, **93**, 389 (1975).
1646.   A. Spencer, *J. Organomet. Chem.*, **124**, 85 (1977).
1647.   A. Spencer, *Eur. Pat. Appl.* 2908 (1979); *C. A.* **92**, 6064 (1980).
1648.   A. Spencer, *J. Organomet. Chem.*, **194**, 113 (1980).
1649.   R. Spogliarich, G. Zassinovich, J. Kaspar and M. Graziani, *J. Mol. Catal.*, **16**, 359 (1982).
1650.   R. Spogliarich, G. Zassinovich, G. Mestroni and M. Graziani, *J. Organomet. Chem.*, **179**, C45 (1979).
1651.   R. Spogliarich, G. Zassinovich, G. Mestroni and M. Graziani, *J. Organomet. Chem.*, **198**, 81 (1980).
1652.   L. Starr, *U.S. Pat.* 3 646 115 (1972); *C. A.* **77**, 6001 (1972).
1653.   G. P. Startseva, B. G. Rogachev, M. L. Khidekel, P. E. Matkovskii and M. P. Garasina, *U.S.S.R. Pat.* 534 244 (1976); *C. A.* **86**, 61 010 (1977).
1654.   G. C. Stathdee and R. M. Given, *J. Catal.*, **30**, 30 (1973).
1655.   A. Stefani, G. Consiglio, C. Botteghi and P. Pino, *J. Am. Chem. Soc.*, **95**, 6504 (1973).
1656.   A. Stefani, G. Consiglio, C. Botteghi and P. Pino, *J. Am. Chem. Soc.*, **99**, 1058 (1977.).
1657.   A. Stefani, D. Tatone and P. Pino, *Helv. Chim. Acta*, **59**, 1639 (1976).
1658.   R. Stern, Y. Chevallier and L. Sajus, *C.R. Acad. Sci., Ser. C*, **264**, 1740 (1967).
1659.   R. Stern, A. Hirschauer and L. Sajus, *Tetrahedron Lett.*, 3247 (1973).
1660.   J. K. Stille, *Pure Appl. Chem.*, **50**, 273 (1978).
1661.   J. K. Stille and Y. Becker, *J. Org. Chem.*, **45**, 2139 (1980).
1662.   J. K. Stille and R. W. Fries, *J. Am. Chem. Soc.*, **96**, 1514 (1974).

1663. J. K. Stille, S. J. Fritschel, N. Takaishi, T. Masuda, H. Imai and C. A. Bertelo, *Ann. N.Y. Acad. Sci.*, **333**, 35 (1980).
1664. J. K. Stille, F. Huang and M. T. Regan, *J. Am. Chem. Soc.*, **96**, 1518 (1974).
1665. J. K. Stille and M. T. Regan, *J. Am. Chem. Soc.*, **96**, 1508 (1974).
1666. J. K. Stille, M. T. Regan, R. W. Fries, F. Huang and T. McCarley, *Adv. Chem. Ser.*, **132**, 181 (1974).
1667. S. H. Strauss and D. F. Shriver, *Inorg. Chem.*, **17**, 3069 (1978).
1668. S. H. Strauss, K. H. Whitmire and D. F. Shriver, *J. Organomet. Chem.*, **174**, C59 (1979).
1669. W. Strohmeier, *J. Organomet. Chem.*, **32**, 137 (1971).
1670. W. Strohmeier, *J. Organomet. Chem.*, **60**, C60 (1973).
1671. W. Strohmeier and G. Csontos, *J. Organomet. Chem.*, **67**, C27 (1974).
1672. W. Strohmeier and G. Csontos, *J. Organomet. Chem.*, **72**, 277 (1974).
1673. W. Strohmeier and W. Diehl, *Z. Naturforsch., Teil B*, **28**, 207 (1973).
1674. W. Strohmeier and E. Eder, *Z. Naturforsch., Teil B*, **29**, 280 (1974).
1675. W. Strohmeier and E. Eder, *J. Organomet. Chem.*, **94**, C14 (1975).
1676. W. Strohmeier and R. Endres, *Z. Naturforsch., Teil B*, **25**, 1068 (1970).
1677. W. Strohmeier and R. Endres, *Z. Naturforsch., Teil B*, **26**, 730 (1971).
1678. W. Strohmeier and R. Endres, *Z. Naturforsch., Teil B*, **27**, 1415 (1972).
1679. W. Strohmeier and R. Fleischmann, *Z. Naturforsch., Teil B*, **24**, 1217 (1969).
1680. W. Strohmeier and R. Fleischmann, *J. Organomet. Chem.*, **29**, C39 (1971).
1681. W. Strohmeier and R. Fleischmann, *J. Organomet. Chem.*, **42**, 163 (1972).
1682. W. Strohmeier, R. Fleischmann and T. Onoda, *J. Organomet. Chem.*, **28**, 281 (1971).
1683. W. Strohmeier, R. Fleischmann and W. Rehder-Stirnweiss, *J. Organomet. Chem.*, **47**, C37 (1973).
1684. W. Strohmeier, B. Graser, R. Marcec and K. Holke, *J. Mol. Catal.*, **11**, 257 (1981).
1685. W. Strohmeier and K. Gruenter, *J. Organomet. Chem.*, **90**, C45 (1975).
1686. W. Strohmeier and K. Gruenter, *J. Organomet. Chem.*, **90**, C48 (1975).
1687. W. Strohmeier and E. Hitzel, *J. Organomet. Chem.*, **87**, 353 (1975).
1688. W. Strohmeier and E. Hitzel, *J. Organomet. Chem.*, **91**, 373 (1975).
1689. W. Strohmeier and E. Hitzel, *J. Organomet. Chem.*, **110**, C22 (1976).
1690. W. Strohmeier and E. Hitzel, *J. Organomet. Chem.*, **110**, 389 (1976).
1691. W. Strohmeier and S. Hohmann, *Z. Naturforsch., Teil B*, **25**, 1309 (1970).
1692. W. Strohmeier and A. Kühn, *J. Organomet. Chem.*, **110**, 265 (1976).
1693. W. Strohmeier and M. Lukács, *J. Organomet. Chem.*, **129**, 331 (1977).
1694. W. Strohmeier and M. Lukács, *J. Organomet. Chem.*, **133**, C47 (1977).
1695. W. Strohmeier, R. Marcec and B. Graser, *J. Organomet. Chem.*, **221**, 361 (1981).
1696. W. Strohmeier and M. Michel, *J. Catal.*, **69**, 209 (1981).
1697. W. Strohmeier and M. Michel, *Z. Phys. Chem. (Weisbaden)*, **124**, 23 (1981).
1698. W. Strohmeier, M. Michel and L. Weigelt, *Z. Naturforsch., Teil B*, **35**, 648 (1980).
1699. W. Strohmeier and F. J. Müller, *Z. Naturforsch., Teil B*, **24**, 931 (1969).
1700. W. Strohmeier and T. Onoda, *Z. Naturforsch., Teil B*, **23**, 1377 (1968).
1701. W. Strohmeier and T. Onoda, *Z. Naturforsch., Teil B*, **23**, 1527 (1968).
1702. W. Strohmeier and T. Onoda, *Z. Naturforsch., Teil B*, **24**, 461 (1969).
1703. W. Strohmeier and T. Onoda, *Z. Naturforsch., Teil B*, **24**, 515 (1969).
1704. W. Strohmeier and T. Onoda, *Z. Naturforsch., Teil B*, **24**, 1493 (1969).
1705. W. Strohmeier and P. Pfoehler, *J. Organomet. Chem.*, **108**, 393 (1976).
1706. W. Strohmeier and W. Rehder-Stirnweiss, *J. Organomet. Chem.*, **18**, P28 (1969).

1707. W. Strohmeier and W. Rehder-Stirnweiss, *J. Organomet. Chem.*, **19**, 417 (1969).
1708. W. Strohmeier and W. Rehder-Stirnweiss, *Z. Naturforsch., Teil B*, **24**, 1219 (1969).
1709. W. Strohmeier and W. Rehder-Stirnweiss, *J. Organomet. Chem.*, **22**, C27 (1970).
1710. W. Strohmeier and W. Rehder-Stirnweiss, *Z. Naturforsch., Teil B*, **25**, 549 (1970).
1711. W. Strohmeier and W. Rehder-Stirnweiss, *Z. Naturforsch., Teil B*, **26**, 61 (1971).
1712. W. Strohmeier and W. Rehder-Stirnweiss, *Z. Naturforsch., Teil B*, **26**, 193 (1971).
1713. W. Strohmeier, W. Rehder-Stirnweiss and R. Fleischmann, *Z. Naturforsch., Teil B*, **25**, 1480 (1970).
1714. W. Strohmeier, W. Rehder-Stirnweiss and R. Fleischmann, *Z. Naturforsch., Teil B*, **25**, 1481 (1970).
1715. W. Strohmeier and J. P. Stasch, *Z. Naturforsch., Teil B*, **34**, 755 (1979).
1716. W. Strohmeier, H. Steigerwald and M. Lukacs, *J. Organomet. Chem.*, **144**, 135 (1978).
1717. W. Strohmeier, H. Steigerwald and L. Weigelt, *J. Organomet. Chem.*, **129**, 243 (1977).
1718. W. Strohmeier and L. Weigelt, *J. Organomet. Chem.*, **82**, 417 (1974).
1719. W. Strohmeier and L. Weigelt, *J. Organomet. Chem.*, **86**, C17 (1975).
1720. W. Strohmeier and L. Weigelt, *J. Organomet. Chem.*, **125**, C40 (1977).
1721. W. Strohmeier and L. Weigelt, *J. Organomet. Chem.*, **129**, C47 (1977).
1722. W. Strohmeier and L. Weigelt, *J. Organomet. Chem.*, **133**, C43 (1977).
1723. G. Strukul, M. Bonivento, M. Graziani, E. Cernia and N. Palladino, *Inorg. Chim. Acta*, **12**, 15 (1975).
1724. G. Strukul, P. D'Olimpio, M. Bonivento, F. Pina and M. Graziani, *J. Mol. Catal.*, **2**, 179 (1977).
1725. D. W. Studer and G. L. Schrader, *J. Mol. Catal.*, **9**, 169 (1980).
1726. H. O. Stuehler, *Z. Naturforsch., Teil B*, **35**, 340 (1980).
1727. A. C. L. Su, *Adv. Organomet. Chem.*, **17**, 269 (1978).
1728. A. C. L. Su, *Ger. Pat.* 2 015 077 (1970); *C. A.* **73**, 124 012 (1970).
1729. A. C. L. Su and J. W. Collette, *J. Organomet. Chem.*, **46**, 369 (1972).
1730. A. C. L. Su and J. W. Collette, *J. Organomet. Chem.*, **90**, 227 (1975).
1731. J. W. Suggs, *J. Am. Chem. Soc.*, **101**, 489 (1979).
1732. Y. Sugihara, N. Morokoshi and I. Murata, *Chem. Lett.*, 745 (1979).
1733. H. Sugiyama, S. Tsuchiya, S. Seto, Y. Senda and S. Imaizumi, *Tetrahedron Lett.*, 3291 (1974).
1734. P. Svoboda, M. Capka, V. Chvalovsky, V. Bazant, J. Hetflejs, H. Jahr and H. Pracejus: *Z. Chem.*, **12**, 153 (1972).
1735. P. Svoboda, M. Capka and J. Hetflejs, *Collect. Czech. Chem. Commun.*, **37**, 3059 (1972).
1736. P. Svoboda, M. Capka and J. Hetflejs, *Collect. Czech. Chem. Commun.*, **38**, 1235 (1973).
1737. P. Svoboda, M. Capka, J. Hetflejs and V. Chvalovsky, *Collect. Czech. Chem. Commun.*, **37**, 1585 (1972).
1738. P. Svoboda and J. Hetflejs, *Collect. Czech. Chem. Commun.*, **42**, 2177 (1977).
1739. P. Svoboda, J. Soucek and J. Hetflejs, *Czech. Pat.* 161 177 (1975); *C. A.* **85**, 33 182 (1976).
1740. A. M. Taber, A. I. Rudenkov, V. I. Pyatiletov, S. P. Chernykh and I. V. Kalechits, *Khim. Prom-st. (Moscow)*, 7, 404 (1981); *C. A.* **96**, 20 526 (1982).
1741. Y. Takagi, S. Takahashi, J. Nakayama and K. Tanaka, *J. Mol. Catal.*, **10**, 3 (1980).
1742. N. Takahashi, I. Okura and T. Keii, *J. Mol. Catal.*, **4**, 65 (1978).

1743. N. Takaishi, H. Imai, C. A. Bertelo and J. K. Stille, *J. Am. Chem. Soc.*, **100**, 264 (1978).
1744. N. Takaishi, Y. Inamoto and M. Matsukane, *Ger. Pat.* 2 918 107 (1979); *C. A.* **92**, 93 968 (1980).
1745. K. Takao, H. Azuma, Y. Fujiwara, T. Imanaka and S. Teranishi, *Bull. Chem. Soc. Jpn.,* **45**, 2003 (1972).
1746. K. Takao, Y. Fujiwara, T. Imanaka and S. Teranishi, *Bull. Chem. Soc. Jpn.*, **43**, 1153 (1970).
1747. K. Takao, M. Wayaku, Y. Fujiwara, T. Imanaka and S. Teranishi, *Bull. Chem. Soc. Jpn.,* **45**, 1505 (1972).
1748. M. Takeda, H. Iwane and T. Hashimoto, *Jpn. Pat.* 27 165 (1980); *C. A.* **94**, 65 899 (1981).
1749. M. Takeda, H. Iwane and T. Hashimoto, *Jpn. Pat.* 28 969 (1980); *C. A.* **93**, 114 041 (1980).
1750. M. Takeda, S. Mori, E. Taniyama and H. Iwane, *Jpn. Pat.* 24 843 (1979); *C. A.* **90**, 203 686 (1979).
1751. Y. Takegami, Y. Watanabe and H. Masada, *Bull. Chem. Soc. Jpn.,* **40**, 1459 (1967).
1752. M. Takesada and H. Wakamatsu, *Bull. Chem. Soc. Jpn.,* **43**, 2192 (1970).
1753. M. Takesada, H. Yamazaki and N. Hagihara, *J. Chem. Soc. Jpn.,* **89**, 1126 (1968).
1754. W. H. Tamblyn, S. R. Hoffmann and M. P. Doyle, *J. Organomet. Chem.,* **216**, C64 (1981).
1755. K. Tanaka, K. L. Watters and R. F. Howe, *J. Catal.,* **75**, 23 (1982).
1756. M. Tanaka, T. Hayashi and I. Ogata, *Bull. Chem. Soc. Jpn.,* **50**, 2351 (1977).
1757. M. Tanaka and I. Ogata, *Chem. Commun.,* 735 (1975).
1758. M. Tanaka, K. Saeki and N. Kihara, *Jpn. Pat.* 83 311 (1977); *C. A.* **87**, 200 792 (1977).
1759. M. Tanaka, Y. Watanabe, T. Mitsudo, H. Iwane and Y. Takegami, *Chem. Lett.,* 239 (1973).
1760. M. Tanaka, Y. Watanabe, T. Mitsudo, K. Yamamoto and Y. Takegami, *Chem. Lett.,* 483 (1972).
1761. M. Tanaka, Y. Watanabe, T. Mitsudo, Y. Yasunori and Y. Takegami, *Chem. Lett.,* 137 (1974).
1762. R. Tang, F. Mares, N. Neary and D. E. Smith, *Chem. Commun.,* 274 (1979).
1763. S. C. Tang, T. E. Paxson and L. Kim, *J. Mol. Catal.,* **9**, 313 (1980).
1764. K. Tani, K. Suwa, E. Tanigawa, T. Yoshida, T. Okano and S. Otsuka, *Chem. Lett.,* 261 (1982).
1765. K. Tani and S. Yaguchi, *Jpn. Pat.* 28 447 (1968); *C. A.* **70**, 77 283 (1969).
1766. K. Tani and S. Yaguchi, *Jpn. Pat.* 08 965 (1970); *C. A.* **73**, 14 129 (1970).
1767. K. Tani, T. Yamagata, S. Otsuka, S. Akutagawa, H. Kumobayashi, T. Taketomi, H. Takaya, A. Miyashita and R. Noyori, *Chem. Commun.,* 600 (1982).
1768. R. Tarao, S. Sakano, K. Miyazaki, K. Fujita, C. Saito and S. Miyajima, *Jpn. Pat.* 43 799 (1973); *C. A.* **81**, 82 836 (1974).
1769. R. B. Taylor and P. W. Jennings, *Inorg. Chem.,* **20**, 3997 (1981).
1770. A. T. Teleshev, G. M. Grishini and E. E. Nifantev, *Zh. Obsch. Khim.,* **52**, 533 (1982); *C. A.* **96**, 223 895 (1982).
1771. Ph. Teyssie, *C.R. Acad. Sci.,* **256**, 2846 (1963).
1772. Ph. Teyssie and R. Dauby, *J. Polym. Sci., Part B,* **2**, 413 (1964).
1773. Ph. Teyssie and R. Dauby, *Bull. Soc. Chim. Fr.,* 2842 (1965).

1774. Ph. Teyssie, F. Tripier, A. Isfendiyaroglu, B. Francois, V. Sinn and J. Parrod, *C.R. Acad. Sci.*, **261**, 997 (1965).
1775. I. Tkatchenko, *C.R. Acad. Sci., Ser. C*, **282**, 229 (1976).
1776. A. Theolier, A. K. Smith, M. Leconte, J. M. Basset, G. M. Zanderichi, R. Psaro and R. Ugo, *J. Organomet. Chem.*, **191**, 415 (1980).
1777. J. A. Thomas, *Br. Pat.* 1 367 623 (1974); *C. A.* **82**, 142 591 (1975).
1778. M. G. Thomas, B. F. Beier and E. L. Muetterties, *J. Am. Chem. Soc.*, **98**, 1296 (1976).
1779. H. W. Thompson and E. McPherson, *J. Am. Chem. Soc.*, **96**, 6232 (1974).
1780. E. W. Thornton, H. Knoezinger, B. Tesche, J. J. Rafalko and B. C. Gates, *J. Catal.*, **62**, 117 (1980).
1781. B. I. Tikhomirov, I. A. Klopotova and A. I. Yakubchik, *U.S.S.R. Pat.* 265 432 (1970); *C. A.* **73**, 26 397 (1970).
1782. H. B. Tinker and D. E. Morris, *J. Organomet. Chem.*, **52**, C55 (1973).
1783. H. B. Tinker and A. J. Solodar, *U.S. Pat.* 4 268 688 (1981); *C. A.* **95**, 114 798 (1981).
1784. C. A. Tolman, *Chem. Soc. Rev.*, **1**, 337 (1972).
1785. C. A. Tolman, P. Z. Meakin, D. I. Lindner and J. P. Jesson, *J. Am. Chem. Soc.*, **96**, 2762 (1974).
1786. S. Toros, *Magy. Kem. Lapja*, **29**, 543 (1974); *C. A.* **83**, 177 870 (1975).
1787. S. Toros, B. Heil and L. Marko, *J. Organomet. Chem.*, **159**, 401 (1978).
1788. S. Toros, B. Heil, L. Kollar and L. Marko, *J. Organomet. Chem.*, **197**, 85 (1980).
1789. S. Toros, L. Kollar, B. Heil and L. Marko, *J. Organomet. Chem.*, **232**, C17 (1982).
1790. Toshiba Silicone Co., Ltd., *Jpn. Pat.* 122 390 (1981); *C. A.* **96**, 162 933 (1982).
1791. J. M. Townsend, J. F. Blount, R. C. Sun, S. Zawoiski and D. Valentine, *J. Org. Chem.*, **45**, 2995 (1980).
1792. J. M. Townsend and D. H. Valentine, *U.S. Pat.* 4 120 870 (1978); *C. A.* **90**, 87 660 (1979).
1793. J. M. Townsend and D. H. Valentine, *U.S. Pat.* 4 187 241 (1980); *C. A.* **93**, 8289 (1980).
1794. F. Tripier, B. Francois, V. Sinn and J. Parrod, *C.R. Acad. Sci., Ser. C*, **267**, 1017 (1968).
1795. J. Tsuji, *Jpn. Pat.* 08 442 (1968); *C. A.* **70**, 11 368 (1969).
1796. J. Tsuji, M. Morikawa and J. Kiji, *Tetrahedron Lett.*, 1437 (1963).
1797. J. Tsuji and K. Ohno, *Tetrahedron Lett.*, 3969 (1965).
1798. J. Tsuji and K. Ohno, *J. Am. Chem. Soc.*, **88**, 3452 (1966).
1799. J. Tsuji and K. Ohno, *Tetrahedron Lett.*, 4713 (1966).
1800. J. Tsuji and K. Ohno, *Tetrahedron Lett.*, 2173 (1967).
1801. J. Tsuji and K. Ohno, *Synthesis*, 157 (1969).
1802. J. Tsuji and K. Ohno, *Jpn. Pat.* 17 128 (1969); *C. A.* **71**, 101 570 (1969).
1803. J. Tsuji and K. Ohno, *Jpn. Pat.* 10 922 (1970); *C. A.* **73**, 14 140 (1970).
1804. J. Tsuji and K. Ohno, *Jpn. Pat.* 32 402 (1970); *C. A.* **74**, P63 593 (1971).
1805. J. Tsuji and K. Ohno, *Jpn. Pat.* 21 603 (1971); *C. A.* **75**, 129 529 (1971).
1806. Y. Tsutsumi, M. Ohshio and Y. Fukuda, *Jpn. Pat.* 91 095 (1974); *C. A.* **83**, 153 192 (1975).
1807. M. Tuner, J. von Jouanne, H. D. Brauer and H. Kelm, *J. Mol. Catal.*, **5**, 425 (1979).
1808. M. Tuner, J. von Jouanne, H. D. Brauer and H. Kelm, *J. Mol. Catal.*, **5**, 433 (1979).
1809. H. S. Tung and C. Brubaker, *J. Organomet. Chem.*, **216**, 129 (1981).

1810. J. O. Turner and J. E. Lyons, *Ger. Pat.* 2 231 678 (1973); *C. A.* **78**, 111 925 (1973).
1811. L. D. Tyutchenkova, L. G. Privalova, Z. K. Maizus, N. F. Gol'dshleger, M. L. Khidekel, I. V. Kalechits and N. M. Emanuel, *Dokl. Akad. Nauk SSSR,* **199**, 872 (1971); *C. A.* **75**, 140 031 (1971).
1812. L. D. Tyutchenkova, V. G. Vinogradova and Z. K. Maizus, *Izv. Akad. Nauk SSSR, Ser. Khim.*, 773 (1978); *Bull. Acad. Sci. USSR (Engl. Transl.)*, **27**, 666 (1978).
1813. T. Uematsu, T. Kawakami, F. Saitho, M. Miura and H. Hashimoto, *J. Mol. Catal.*, **12**, 11 (1981).
1814. S. Uemura and S. R. Patil, *Tetrahedron Lett.*, 4353 (1982).
1815. H. Umezaki, Y. Fujiwara, K. Sawara and S. Teranishi, *Bull. Chem. Soc. Jpn.*, **46**, 2230 (1973).
1816. Uniroyal, Inc., *Br. Pat.* 1 131 160 (1968); *C. A.* **70**, 20 488 (1969).
1817. United States Rubber Co., *Belg. Pat.* 656 793 (1965); *C. A.* **64**, 17 742 (1966).
1818. J. D. Unruh, *Ger. Pat.* 2 834 742 (1979); *C. A.* **90**, 203 487 (1979).
1819. J. D. Unruh and J. R. Christenson, *J. Mol. Catal.*, **14**, 19 (1982).
1820. J. D. Unruh and L. E. Wade, *Ger. Pat.* 2 813 963 (1978); *C. A.* **90**, 22 315 (1979).
1821. J. D. Unruh and L. E. Wade, *Ger. Pat.* 2 813 963 (1978).
1822. J. D. Unruh and W. J. Wells, *Ger. Pat.* 2 617 306 (1975); *C. A.* **86**, 22 369 (1977).
1823. K. Unverferth, R. Hoentsch and K. Schwetlick, *J. Prakt. Chem.*, **321**, 86 (1979).
1824. K. Unverferth, R. Hoentsch and K. Schwetlick, *J. Prakt. Chem.*, **321**, 928 (1979).
1825. K. Unverferth, J. Pelz, F. Flemming and K. Schwetlick, *Z. Chem.*, **14**, 304 (1974).
1826. K. Unverferth, C. Rueger and K. Schwetlick, *J. Prakt. Chem.*, **319**, 841 (1977).
1827. UOP Inc., *Fr. Pat.* 2 431 481 (1980); *C. A.* **93**, 149 800 (1980).
1828. V. M. Ushakov, D. B. Kazarnovskaya, L. L. Klinova and V. M. Egorova, *Neftekimiya*, **18**, 920 (1978); *C. A.* **90**, 86 582 (1979).
1829. V. M. Ushakov, D. B. Kazarnovskaya, L. L. Klinova and N. M. Malygina, *Neftekimiya*, **19**, 62 (1979); *C. A.* **90**, 186 438 (1979).
1830. V. M. Ushakov, D. B. Kazarnovskaya, L. L. Klinova and Z. V. Prilepskaya, *Neftekimiya*, **18**, 739 (1978); *C. A.* **90**, 22 002 (1979).
1831. R. Uson, P. Lahuerta, L. A. Carmona, L. A. Oro and K. Hildenbrand, *J. Organomet. Chem.*, **157**, 63 (1978).
1832. R. Uson, L. A. Oro, C. Claver, M. A. Garralda and J. M. Moreto, *J. Mol. Catal.*, **4**, 231 (1978).
1833. R. Uson, L. A. Oro, M. J. Fernandez and M. T. Pinillos, *Inorg. Chim. Acta*, **39**, 57 (1980).
1834. R. Uson, L. A. Oro, M. A. Garralda, C. M. Carmen and P. Lahuerta, *Transition Met. Chem.*, **4**, 55 (1979).
1835. R. Uson, L. A. Oro, M. J. Pinillos, M. Royo and E. Pastor, *J. Mol. Catal.*, **14**, 375 (1982).
1836. R. Uson, L. A. Oro, R. Sariego and M. A. Esteruelas, *J. Organomet. Chem.*, **214**, 399 (1981).
1837. D. Valentine, *Ger. Pat.* 2 548 884 (1976); *C. A.* **85**, 94 550 (1976).
1838. D. Valentine, *U.S. Pat.* 4 115 417 (1978); *C. A.* **90**, 104 171 (1979).
1839. D. Valentine, K. K. Johnson, W. Priester, R. C. Sun, K. Toth and G. Saucy, *J. Org. Chem.*, **45**, 3698 (1980).
1840. D. Valentine and J. W. Scott, *Synthesis*, 329 (1978).
1841. G. Valentini, G. Sbrana and G. Braca, *J. Mol. Catal.*, **11**, 383 (1981).
1842. H. van Bekkum, F. van Rantwijk and T. van de Putte, *Tetrahedron Lett.*, 1 (1969).
1843. A. van der Ent and H. G. Cuppers, *Neth. Pat. Appl.* 01 018 (1970); *C. A.* **75**, 151 336.

1844. P. van der Plank, A. van der Ent, A. L. Onderdehnden and H. J. van Oosten, *J. Am. Oil Chem. Soc.,* **57**, 343 (1980).

1845. H. van Gaal, H. G. A. M. Cuppers and A. van der Ent, *Chem. Commun.*, 1694 (1970).

1846. P. W. N. M. van Leeuwen and C. F. Roobeek, *Eur. Pat. Appl.* 33 554 (1981); *C. A.* **96**, 6174 (1982).

1847. F. van Rantwijk, Th. G. Spek and H. van Bekkum, *Recl. Trav. Chim. Pays-Bas,* **91**, 1057 (1972).

1848. B. H. van Vugt, N. J. Koole, W. Drenth and F. P. J. Kuijpers, *Recl. Trav. Chim. Pays-Bas,* **92**, 1321 (1973).

1849. Y. S. Varshavskii, E. P. Shestakova, N. V. Kiseleva, T. G. Cherkasova, N. A. Buzina, L. S. Bresler and V. A. Kormer, *J. Organomet. Chem.,* **170**, 81 (1979).

1850. C. G. Vasiyarov, M. L. Kalinkin, S. N. Ananchenko, G. D. Kolomnikova, I. V. Torgov, Z. N. Parnes and D. N. Kursanov, *Izv. Akad. Nauk SSSR, Ser. Khim.,* 2144 (1981); *C. A.* **96**, 123 081 (1982).

1851. L. Vaska and R. E. Rhodes, *J. Am. Chem. Soc.,* **87**, 4970 (1965).

1852. L. Vaska and M. F. Werneke, *Trans. N. Y. Acad. Sci.,* **33**, 70 (1971).

1853. S. Vastag, B. Heil, S. Toros and L. Marko, *Transition Met. Chem.,* **2**, 58 (1977).

1854. V. M. Vdovin, V. E. Federov, N. A. Pritula and G. K. Fedorova, *Izv. Akad. Nauk SSSR, Ser. Khim.*, 181 (1982); *Bull. Acad. Sci. USSR (Engl. Transl.)*, **31**, 170 (1982).

1855. L. Verdet and J. K. Stille, *Organometallics,* **1**, 380 (1982).

1856. B. Veruovic, J. Zachoval and S. Bittner, *Collect. Czech. Chem. Commun.,* **33**, 3026 (1968).

1857. J. L. Vidal, Z. C. Mester and W. E. Walker, *U.S. Pat.* 4 115 428 (1978); *C. A.* **90**, 103 384 (1979).

1858. J. L. Vidal, Z. C. Mester and W. E. Walker, *Ger. Pat.* 2 842 307 (1979); *C. A.* **91**, 74 202 (1979).

1859. J. L. Vidal and W. E. Walker, *Ger. Pat.* 2 906 683 (1979); *C. A.* **92**, 8741 (1980).

1860. J. L. Vidal and W. E. Walker, *Inorg. Chem.,* **19**, 896 (1980).

1861. J. L. Vidal and W. E. Walker, *Inorg. Chem.,* **20**, 249 (1981).

1862. J. L. Vidal, W. E. Walker, R. L. Pruett, R. C. Schoening and R. A. Fiato, *Fundam, Res. Homogeneous Catal.,* **3**, 499 (1979).

1863. J. Vilim and J. Hetflejs, *Chem. Prum.,* **28**, 135 (1978); *C. A.* **89**, 180 317 (1978).

1864. J. Vilim and J. Hetflejs, *Czech. Pat.* 185 974 (1980); *C. A.* **95**, 169 785 (1981).

1865. B. D. Vineyard, W. S. Knowles, M. J. Sabacky, G. L. Bachman and O. J. Weinkauff, *J. Am. Chem. Soc.,* **99**, 5946 (1977).

1866. G. Vitulli, P. Salvadori, A. Raffaelli, P. A. Costantino and R. Lazzaroni, *J. Organomet. Chem.,* **239**, C23 (1982).

1867. W. Voelter and C. Djerassi, *Chem. Ber.,* **101**, 1154 (1968).

1868. H. W. Voigt and J. A. Roth, *J. Catal.,* **33**, 91 (1974).

1869. H. C. Volger and M. M. P. Gaasbeek, *Recl. Trav. Chim. Pay-Bas,* **87**, 1290 (1968).

1870. H. C. Volger and H. Hogeveen, *Recl. Trav. Chim. Pay-Bas,* **86**, 830 (1967).

1871. H. C. Volger, H. Hogeveen and M. M. P. Gaasbeek, *J. Am. Chem. Soc.,* **91**, 218 (1969).

1872. H. C. Volger, H. Hogeveen and M. M. P. Gaasbeek, *J. Am. Chem. Soc.,* **91**, 2137 (1969).

1873. M. E. Vol'pin, V. P. Kukolev, V. O. Chernyshev and I. S. Kolomnikov, *Tetrahedron Lett.*, 4435 (1971).

1874. M. E. Vol'pin, A. M. Yurkevich, L. G. Volkova, E. G. Chauser, I. P. Rudakova,

I. Ya. Levitin, E. M. Tachkova and T. M. Ushakova, *Zh. Obshch. Khim.*, **45**, 164 (1975); *C. A.* **83**, 27 147 (1975).

1875. D. A. von Bezard, G. Consiglio, F. Morandini and P. Pino, *J. Mol. Catal.*, 7, 431 (1980).

1876. N. von Kuterpow, *Ger. Pat.* 2 037 782 (1972); *C. A.* **76**, 154 404 (1972).

1877. N. von Kuterpow, F. J. Mueller and P. Reuter, *Ger. Pat.* 2 262 852 (1973); *C. A.* **79**, 83 962 (1973).

1878. K. P. Vora, C. F. Lochow and R. G. Miller, *J. Organomet. Chem.*, **192**, 257 (1980).

1879. M. G. Voronkov, V. Chvalovsky, S. V. Kirpichenko, N. A. Vlasova, S. T. Bolshakov, G. Kuncova, V. V. Kieko and E. O. Tsetlina, *Collect. Czech. Chem. Commun.*, **44**, 742 (1979).

1880. M. G. Voronkov, S. V. Kirpichenko, V. V. Kieko and E. O. Tseitlina, *Izv. Akad. Nauk SSSR, Ser. Khim.*, 174 (1981); *C. A.* **95**, 25 192 (1981).

1881. M. G. Voronkov, S. V. Kirpichenko, E. N. Suslova and V. V. Keiko, *J. Organomet. Chem.*, **204**, 13 (1981).

1882. M. G. Voronkov, V. B. Pukhnarevich, I. I. Tsykhanskaya and Y. S. Varshavskii, *Dokl. Akad. Nauk SSSR*, **254**, 887 (1980); *C. A.* **94**, 192 397 (1981).

1883. M. G. Voronkov, N. N. Vlasova, S. A. Bol'shakova and S. V. Kirpichenko, *J. Organomet. Chem.*, **190**, 335 (1980).

1884. M. G. Voronkov, N. N. Vlasova, S. V. Kirpichenko, S. A. Bol'shakova and V. V. Keiko, *Izv. Akad. Nauk SSSR, Ser. Khim.*, 422 (1979); *C. A.* **90**, 168 669 (1979).

1885. M. G. Voronkov, N. N. Vlasova, S. V. Kirpichenko, S. A. Bol'shakova and A. E. Pestunovich, *Katal. Sint. Org. Soedin. Sery*, 7 (1979); *C. A.* **93**, 239 514 (1980).

1886. K. Vrieze, J. P. Collman, C. T. Sears and M. Kubota, *Inorg. Synth.*, **11**, 101 (1968).

1887. H. M. Walborsky and L. E. Allen, *Tetrahedron Lett.*, 823 (1970).

1888. H. M. Walborsky and L. E. Allen, *J. Am. Chem. Soc.*, **93**, 5465 (1971).

1889. W. E. Walker, D. R. Bryant and E. S. Brown, *Ger. Pat.* 2 531 149 (1976); *C. A.* **85**, 35 400 (1976).

1890. H. K. Wang, H. W. Choi and E. L. Muetterties, *Inorg. Chem.*, **20**, 2661 (1981).

1891. M. D. Ward and J. Schwartz, *J. Am. Chem. Soc.*, **103**, 5253 (1981).

1892. M. D. Ward and J. Schwartz, *J. Mol. Catal.*, **11**, 397 (1981).

1893. M. D. Ward and J. Schwartz, *Organometallics*, **1**, 1030 (1982).

1894. H. Watanabe, M. Aoki, N. Sakurai, K. Watanabe and Y. Nagai, *J. Organomet. Chem.*, **160**, C1 (1978).

1895. H. Watanabe, M. Asami and Y. Nagai, *J. Organomet. Chem.*, **195**, 363 (1980).

1896. H. Watanabe, T. Kitahara, T. Motegi and Y. Nagai, *J. Organomet. Chem.*, **139**, 215 (1977).

1897. Y. Watanabe, T. Mitsudo, Y. Yasunori, J. Kiguchi and Y. Takegami, *Bull. Chem. Soc. Jpn.*, **52**, 2735 (1979).

1898. Y. Watanabe, S. C. Shim and T. Mitsudo, *Bull. Chem. Soc. Jpn.*, **54**, 3460 (1981).

1899. Y. Watanabe, Y. Shimizu, K. Takatsuki and Y. Takegami, *Chem. Lett.*, 215 (1978).

1900. Y. Watanabe, N. Suzuki, Y. Tsuji, S. C. Shim and T. Mitsudo, *Bull. Chem. Soc. Jpn.*, **55**, 1116 (1982).

1901. Y. Watanabe, M. Yamamoto, T. Mitsudo and Y. Takegami, *Tetrahedron Lett.*, 1289 (1978).

1902. K. L. Watters, R. F. Howe, T. P. Chojnacki, C. M. Fu, R. L. Schneider and N. B. Wong, *J. Catal.*, **66**, 424 (1980).

1903. M. Wayaku, K. Kaneda, T. Imanaka and S. Teranishi, *Bull. Chem. Soc. Jpn.*, **48**, 1957 (1975).

1904. K. M. Webber, B. C. Gates and W. Drenth, *J. Mol. Catal.*, **3**, 1 (1977).
1905. D. E. Webster, *Adv. Organomet. Chem.*, **15**, 147 (1976).
1906. G. E. Weismantel, *Chem. Eng. (London)*, 47 (Jan. 12th, 1981).
1907. D. J. Westlake and M. J. Wriglesworth, *Br. Pat.* 1 362 997 (1974); *C. A.* **82**, 3805 (1975).
1908. C. White, D. S. Gill, J. W. Kang, H. B. Lee and P. M. Maitlis, *Chem. Commun.*, 734 (1971).
1909. R. Whyman, *J. Organomet. Chem.*, **94**, 303 (1975).
1910. R. Whyman, *Eur. Pat. Appl.* 33 425 (1981); *C. A.* **95**, 219 767 (1981).
1911. K. B. Wiberg, *Adv. Alicyclic Chem.*, **2**, 185 (1968).
1912. K. B. Wiberg and K. C. Bishop, *Tetrahedron Lett.*, 2727 (1973).
1913. K. B. Wiberg and H. A. Connon, *J. Am. Chem. Soc.*, **98**, 5411 (1976).
1914. P. Wilhelmus, N. M. van Leeuwen and C. F. Roobeek, *Br. Pat. Appl.* 2 068 377 (1982); *C. A.* **96**, 217 237 (1982).
1915. G. Wilkinson, *Fr. Pat.* 1 459 643 (1966); *C. A.* **67**, 53 652 (1967).
1916. G. Wilkinson, *Ger. Pat.* 1 816 063 (1969); *C. A.* **71**, 90 766 (1969).
1917. G. Wilkinson, *Ger. Pat.* 1 939 322 (1970); *C. A.* **73**, 14 195 (1970).
1918. G. Wilkinson, *Ger. Pat.* 2 064 471 (1971); *C. A.* **75**, 109 848 (1971).
1919. G. Wilkinson, *Fr. Pat.* 2 072 146 (1971); *C. A.* **77**, 4914 (1972).
1920. G. Wilkinson, *Ger. Pat.* 2 136 470 (1972); *C. A.* **76**, 145 404 (1972).
1921. G. Wilkinson, *Br. Pat.* 1 368 432 (1974); *C. A.* **82**, 30 930 (1975).
1922. G. Wilkinson, *Ger. Pat.* 2 064 471 (1971); *C. A.* **75**, 109 848 (1971).
1923. G. Wilkinson, R. A. Schunn and W. G. Peet, *Inorg. Synth.*, **13**, 126 (1971).
1924. G. Wilkinson, F. G. A. Stone and E. W. Abel, *Comprehensive Organometallic Chemistry*, Pergamon, New York, 1982.
1925. R. C. Williamson and T. P. Kobylinski, *U.S. Pat.* 4 170 606 (1979); *C. A.* **92**, 25 342 (1980).
1926. M. J. Wrigelsworth and D. J. Westlake, *Br. Pat.* 1 363 961 (1974); *C. A.* **82**, 3797 (1975).
1927. Y. Wu and E. A. Zuech, *U.S. Pat.* 4 246 177 (1981); *C. A.* **94**, 139 621 (1981).
1928. G. Yagupski and G. Wilkinson, *J. Chem. Soc. A*, 725 (1969).
1929. M. Yagupski, C. K. Brown, G. Yagupski and G. Wilkinson, *J. Chem. Soc. A*, 937 (1970).
1930. M. Yagupski and G. Wilkinson, *J. Chem. Soc. A*, 941 (1970).
1931. M. Yamada, M. Yamashita and S. Inokawa, *Carbohydr. Res.*, **95**, C9 (1981).
1932. M. Yamaguchi, *J. Chem. Soc. Jpn.*, **70**, 675 (1967).
1933. M. Yamaguchi, *J. Chem. Soc. Jpn.*, **72**, 671 (1969).
1934. M. Yamaguchi, *Shokubai*, **11**, 179 (1969); *C. A.* **73**, 13 787 (1970).
1935. K. Yamamoto, T. Hayashi and M. Kumada, *J. Organomet. Chem.*, **54**, C45 (1973).
1936. K. Yamamoto, J. Wakatsuki and R. Sugimoto, *Bull. Chem. Soc. Jpn.*, **53**, 1132 (1980).
1937. H. C. Yao and W. G. Rothschild, *J. Chem. Phys.*, **68**, 4774 (1978).
1938. Yu: I. Yermakov and V. A. Zakharov, in *Studies in Surface Science and Catalysis*, Vol. 8, 1981.
1939. C. C. Yin and A. J. Deeming, *J. Organomet. Chem.*, **133**, 123 (1977).
1940. T. Yoshida, T. Okano and S. Otsuka, *Chem. Commun.*, 870 (1979).
1941. T. Yoshida, T. Okano and S. Otsuka, *J. Am. Chem. Soc.*, **102**, 5966 (1980).
1942. T. Yoshida, T. Okano, K. Saito and S. Otsuka, *Inorg. Chim. Acta*, **44**, L135 (1980).

1943. T. Yoshida, T. Okano, Y. Ueda and S. Otsuka, *J. Am. Chem. Soc.*, **103**, 3411 (1981).
1944. S. Yoshikawa, J. Kiji and J. Furukawa, *Makromol. Chem.*, **178**, 1077 (1977).
1945. T. Yoshikuni and J. C. Bailar, *Inorg. Chem.*, **21**, 2129 (1982).
1946. A. Yoshioka, T. Sugimura and M. Takahashi, *Jpn. Pat.* 110 293 (1977); *C. A.* **88**, 24 103 (1978).
1947. J. F. Young, J. A. Osborn, F. H. Jardine and G. Wilkinson, *Chem. Commun.*, 131 (1965).
1948. H. Yukimasa, H. Sawai and T. Takizawa, *Makromol. Chem., Rapid Commun.*, **1**, 579 (1980).
1949. E. N. Yurchenko, *Kinet. Katal.*, **14**, 515 (1973); *C. A.* **79**, 4701 (1973).
1950. E. N. Yurchenko, A. D. Troitskaya, L. S. Gracheva and L. Ya. A'lt, *Izv. Akad. Nauk SSSR, Ser. Khim.*, 2153 (1975); *Bull. Acad. Sci. USSR (Engl. Transl.)*, **24**, 2041 (1975).
1951. J. Zachoval, J. Krepelka and M. Klimova, *Collect. Czech. Chem. Commun.*, **37**, 327 (1972).
1952. J. Zachoval, F. Mikes, J. Krepelka, O. Prouzova and O. Pradova, *Eur. Polym. J.*, **8**, 397 (1972).
1953. J. Zachoval and B. Veruovic, *J. Polym. Sci., Part B*, **4**, 965 (1966).
1954. J. Zachoval and B. Veruovic, *Czech. Pat.* 126 437 (1968); *C. A.* **70**, 48 422 (1969).
1955. J. B. Zachry and C. L. Aldridge, *U.S. Pat.* 3 161 672 (1964); *C. A.* **62**, 9018 (1965).
1956. L. I. Zakharkin and T. B. Agakhanova, *Izv. Akad. Nauk SSSR, Ser. Khim.*, 2632 (1977); *Bull. Acad. Sci. USSR (Engl. Transl.)*, **26**, 2436 (1977).
1957. L. I. Zakharkin and T. B. Agakhanova, *J. Gen. Chem. USSR (Engl. Transl.)*, **47**, 2191 (1977).
1958. L. I. Zakharkin and T. B. Agakhanova, *Izv. Akad. Nauk SSSR, Ser. Khim.*, 2151 (1978); *C. A.* **90**, 6458 (1979).
1959. L. I. Zakharkin and T. B. Agakhanova, *Izv. Akad. Nauk SSSR, Ser. Khim.*, 2833 (1978); *C. A.* **90**, 120 948 (1979).
1960. L. I. Zakharkin and T. B. Agakhanova, *Izv. Akad. Nauk SSSR, Ser. Khim.*, 1208 (1980); *C. A.* **93**, 132 547 (1980).
1961. L. I. Zakharkin, I. V. Pisareva and T. B. Agakhanova, *Izv. Akad. Nauk SSSR, Ser. Khim.*, 2389 (1977); *Bull. Acad. Sci. USSR (Engl. Transl.)*, **26**, 2222 (1977).
1962. R. Zanella, F. Canziani, R. Ros and M. Graziani, *J. Organomet. Chem.*, **67**, 449 (1974).
1963. G. Zassinovich, A. Camus and G. Mestroni, *J. Mol. Catal.*, **9**, 345 (1980).
1964. G. Zassinovich, C. Del Bianco and G. Mestroni, *J. Organomet. Chem.*, **222**, 323 (1981).
1965. G. Zassinovich, G. Mestroni and A. Camus, *J. Organomet. Chem.*, **168**, C37 (1979).
1966. G. Zassinovich, G. Mestroni and A. Camus, *J. Mol. Catal.*, **2**, 63 (1977).
1967. S. Y. Zhang, S. Yemul, H. B. Kagan, R. Stern, D. Commereuc and Y. Chauvin, *Tetrahedron Lett.*, 3955 (1981).
1968. Yu. M. Zhorov, G. M. Panchenkov, A. V. Selkov and G. V. Demidovich, *Kinet. Katal.*, **14**, 809 (1973); *C. A.* **79**, 77 962 (1973).
1969. M. Zuber, B. Banas and F. Pruchnik, *J. Mol. Catal.*, **10**, 143 (1981).
1970. M. Zuber and F. Pruchnik, *Rocz. Chem.*, **49**, 1375 (1975); *C. A.* **84**, 16 524 (1976).
1971. E. A. Zuech, *U.S. Pat.* 3 956 177 (1976); *C. A.* **85**, 37 669 (1976).

INDEX

264INDEX

Azirines
carbonylation, 136
Azobenzene
reduction, 70, 74
Azoxybenzene
reduction, 74

Barbaralane
from tricyclononatriene, 33
'BDPCH' (*see* Chiral ligands)
'BDPCP' (*see* Chiral ligands)
Benzaldehyde
oxidation, 165
from styrene, 160
Benzaldimines
reduction, 74
Benzamides, *N*-aryl
formation, from nitroarenes, 135
Benzene
alkylation, 187
conversion to cyclohexane, 80
H/D exchange, 11
reduction, 80, 82, 85
Benzoic anhydride
conversion to fluorenone, 173
Benzoin
reduction, 70
Benzonitrile
from *N*-benzylbenzamide, 173
Benzophenone
from diphenylmethane, 165
reduction, 70
Benzothiepin
from thiatetracycloundecatriene, 33
Benzyl alcohol
carbonylation, 132
Benzylamine
dehydrogenation, 11
Benzyl chloride
carbonylation, 134
Benzylic alcohols (*see also* Benzyl alcohol)
dehydrogenation, 11
Benzylidene aniline
as chiral Schiff base ligand, 89–90
reduction, 70, 74
Biaryls
from aryl-mercurials, 194
Bicyclic compounds (*see* specific compounds)

Bicyclo[1.1.0]butanes
ring opening, 24–26
Bicyclo[4.1.0]hept-2-ene
ring opening, 28
Bicyclo[2.2.0]hexane
valence isomerization, 33
Bicyclo[3.1.0]hex-2-enes
ring opening, 27
Bicyclononatrienes
formation, in ring opening reactions, 32, 33
Bicyclo[6.1.0]nona-2,4,6-triene
rearrangement, 28, 29
Bicyclo[6.1.0]non-2-ene
rearrangement, 28
Bicyclo[2.2.0]oct-2-ene
hydroformylation, 147
Bicyclo[2.1.0]pentane
ring opening, 26, 27
'BINAP' (*see also* Chiral ligands)
in asymmetric isomerization catalyst, 16
crystal structure of rhodium complex, 98
'BINOR-S'
from norbornadiene, 183
Biphenylene
dimerization, 183
Bishomocubane
valence isomerization, 34, 35
Bishomocycloheptatriene
valence isomerization, 32, 33
'BMPP' (*see* Chiral ligands)
'BPPFA' (*see* Chiral ligands)
'BPPM' (*see also* Chiral ligands)
use in asymmetric reduction of ketones, 107, 108
Butadiene
from bicyclo[1.1.0]butane, 24
carboxylation, 157
co-oligomerization with allyl alcohol, 182
co-oligomerization with ethylene, 181
hydrogenation, 46, 50, 59, 65
hydrogenation, with supported catalyst, 86
hydroformylation, 140, 144, 149
hydrosilation, 112
oligomerization, 178

DATE DUE

MAY 2 1986			
NO 27 '91			

DEMCO 38-297